LEGO® MINDSTORMS™ NXT™
POWER PROGRAMMING

LEGO® MINDSTORMS™ NXT™
POWER PROGRAMMING

JOHN C. HANSEN

VARIANT PRESS

Designed by Relish Design Studio Ltd.
Instructions created with LEGO Digital Designer
Printed in Canada

Library and Archives Canada Cataloguing in Publication

Hansen, John C., 1964-
     Lego NXT power programming : robotics in C / John C. Hansen.

Includes index.

ISBN 978-0-9738649-2-2

     1. Robotics--Popular works. 2. LEGO toys. 3. Robots—Programming.
4. Robots—Design and construction. I. Title.

TJ211.15.H365 2007   629.8'92

C2007-905721-7

VARIANT PRESS
143 Goldthorpe Crescent
Winnipeg, Manitoba
R2N 3E6

Acknowledgements

Thanks to Michael Barrett Andersen for his substantial contributions to the development of NBC and NXC and for Pong. Thanks to Philippe Hurbain for his generous help with LEGO Digital Designer and LDraw issues and for being a constant source of inspiration. Thanks to Ross Crawford for the NBC Tutorial and for tic-tac-toe. Thanks to my brother, Karl Hansen, for diving into NXC and coming up with some real pearls. Thanks to Andreas Dreier for his amazing nxtRICedit tool and his NXT pictures. Thanks to Danny Silver for reading the early chapters and providing helpful feedback. Thanks to Brian Bagnall for all his hard work and encouragement. Most of all I thank my lovely wife, Carol, and my four beautiful children, Juliet, Elisabeth, Kristina, and Nikolas for their patience, love, and support throughout the entire process.

—John Hansen

Preface

My goal is to turn you into an NXT power programmer. The more you know about NXT and its firmware, the better you can harness its power to build amazing creations. To that end, I've tried to include all the relevant knowledge I can about this fantastic little brick.

The NXT set includes electronic parts that vary in complexity from the simplest to the most advanced. To help you realize the full power of these devices, we will explore them using Versa, a versatile mobile robot featuring plug-and-play attachments. We'll also examine a wide assortment of third-party sensors from companies like HiTechnic and mindsensors.com.

To become a power programmer, you need to understand all of the tools at your disposal. This book introduces *Not eXactly C* (NXC) and *NeXT Byte Codes* (NBC), two powerful text-based programming languages that are designed to work with the NXT right out of the box. I'll teach you how to write programs that are many times smaller, faster and more manageable than equivalent programs written using NXT-G.

Although it's easy to program NXC and NBC code using a notepad and command line, it's even easier when using **Bricx Command Center** (BricxCC). This application gives you many of the features found in professional software development systems, such as Microsoft Visual Studio. I'll also present a wide array of NXT utilities that will further enhance your programming prowess.

The most important thing you can do as you begin your journey is to try something new every day. Ask questions. Experiment. Freely share your gifts with fellow travelers. These characteristics will lead you to greatness not just with LEGO robots but in other aspects of your life. Let's take that first big step together.

Contents

Getting Started

Topics in this Chapter

- The Basics
- Programming Languages
- Quick Start
- Starting to Program

Chapter **1**

W elcome to NXT Power Programming. Together, we'll examine an exciting entry in the world of LEGO MINDSTORMS, the NXT programmable brick. This will be a journey of discovery where we will explore the technology in increasing detail as you progress through the chapters of this book. We'll start off with the basics and take things slowly for a while. Later chapters will introduce more advanced subjects and provide additional depth. By the end of our journey, you will be able to create your own amazing robots.

The Basics

Before we begin, let's become familiar with how to communicate with the NXT from a computer. The NXT set comes with the LEGO MINDSTORMS NXT software, which is a graphical environment that includes an easy to use, icon-based, drag-and-drop programming language called NXT-G. This software package also includes a huge amount of instructional material in the Robo Center, where you can find building instructions for four very interesting robots: TriBot, RoboArm T-56, Spike, and Alpha Rex. Working through these examples will lay the ground work for what we will try later.

Learning the Ropes
Make sure the NXT software is installed on your computer and that you can successfully connect to the NXT before continuing. You can skip this step if you are an advanced user or if you are using the NXT on Linux or another operating system that is not supported by the LEGO MINDSTORMS NXT software. However, the instructional material in the Robo Center is written as a set of Shockwave Flash movies and HTML pages, so even if you are using an operating system other than Windows or Mac OSX you can still use the material. The building instructions for the four standard robots can also be downloaded from the web in PDF format via the link below.

 WEBSITE: http://www.active-robots.com/products/
mindstorms4schools/building-instructions.shtml

For Windows and Mac OSX users it is important to download and install the latest version of the NXT Driver software. This update is available from the LEGO MINDSTORMS support web page.

 WEBSITE: *http://mindstorms.lego.com/Support/Updates/*

Download the NXT Driver update from this page and install it. This driver provides what is known as the Fantom SDK, which enables programs running on a PC or Mac to communicate with the NXT either via USB or Bluetooth. The software described in the next section requires the Fantom SDK so it is important to have this software installed and working correctly.

Programming Languages

Whenever you write a program using a programming language (graphical or text-based) the code that you write will need to be processed by a utility known as a compiler before the program can run on the NXT. A compiler converts the graphical or text representation of a program into the executable file format expected by the device that you want to run the program on. In our case, the device happens to be the NXT programmable brick. This process is referred to as either assembling or compiling the code.

The NXT Operating System

The NXT is a very small computer, and just like any other computer, it can't do much without an operating system running on it. The NXT operating system (OS) is also commonly known as the standard NXT firmware. The word firmware is the term used by computing professionals for the low-level operating system or software that runs on computer devices such as the NXT.

Just as Windows provides PC programmers with a large set of built-in features and functions, the standard NXT firmware provides NXT programmers with a set of commands and functions which let you interact with the motors, sensors, sound system, buttons, and LCD screen. The set of commands and functions provided by the firmware for the programmer is called the application programming interface or API.

The NXT firmware also defines the required structure for executable files. A compiler is needed to translate our graphical or text-based programs into the right format so that they can run on the NXT firmware just like your favorite game runs on the Windows or Macintosh operating system.

In this book, we'll look at a few simple programs written for the NXT in the NXT-G programming language. NXT-G gets its name from the programming language used within the National Instruments

LabVIEW product, which is simply called G. LabVIEW stands for Laboratory Virtual Instrumentation Engineering Workbench. The LabVIEW software is a powerful development platform commonly used for data acquisition and for controlling various types of laboratory instruments.

The name "G" comes from the fact that the programming language is graphical in nature. Programs written in G are drawn using a palette of blocks called virtual instruments that you can place on a workspace and wire together into a sequence or network of blocks in a manner similar to drawing a flowchart. G is a type of dataflow programming language. This means a block of code can execute at any moment as soon as data is available on all of the block's inputs.

LEGO and National Instruments worked together to develop the NXT firmware and the LEGO MINDSTORMS NXT software. As a result of this collaboration, the NXT-G programming language is much more complete and powerful than the graphical programming language provided with earlier generations of LEGO MINDSTORMS. National Instruments even released a special LabVIEW toolkit for the NXT that enables LabVIEW users to write lower-level code directly in G that can run faster and use features of the NXT firmware that are not exposed from within NXT-G.

 WEBSITE: *The LabVIEW NXT toolkit can be downloaded from the National Instruments website:* http://www.ni.com/academic/mindstorms/

The NXT-G compiler is built into the software that comes with the NXT set. It will run automatically any time you try to download your NXT-G program to the NXT. However, while we will see some NXT-G in this book, it is not the main programming language that we will use.

The NBC Compiler

Our primary focus will be on text-based programming languages rather than on NXT-G, so most of the programs in this book will use a third-party alternate programming tool for the NXT called NBC. NBC is a compiler that can create executable programs for the standard NXT firmware. The NBC compiler can compile a few different text-based programming languages into the standard NXT executable file format.

The NBC compiler is available for multiple platforms, including Windows, Mac OSX, Linux, and FreeBSD. Regardless of whether you write a program in NBC on Linux using Emacs as your editor, on Mac OSX using XCode, or on Windows using the Bricx Command Center, in every case the language stays the same. The NBC compiler supports four main programming languages: NBC, NXC, NPG, and RICScript.

NeXT Byte Codes (NBC) is a low-level assembly language that can be used to produce the smallest and fastest executables possible for the standard NXT firmware. Because it is an assembly language, it does not provide some of the language constructs that are supported by high-level programming languages that you may be familiar with, such as Basic, Pascal, or C/C++. It's a lower level, pedal-to-the-metal programming solution. We'll explore how to take advantage of programming in assembly when the situation demands it. We'll also learn how to use the NBC code from within programs written in higher-level languages.

Not eXactly C (NXC) is a C-like programming language supported by the NBC compiler. NXC is very similar to Not Quite C (NQC), a programming language developed by Dave Baum for the RCX programmable brick. You could say that NXC is NQC for the NXT! Both languages implement a subset of the standard C programming language, with some limitations and differences that reflect the capabilities of the RCX and NXT bricks.

The set of constants and functions (the API) which can be used within an NXC program has been designed to follow those available in NQC, where possible, so that users of NQC can easily transition to using NXC for the NXT brick. We'll look closely at the similarities and differences between NXC and NQC.

The third programming language that the NBC compiler supports is a very simple language called NPG that can be used to create short 5-step scripts. These scripts can be executed on the NXT via a firmware feature known as On Brick Programming or NXT Programs. In the next chapter we'll explore the NXT Programs portion of the menu system. In addition to creating these simple programs via the NXT menu interface, you can also use the NBC compiler to create these simple scripts.

 WARNING: *These scripts are stored as RPG files. These files require a special system program called RPGReader.sys installed on the NXT brick. By default this will be the case but if you delete all the files off your NXT to make room for your own programs then the RPG programs will no longer execute. You can download a smaller version of the RPGReader.sys program from the NBC website.*

NBC also supports compiling NXT picture files with a .ric file extension using the RICScript programming language. RICScript is also a very simple language much like NPG. There aren't any conditional statements or looping constructs. RICScript supports nine RIC drawing functions. You can construct a RICScript program using those nine functions in any order and repeated as often as you like. We'll look more closely at RICScript and RIC files along with details about the supported drawing functions in Chapter 6 and Chapter 13.

We will be using NXC and NBC to write most of the programs found throughout this book. One of the major purposes behind this book is to fully examine these two programming languages along with their associated APIs and learn how to use them to maximize the power of the NXT brick. Before we get there you will need to install the tools that we'll be using throughout the rest of the book.

NXC & NBC Quick Start

We'll use two different methods for compiling programs written in NBC and NXC. The more powerful of the two is to use the integrated development environment (IDE) called BricxCC. An IDE is a programming tool that combines or integrates a full-featured editor with a compiler and other tools that simplify the process of developing programs. The BricxCC IDE will be described in more detail in Chapter 4. This option is only available for Windows so if you are using another operating system you will use the NBC compiler directly from a command shell or terminal window. Each of these methods is described below.

Installing and Using BricxCC

Before you can install BricxCC you will need to download the latest version from the website below.

 WEBSITE: http://bricxcc.sourceforge.net/

The download link is in the Downloading section of the BricxCC page and is labeled "BricxCC latest version". The steps outlined below should get you up and running.

1. Run the BricxCC setup executable after downloading it to your computer. Select the default values and everything will be installed automatically.
2. Turn on your NXT brick and connect it to your computer using a USB cable.
3. Run BricxCC from the Start menu. You'll be prompted to select a brick type and a port.
4. Choose NXT as the brick type.
5. Choose "usb" as the port.
6. Click OK or press Enter on your keyboard.
7. If everything works properly, BricxCC should successfully connect to the NXT.

Trouble Shooting:

If BricxCC tells you it is unable to connect to the NXT there are a couple of things you should double check. First, make sure the Fantom drivers mentioned above are properly installed and that the NXT is recognized as a LEGO Device rather than an unknown USB device. The NXT Fantom Driver Help document on the web will help make sure the required driver is installed correctly.

WEBSITE: *http://bricxcc.sourceforge.net/NXTFantomDriverHelp.pdf*

Also make sure that you selected the right brick type (NXT) and that the Firmware selection is set to Standard. Drop down the list of known ports and check whether the list contains an entry with your NXT's brick name in it. If you haven't changed your brick's name then it will default to NXT. Try selecting that port rather than "usb" if you still have problems connecting. You may also want to try selecting the long entry that ends with "RAW". It should look something like the following:

```
USB0::0X0694::0X0002::00165301E2CB::RAW
```

If BricxCC successfully connected to the NXT then select File|New or press the New file button (or Ctrl+N) to create a new empty editor window. Enter the following source code text:

```
thread main
  TextOut(0, LCD_LINE1, 'Hello World')
  wait 10000
endt
```

Save this program as "helloworld.nbc" using the File|Save menu or by pressing the Save file button (or Ctrl+S). The .nbc file extension tells the NBC compiler that it should interpret the source code as NeXT Byte Codes rather than Not eXactly C code or some other programming language supported within BricxCC.

Compile the code by clicking the Compile button, by selecting the Compile|Compile menu item, or by pressing F5. To download it, use F6, Compile|Download, or press the Download button on the BricxCC toolbar. Now run the program with F7, Compile|Run, or press the Run button on the toolbar. You should see something like the image shown in Figure 1-1 on the NXT screen.

Figure 1-1. Helloworld program running on the NXT

7

BricxCC can either execute the NBC compiler as a separate process using the stand-alone NBC executable or it can use a built-in version of the compiler. This option is user configurable via the Compiler|NBC/NXC tab in the BricxCC Preferences dialog. If you intend to connect to your NXT using Bluetooth then you will definitely want to use the internal NBC compiler from within BricxCC. This is because only one process can communicate with the NXT brick via Bluetooth at any given moment, so BricxCC has to disconnect and reconnect when it launches the NBC external compiler in order for the external compiler process to successfully make its own connection to the NXT via Bluetooth.

The delay caused by the two processes (BricxCC and NBC) connecting and disconnecting the Bluetooth link to the NXT makes compiling and downloading via the external compiler far too slow when you are connected via Bluetooth. If you are connecting to the NXT using a USB connection then either the internal compiler or the external compiler will be extremely fast so it really doesn't matter which option you choose.

If you plan on only using the BricxCC IDE then you can skip the remaining sub-sections that describe the command-line compiler method. You can skip to the Starting to Program section or you can read on and learn how to use the compiler from a command prompt in case you ever need to do so.

Installing and Using NBC for Windows

To install the command line NBC compiler, just extract the files from the latest NBC release zip that you can download via the link below.

WEBSITE: http://bricxcc.sourceforge.net/nbc/

Now open a command prompt window by selecting Start|Run, typing "cmd" into the edit window, and pressing OK. Change to the directory where you extracted the contents of the NBC zip.

```
c:
cd \nbc
```

Use notepad to create an NXC program by typing the following command from the prompt. Answer yes when you are asked if you want to create the file. Since this file has a .nxc file extension we will be writing NXC code rather than NBC code.

```
notepad helloworld.nxc
```

In notepad, enter the short program as written below and save it using Ctrl+S or File|Save. Then exit notepad.

```
task main()
{
  TextOut(0, LCD_LINE1, "Hello World");
  Wait(10000);
}
```

To compile and download this program, enter the following command:

```
nbc -S=usb -d helloworld.nxc
```

You should hear the NXT make a short beep indicating that the program downloaded successfully. You can now run this program by navigating through the NXT menu system using the orange enter button and the left and right arrow buttons as shown in the screenshots in Figure 1-2. After you drill down into the Software files section, use the left and right buttons to scroll through the various programs on the NXT and select the helloworld program. When it is active, press the orange enter button twice to select it and run it. You can abort a running program by pressing the dark grey escape key.

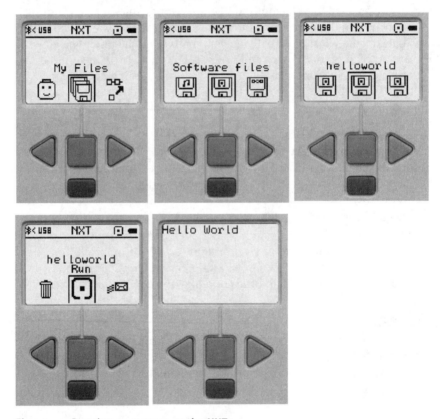

Figure 1-2. Running a program on the NXT

Installing and Using NBC for Mac OS X

To install the command line NBC compiler for Mac OSX, follow the instructions listed below. You should be familiar with using Finder to create directories and move files from one location to another. You will need to be familiar with extracting files from a compressed archive file with a .tgz file extension. If you want to put the compiler in a different location than the one described below, you will have to modify the application paths appropriately.

1. Using Safari or the browser of your choice, download the latest NBC release from the NBC website. The latest release will be described on the main welcome page here: http://bricxcc.sourceforge.net/nbc/

2. If the browser doesn't automatically extract the files from the archive for you, use Finder to locate the archive and double click it to extract its contents. The compiler, the nxtcom download utility written by Dave Baum, three Applescript scripts (saved as applications), and three text files should wind up in an "nxt" folder – probably on your desktop.

3. Use Finder to create a new folder in the Applications folder that is called Mindstorms. Open that folder and drag the nxt folder from your desktop into the Mindstorms folder. If the folder is copied successfully to the Mindstorms folder then you can delete the nxt folder off your desktop.

4. If you chose to use a directory other than /Applications/ Mindstorms/nxt then you will need to edit the Applescript scripts and modify the paths that are used in those files. They need to point to the correct location of nbc and nxtcom.

5. To make compiling and downloading via USB simple, you can drag the NBC Compile and Download via USB Applescript to the Dock. Once you have done that you will be able to just drag and drop any NBC, NXC, NPG, or RICScript program to the script in your Dock and it will compile the file and download the resulting program to your NXT.

Now you should have the compiler and a nice download utility installed and ready to use. Open a text editor to create an NXC program. Enter the short program as written below and save it to a file called helloworld.nxc. Make sure the file is saved in plain text format. Then close your editor program.

```
task main()
{
  TextOut(0, LCD_LINE1, "Hello World");
  Wait(10000);
}
```

To compile this program and download it to the NXT just drag it to the Applescript we put in your Dock. This will compile your program and transfer it to your NXT as helloworld.rxe.

You can also execute these commands from a terminal window if you are familiar with that technique. For those who prefer a command prompt you can start by opening a terminal window and entering the commands below. Be sure to substitute your actual sourcecode path in the first command. The lines may have wrapped here but make sure you enter each of the four commands on its own line.

```
cd ~/sourcecode/path
/Applications/Mindstorms/nxt/nbc -Z1 helloworld.nxc -
O=helloworld.rxe
/Applications/Mindstorms/nxt/nxtcom helloworld.rxe
rm helloworld.rxe
```

On the Macintosh platform you also have the option of running the LEGO MINDSTORMS NXT software to download the compiled files to the NXT. Alternatively, if you are running OSX 10.4 you can install the NXTBrowser utility, which supports Bluetooth communication on both Power PC and Intel Macs.

WEBSITE: http://web.mac.com/carstenm/iWeb/Lego/NXT/
F2F73940-D837-4038-9011-2968725A2872_files/NXTBrowser.dmg

Installing and Using NBC on Linux

To install the command line NBC compiler on an i386 Intel platform, just extract the files from the latest NBC i386 Linux binary compressed tar archive that you can download from the official NBC website:

http://bricxcc.sourceforge.net/nbc/

Now open a terminal window. Change to the directory where you extracted the contents of the NBC archive.

```
cd ..\nbc
```

Use a text editor to create an NXC program. Enter the short program as written below and save it to a file called helloworld.nxc. Then close your editor program.

```
task main()
{
  TextOut(0, LCD_LINE1, "Hello World");
  Wait(10000);
}
```

To compile this program, enter the following command.

```
nbc helloworld.nxc —O=helloworld.rxe
```

This will compile your program and save the executable output to a file in the same directory with the name helloworld.rxe. On Linux platforms you can use linxt to download your programs to the NXT brick after you have compiled.

Starting to Program

Now we will write a more complex but still very simple program using both NXT-G and our text-based programming languages, NXC and NBC. This will allow us to compare the languages in terms of source code size and executable file size. The desired outcome of the program is to run forever with the motor attached to output A turning forward continuously except while the touch sensor attached to input 1 is pressed. While the touch sensor is pressed the motor should reverse its direction. Before we write the program, attach a motor to output A and attach the touch sensor to input 1.

In NXT-G we can drag and drop a few icons onto the grid to write a simple program that controls a single motor based on the input from a single touch sensor. The screenshot in Figure 1-3 shows one way you can write the program using NXT-G so that the motor does exactly what we want it to do as we press and release the touch sensor.

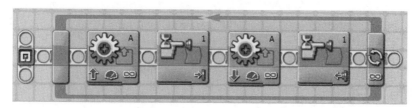

Figure 1-3. Starting to program in NXT-G

Save this to a file called start.rbt. It should save to a file that is about 321 kilobytes in its binary source form. Then compile and download the program to the NXT using the download button (that looks like a down arrow) on the controller button panel as shown in Figure 1-4.

Figure 1-4. The NXT software's controller buttons

The compiled executable on the NXT (start.rxe) is 5628 bytes. If you use the new Motor Mini blocks instead it compiles to 3298 bytes.

On the NXT, press the orange enter button four times to drill down into the menu system and execute the start program. Alternatively, you can press the download and run button on the controller, which is the button in the center with a triangle on it.

Let's try to write the same simple program in both NBC and NXC to see how it compares. Both NBC and NXC use a set of API functions similar to those included in NQC for controlling motors and sensors. We'll try to use those in both languages so that the resulting program is similar to what you might see in NQC.

First, let's see the contents of the text file that we can create using BricxCC or any other text editor we choose. This code is NBC source so make sure you save the file as "start.nbc".

```
thread main
  dseg segment
    sensorValue word
  dseg ends

  // configure sensor 1 as a touch sensor
  SetSensorTouch(IN_1)

  Forever:
    OnFwd(OUT_A, 75)
    untilPressed:
      ReadSensor(IN_1, sensorValue)
    brtst NEQ, untilPressed, sensorValue
    OnRev(OUT_A, 75)
    untilReleased:
      ReadSensor(IN_1, sensorValue)
    brtst EQ, untilReleased, sensorValue
  jmp Forever
endt
```

The plain text source code is only 439 bytes long. It compiles into an executable that is 288 bytes long.

Now we can write the same program in NXC. Save this text to a file called "start.nxc".

```
task main()
{
  // configure sensor 1 as a touch sensor
  SetSensorTouch(IN_1);

  while(true)
  {
    OnFwd(OUT_A, 75);
    until(SENSOR_1);
    OnRev(OUT_A, 75);
```

```
    until(!SENSOR_1);
  }
}
```

This source code file is only 233 bytes long and it compiles into an executable that is just a little bigger than the NBC program: 354 bytes. Table 1-1 contains a summary of the resulting source code size and compiled executable size of the three programs that we just wrote.

Programming Language	Source Size	Executable Size
NXT-G	321 kilobytes	5628 bytes
NBC	439 bytes	238 bytes
NXC	233 bytes	354 bytes

Table 1-1. Program sizes

As this table suggests, programs written using a text-based programming language such as NBC and NXC will be much smaller, both in the size of the source code and in the size of the executable program that you download to the NXT. The 321 kilobytes of NXT-G source code also can only be viewed from within the LEGO MINDSTORMS NXT software or as pictures stored in some image format such as JPG or PNG, whereas the much smaller text files for languages such as NBC and NXC are easily shared using any sort of text editor or viewer. The code can even be presented on the web in HTML format with syntax highlighting if you want to.

The smaller size of the resulting executables means you can fit many more of your programs in the NXT flash memory, making them more readily accessible for execution. But even better than the benefit of fitting more files on your NXT is the benefit that is implied by the size of the executable. Programs that you create using the NBC compiler are smaller because the executable code in the compiled files is much more optimal than the code in a comparable program written using NXT-G. The more optimal your executable code, the faster your program will run on the NXT. You'll find that programs written in NBC or NXC will often be as much as *ten times* faster than the same program written using the standard NXT-G programming language.

To make sure you get the smallest and fastest executable code when compiling NBC and NXC programs, you should enable compiler optimizations. This will help keep the executable file size smaller and it will help improve the runtime speed of your programs by removing unneeded variables and unused code. You can enable optimization by adding -Z1 to the NBC command line or via the Compiler|NBC/NXC|Optimization level option on the BricxCC Preferences dialog as shown in Figure 1-5. You can edit the BricxCC preferences via the Edit|Preferences menu item.

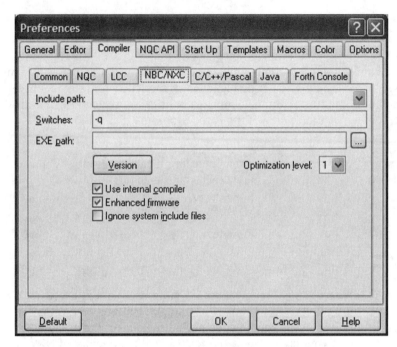

Figure 1-5. Setting the optimization level for NBC and NXC

You may also have noticed in the above programs written in NBC and NXC that even though NBC is lacking in the high level programming constructs such as the while-loop and the until-loop that are supported in NXC, the bulk of the NBC program is very similar to the NXC program. This is because the APIs for both languages are designed to be nearly identical. The similarity between the programming APIs should make it very easy to transition between the two languages in case you need to squeeze a little more speed out of your NXT programs.

The NXT Hardware

Topics in this Chapter

- Introducing the Hardware
- Moving About
- Sensing Surroundings

Chapter **2**

Before we dive deeper into programming, let's first take a closer look at that large white and grey brick we found in our NXT set. What are some of the similarities between the NXT generation of LEGO MINDSTORMS and some of the previous programmable bricks produced by LEGO, such as the RCX, the Scout, or the Cybermaster? What are the important differences between the NXT and its ancestors? In this chapter we will try to answer these questions.

You want to know how to access the new hardware and take advantage of the many new features and functions that it provides. To that end, we will investigate the NXT along with the LCD screen, the sound system, the new servo motors, and the sensors.

Introducing the Hardware

The NXT brick represents a major improvement over previous generations of LEGO MINDSTORMS programmable bricks. The RCX brick that came with the Robotics Invention System (RIS) used an 8-bit Hitachi microprocessor running at 16 MHz and it only had 32 kilobytes of RAM. If you ever removed the batteries from the RCX, the memory would lose not only all of the programs (up to 5) that you had downloaded to the brick but also the brick's operating system.

In comparison, the NXT uses a 32-bit ARM7 Atmel microprocessor running at 48 MHz (see Figure 2-1) with 256 kilobytes of non-volatile flash memory as well as 64 kilobytes of RAM. You can download as

Figure 2-1. Atmel AT91SAM7S chips

many as 64 files to the NXT brick's flash memory. And whether you have batteries in your NXT or not, both the brick's operating system and your programs will remain safely stored in the flash RAM waiting to be used again once power is restored to the brick.

Not only does the NXT have a far more powerful central processing unit (CPU) at its core but it also includes an 8-bit AVR microprocessor. This chip comes with four kilobytes of flash memory and 512 bytes of RAM. It is used by the NXT to control the new interactive servo motors that come with the NXT set.

LCD Display

A major improvement from the RCX to the NXT is the LCD display. With the RCX you had an LCD that used specific segments or images that you could turn on or off. The RCX firmware provided very limited control over exactly what was displayed on the tiny LCD screen. All that has changed in the NXT.

The NXT has a large 100 pixel wide by 64 pixel tall LCD. Each pixel can be individually set or cleared. It has a very fast 17 millisecond refresh cycle, which means you can display animations and write programs such as Chess and Space Invaders (see Figure 2-2). While the LCD only supports setting or clearing pixels, giving only black and white as available colors, you can still do some extraordinary things on the screen.

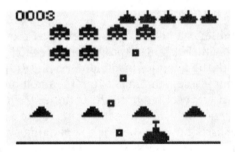

Figure 2-2. Space Invaders on the NXT

As soon as you turn on the NXT, it demonstrates some graphical tricks that you can do in your own programs. The startup sequence is easily replicated using the drawing functions provided by the standard NXT firmware. We'll explore all of the capabilities of the LCD in several upcoming chapters.

 TRY ME: *Try downloading the Space Invaders game and running it on your NXT brick to see the LCD screen in action. The code is in the NBC samples area of the NXC/NBC website.* http://bricxcc.sourceforge.net

Sounds

Previous generations of LEGO MINDSTORMS bricks included support for playing music. A song was built using a series of sine-wave tones that a programmer could control by specifying each tone's frequency and duration. It was rudimentary but it worked.

Shortly after the RCX was released, programmers created various utilities that let you generate all the required NQC code to play a song. You could convert simple MIDI files into somewhat large NQC programs if you wanted to do so. BricxCC, which we will examine closely in chapter 5, comes with a Brick Piano tool that lets you compose your own simple songs for the RCX using a virtual piano keyboard.

The NXT, not surprisingly, also includes support for simple tone generation. As with the RCX there are utilities, including BricxCC, which provide the MIDI file conversion and virtual piano keyboards for the NXT. But LEGO didn't stop there when they designed the NXT.

One of the great new features in the NXT is support for playing sound files that contain standard pulse-code modulation (PCM) samples. PCM is a method for representing an analog waveform digitally by uniformly sampling or recording the magnitude of the analog signal. Included with the NXT software is a huge collection of NXT sound files that are easy to incorporate into your own programs using NXT-G, NXC or NBC.

Sound files supported by the NXT can only use a single channel, not stereo. The samples are limited to 8-bit, with a sampling frequency anywhere from 2k samples per second to 16k samples per second. The files included in the retail set all feature an 8k sampling frequency. Digital audio files with higher sampling rates have a higher sound quality than files using lower sampling rates, so keep that in mind if you want to create your own sound files. However, a higher sampling frequency produces a larger file.

To help solve the problem of large sound files consuming all the flash memory on the NXT, LEGO has added support in version 1.5 of the standard firmware for a form of digital audio compression known as IMA ADPCM. This audio format gets its name from the International Multimedia Association (IMA), which suggested the algorithm. The AD in ADPCM stands for "adaptive differential", which is the technique used to compress the original waveform samples. The initial sample value is left intact but subsequent samples are replaced by 4-bit values that store the difference between the last sample and the current sample. Two 4-bit values are stored in a single byte, so a file containing 8-bit samples is half as large using ADPCM as using uncompressed PCM samples.

In chapter 6 we'll look at the utilities available for making your own sound files. This will include tools for generating NXC and NBC source code for playing simple tones. Another hidden gem in the NXT brick's support for playing sounds will also be revealed and explored in that chapter. Several other chapters will include programs that show off the various types of sound playback supported by the NXT.

Communication

In the RCX days, if you wanted to communicate with the brick from a computer or another programmable brick you had to rely on a very slow infra-red (IR) transfer protocol. It was wireless so you didn't have to plug your robot into the computer but you did have to hook an IR transmission tower to your computer either via a serial port or a USB connection. And you had to make sure that the RCX IR receiver was lined up nicely with the IR transmitter attached to the computer. You also had to worry about too much ambient light interfering with the transmission. Other robots would have to be turned off or their IR receiver covered to prevent accidentally downloading a program to the wrong brick.

With the NXT, you no longer have to worry about ambient light and you don't need to place your robot in close proximity to a proprietary IR tower. The NXT supports either wireless communication via Bluetooth at a very fast 460.8 Kbits per second or a super fast (12 Mbits per second) wired USB communication using a standard USB cable that comes with the set. That's right, your NXT brick has a built-in standard USB 2.0 connection just like the one you might find at the back of your printer.

Wireless Bluetooth communication is very popular in cell phones, with people walking around using Bluetooth headsets stuck in their ears. Your PC may not have built-in Bluetooth capability but you can find very inexpensive Bluetooth dongles at any computer store or over the Internet (see Figure 2-3).

Figure 2-3. A Bluetooth dongle

Check the LEGO NXT compatibility list to make sure you get a Bluetooth dongle that will work with the LEGO MINDSTORMS NXT software. Other devices may work but they have not been certified by LEGO as being compatible.

WEBSITE: *Check the Bluetooth compatibility list at* http://mindstorms.lego.com/Overview/Bluetooth.aspx

While Bluetooth communication is not nearly as fast as wireless technologies such as 802.11n it is certainly a lot faster and better than the old RCX IR wireless option. Because Bluetooth connections between a PC and an NXT or between two NXT bricks are point-to-point, they will only ever affect the specific brick to which the program or message is sent. You no longer have to worry about accidentally overwriting a program on your competitor's robot when you download that last minute change at a FIRST LEGO League (FLL) competition. More importantly, you don't have to worry about someone else overwriting your own programs, either accidentally or maliciously.

The support for wireless Bluetooth communication built into the NXT is similar to the radio frequency (RF) wireless capability that the Cybermaster programmable brick had. With the Cybermaster you could download programs and send commands to the brick from your PC over radio waves, much like you would control a remote control car. But the Cybermaster had to have two long antennae screwed into it and you had to use a proprietary RF tower with yet another pair of long antennae attached.

With the NXT the Bluetooth transmitter and receiver are inside the brick so there is no need to attach an antenna at all. The type of Bluetooth device implemented in the NXT (class II) can send data transmissions as far as 10 meters. And because it is a popular industry standard protocol, either your computer already has Bluetooth capability like a Macintosh or it is cheap to add.

Remote Processing and Control

An important benefit provided by having a reliable wireless data transmission protocol built into the NXT is that you can actually off-load part of your robot's program to your computer, which is far more powerful than the computer in the NXT. By sending data back and forth between the NXT and a computer you can significantly increase the available memory, storage, and data processing capabilities to your program. Of course, this is not without a cost in speed since sending data over Bluetooth and receiving a response from a PC is not as fast as simply executing code on the brick itself. But in many types of robotic applications the benefits far outweigh the costs associated with shared processing between an NXT and a PC.

As long as your NXT is within the range limitations of the Bluetooth protocol, whether it is in the same room or around a corner, you can still pass messages back and forth either between the NXT and a computer or between two or more NXTs. LEGO even has a program that runs on a Bluetooth-enabled cell phone that gives you the power to control the NXT using your phone. This application is available for free from the LEGO MINDSTORMS website via the link below. It supports such phones by Nokia as the 6680 and 3230 models as well as phones by Siemens such as the CX75 and Sony Ericsson, including models W550i, K800i, and Z710i (see Figure 2-4).

Figure 2-4. Sony Ericsson K800i, Nokia 6680, and Siemens CX75

WEBSITE: *For more information about the LEGO NXT Mobile Application visit* http://mindstorms.lego.com/Overview/Mobile%20Application.aspx

Table 2-1 lists some other third-party programs that run on Bluetooth-enabled devices such as Pocket PCs or PALM PDAs.

Name	URL	Platform
Pocket NXT Remote	http://www.norgesgade14.dk/legoSider/ remote/pocket.zip	Pocket PC
OnBrick	http://www.pspwp.pwp.blueyonder.co.uk/ science/robotics/nxt/bin/onbrick_pda.zip	Pocket PC/ Windows Mobile 5
NXT Director	http://www.razix.com/nxtdirector.htm	Palm

Table 2-1. Bluetooth remote control applications

Batteries

The NXT requires 6 AA batteries as its power source, just like the RCX. Running your robot for very long – especially running the motors – will quickly eat up regular alkaline AA batteries. You should plan on investing in a number of high capacity rechargeable NIMH batteries and a rapid charger. Buying at least twelve of them will provide you with a set of 6 to use and a backup set for when the ones in the NXT run low. It's a good idea to mark each set of six in some fashion so that you always use the same set each time rather than mixing batteries that might not all be charged at the same level.

You can also buy a rechargeable lithium-ion battery pack, which fits into the space where you would normally put the six AA batteries. The rechargeable battery pack uses the same AC adapter that could be used with the RCX 1.0 brick and if your robot is stationary you can even run it while it is plugged into the AC adapter recharging the battery pack. The battery pack extends the depth of the NXT by a few millimeters so if you plan on using it you will need to account for that additional thickness in your robot.

One thing to keep in mind with either the rechargeable battery pack or rechargeable NIMH AA batteries is that your robot will be running on a lower voltage than that provided by regular alkaline AA batteries. This means that motors will turn a little slower when you are running on rechargeable batteries.

Moving About

A robot isn't much good without the ability to move things around, including itself. Robots live or die according to their motors. The RCX brick supported up to three motors connected to outputs A, B, and C. Similarly, the NXT also has only three output ports with the same port designations. But that is as far as the similarity goes.

Motors

The old RCX motors were powered via a two-wire connection using pulse-width modulation to vary the speed of the motor. The motors had built-in gearing to decrease the speed and increase the torque of the motors. The motor output was a short axle, whose length equaled the width of a single Technic brick (see Figure 2-5). If you needed to position the motors in a way that required a longer axle connection then you had to use an axle extender of some sort. And even though the RCX supported up to three motors, the Robotics Invention System (RIS) set only came with two. These old motors are commonly referred to as the Technic mini motor and it came in two very simil.. ....es: LEGO part 71427 and LEGO part 43362.

Figure 2-5. LEGO Technic mini motor

With the NXT set you get three of the new NXT interactive servo motors. These motors are much larger than the Technic mini motors. In terms of overall efficiency, they rate below the Technic mini motor. For a detailed comparison of motor characteristics you can visit Phillipe Hurbain's website where he has posted graphs and tables that detail just about anything you ever wanted to know about the LEGO electric motors.

WEBSITE: http://www.philohome.com/motors/motorcomp.htm

Even though the new NXT motors are larger and less efficient than the RCX motors, they far surpass their older cousins via other measures. This is due in part to their flexibility for connecting to your robot but mostly because of the built-in support for precise control that could not easily be obtained using the old mini motors. If you wanted to measure rotations of the mini motor you had to buy a separate rotation sensor and figure out how to build it into your model. The rotation sensor would use up one of the three available input ports. And you would need to repeat this process for each motor that needed to be measured.

The NXT motors rotate at a slower speed and with more torque due to the internal gearing – a drive train – that is built into the new motors (see Figure 2-6). This could mean you don't need to build your own drive train into your robot. The NXT motor also has a built-in rotation sensor that is far more precise than the RCX rotation sensor. This built-in sensor can measure 360 degrees of rotation per revolution. The rotation values are sent back to the NXT via the same cable and port to which the motor is connected.

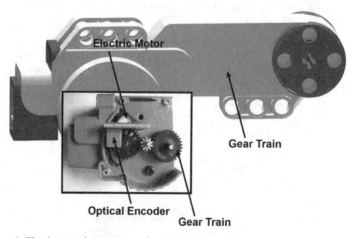

Figure 2-6. The internal structure of an NXT motor

A program running on the NXT can easily access these rotation counter values for use in any way you might imagine. This means that each motor is both an output that can be used to move or manipulate the robot's environment and an input that can provide feedback about the robot's motion and its surroundings to programs that you write. No more using up all your input ports just to measure motor rotations. And no more having to figure out how to insert a rotation sensor into an already complicated gear train.

Another aspect of the new NXT motors is the ability within
the standard NXT firmware to control motors in pairs, in order to
synchronize them. Once again, the built-in rotation sensors help make
this capability possible. Not only can you individually power and control
the motors but you can also write code that tells two motors to rotate
together at a certain speed. If one of them slows down due to some
external cause such as friction, the NXT firmware will automatically
slow down the paired motor. In the RCX you had to manually try to pair
up motors that ran consistently at the same speed given a specific power
level and after all that work if one motor was unexpectedly slowed for
some reason there was no easy way to compensate. With the new NXT
motors, driving in a straight line just got a whole lot easier.

 WEBSITE: *Check out Phillipe Hurbain's exploration of the internals of
the NXT motor at* http://www.philohome.com/nxtmotor/nxtmotor.htm

The built-in rotation sensors provided by the new NXT motors also
allow you to write code that tells the motors to rotate a certain number
of degrees either clockwise or counter-clockwise. With the RCX this
was simply not possible without writing a lot of extra code and even
then it was difficult to do accurately. But with the NXT motors and the
functionality exposed within the standard NXT firmware you have this
power built-in. The same sort of synchronization that was described
above is also available when you control the NXT motors by specifying
the number of degrees you want the motors to rotate.

Another aspect of the new NXT motor that is worth mentioning here
is the flexibility it provides for both anchoring it into your robot and for
connecting it to wheels or other mechanical devices that you want to
move or rotate. There are four separate points to which you can attach
Technic pins for mounting the NXT motor to your robot. Either working
together or individually, these four connection points provide plenty of
ways to fit the motor into your latest creation.

As mentioned earlier, the mini motor only had a short axle to which
you could attach a gear or some other type of axle connector. The NXT
motor provides a cross axle hole instead, which lets you pick the most
appropriate axle length. The axle hole extends completely through the
motor as well so you can connect to an axle from either side of the
motor or even both sides if needed. In addition to the axle hole, there
are four holes into which you can insert any standard Technic pin
connector. Since these holes are offset from the center of rotation you
can even use the motor to produce periodic or offset rotations.

Sensing Surroundings

The sensors included in the NXT set are designed to have a standard sleek appearance with a style and color scheme that matches that of the NXT itself. None of these sensors are small, however, which can make fitting them into tight locations a bit complicated. In comparison, the two touch sensors included in the RIS set were much smaller. They were the same size as a two by three brick. You will need to keep the larger size of sensors in mind when you are designing your robots.

Touching things

Even though the NXT touch sensor is substantially larger than the RCX touch sensor, it has some nice improvements over the design of the old version (see Figure 2-7). It has a cross-axle hole in the button at the front of the sensor, which provides a lot of flexibility for extending the sensor's button or attaching bumpers. It also has a much wider angle of press sensitivity.

Figure 2-7. NXT and RCX touch sensors

One issue with the retail NXT set is that it only comes with one touch sensor, whereas in previous LEGO MINDSTORMS sets you had two or even three touch sensors. When designing robots that interact with their environment it is always nice to have access to more than one touch sensor. The educational version of the NXT set actually comes with a second touch sensor. If you only have one you may want to visit the LEGO shop-at-home website and pick up another one.

WEBSITE: *You can purchase additional touch sensors at* http://shop.lego.com/product/?p = 9843

Hearing things

Unique to the NXT is the new sound sensor (see Figure 2-8). It allows your robots to listen to the surrounding environment and respond to noises that they hear. The sound sensor can be used in normal decibel mode or adjusted decibel mode.

Figure 2-8. NXT sound sensor

When using the sound sensor in adjusted (dBA) mode the sensor is limited to sounds within the frequency range that human ears can detect. In normal mode (dB) it will measure sound level intensity at a wider range of frequencies, including some either too low or too high for humans to hear. You might use this sensor to make your robot start moving anytime it senses a loud sound. Use your imagination!

Seeing things

The NXT light sensor works very much like the light sensor included with the RIS set. Physically it has the same size and shape as the previous two NXT sensors (see Figure 2-9). Its primary use is for detecting different levels of light reflectivity at close range. The NXT light sensor also corrects a problem that the RCX light sensor had by including a barrier between its sensor and its emitter.

Figure 2-9. NXT and RCX light sensors

This sensor uses a small light-emitting diode (LED) to shine a light on close objects along with a light sensor to measure how much light is reflected back by the object. Light surfaces will reflect more light than dark surfaces. This comes in handy when you are designing a robot that can follow a line on the floor. The test pad that comes with the NXT set can be used to see how well your line-following robot can handle a simple oval shape.

The light sensor can also be used with the LED disabled. In this mode it can be used to measure ambient or surrounding light levels. Having your robot measure ambient light can be used to wake it up when the sun comes out in the morning or when somebody turns on a nearby lamp. Or you could design a wandering robot that avoids shadowy areas in your living room.

Measuring distances

A fantastic new addition to the LEGO suite of sensors is the NXT ultrasonic sensor (see Figure 2-10). With the RCX, if you wanted to build a robot that avoided obstacles by sensing how close it was to them you had to buy a third-party sensor. Now this sensor is included in the NXT set.

The ultrasonic sensor uses sound waves like a bat. It sends out a high frequency sound wave ping and calculates the distance to an obstacle by timing how long it takes for the sound to bounce off the object and reflect back to the sensor.

Figure 2-10. Ultrasonic sensor

This sensor acts almost like a pair of eyes for your robot. It lets your robot see how far it is away from things that may be in its path. LEGO seems to have designed this sensor with that thought in mind. It looks a lot like a little robot head with two orange-rimmed eyeballs staring out at you.

One thing to keep in mind when using the ultrasonic sensor is that it uses sound waves. Since it relies on sound wave echoes returning to the sensor, the ultrasonic sensor will have a difficult time measuring distances to objects that have a curved surface or which absorb the sound wave energy, such as a pillow. This sensor will work best for detecting and accurately measuring distances to larger objects with hard surfaces that aren't at sharp angles away from the robot. It will perform well with the walls of a maze.

Another potential pitfall related to using the ultrasonic sensor is that other robots using ultrasonic sensors in the same room can cause interference with each other. Your robot might detect the ping sent out by another robot with the end result being that the sensor reports erroneous distance readings. Fortunately, the ultrasonic sensor has special modes of operation that help to diffuse this situation.

Lowspeed Sensors

The NXT comes with support for a new breed of sensor – the digital I²C sensor. I²C is a standard serial bus designed to allow multiple slave devices to connect to a master device. The name stands for inter-integrated circuit. The ultrasonic sensor that comes with the NXT set is an I²C sensor.

All four of the NXT's input ports support the I²C protocol. Also, it is theoretically possible to hook multiple I²C sensors to the same input port. In that situation all the port-sharing sensors would need to work together in compliance with the NXT's power consumption restrictions and I²C bus termination requirements.

The NXT firmware uses the term "lowspeed" to refer to this type of sensor. This doesn't mean that the sensors are particularly slow. It just means that they use the low-speed I²C bus standard, which runs on the NXT at 9600 bps (bits per second).

Figure 2-11. Third-party I2C sensors from HiTechnic

One fantastic aspect of the NXT's support for the I²C standard is that it is very easy to implement in many different types of sensors (see Figure 2-11). Even before the NXT was publicly available, a number of third-party I²C sensors were under development. You can find compass sensors that tell your robot its orientation with respect to magnetic north, acceleration sensors that report velocity rates of change for three axis, pressure sensors that you can use in conjunction with LEGO pneumatic pumps and cylinders, color sensors which can read RGB color values and infra-red sensors that are like those built into the RCX.

WEBSITE: *For several interesting home-brew I²C projects check out Sivan Toledo's website:* www.cs.tau.ac.il/~stoledo/lego/

Companies such as HiTechnic and MindSensors offer a wide range of different NXT-compatible I²C sensors. You can even buy HiTechnic sensors off the official LEGO shop-at-home website. The HiTechnic sensors are enclosed in the official LEGO plastic containers, so they nicely fit the standard NXT look-and-feel. MindSensors produces I²C sensors with a less polished look. The circuitry in all its raw glory is not wrapped in a fancy plastic shell. As a result their prices seem to be a little lower and they currently have a wider range of selection than other vendors.

I recommend you check out the sensors from both companies. Both have high quality products and great customer support. Why not pick out a sensor that can add a spark to your robot's life by incorporating a fancy new way for it to sense its surroundings?

Mixing old and new

One more thing that needs to be mentioned about using sensors with the NXT is that you had better not get rid of your RCX sensors just yet. They all can work fine with the NXT. You may need to purchase a few RCX sensor converter cables before you can use them, though (see Figure 2-12).

The educational version of the NXT set includes three of these cables. You can also buy them in packs of three from the LEGO shop-at-home website. And for the more adventurous among us, there are instructions for making your own converter cables on Phillipe Hurbain's website.

Figure 2-12. Converter cable

 WEBSITE: *You can make your own converter cables by following the instructions found here:* www.philohome.com/nxtcables/nxtcable.htm

Now that we have examined the NXT hardware it's time to shift our focus to the OS on the NXT: the NXT firmware. The next chapter will give you an overview of the clever menu interface built into the NXT brick.

The NXT Firmware

Topics in this Chapter

- The NXT Menu Interface
- Introducing the Firmware

Chapter 3

Having examined the NXT hardware in the previous chapter, we can now explore the low-level firmware that exposes all this new power for you to effortlessly use in your own code. It has been said that a little knowledge is a dangerous thing. In this chapter you will receive a lot of knowledge. However, given that the NXT brick uses six AA batteries, the danger is probably minimal.

The Menu Interface

Interacting with the NXT is not limited to just writing programs and running them on the brick. The NXT firmware includes an extensive menu system that lets you control many aspects of the NXT without needing to write code. We'll take a quick look at some of the things you can do with your NXT by using the firmware's built-in menus.

You navigate the NXT menus using the four buttons on the brick. The left and right arrow keys let you scroll through the current menu level to select the desired option. Select an option by centering it on the LCD screen using the left and right buttons. At any point you can back up a level using the dark gray exit button. Similarly, you can dive down a level into the selected option by using the orange enter button.

When you first turn on the NXT using the enter button, My Files is selected. At the top of the LCD screen are several icons that tell you about the state of the NXT. See the NXT user guide on page 10 for pictures of these icons.

To the far left is the Bluetooth icon, which can have four different states: blank, meaning that Bluetooth support is turned off, Bluetooth on, Bluetooth on and visible to other devices, and Bluetooth on and connected to another device. To the right of the Bluetooth icon is the USB icon. This icon has three states: blank, meaning the USB cable is not connected to a computer, USB connected and working, and USB connected but not working.

Centered at the top of the LCD is the name of your NXT brick, which defaults to NXT. You can change the name of your NXT to any eight-character string via BricxCC's NeXT Screen tool or by using the LEGO MINDSTORMS NXT software. Your brick's name is an important part of making Bluetooth connections to other NXT bricks.

To the right of the brick name are two icons. The first is the running icon, which appears to spin around in circles while the NXT is running. The last icon is the battery level indicator. As your batteries run down, this icon will progressively change from looking completely full to looking more and more empty. Eventually the icon will start flashing on and off. If it gets low enough you will see a popup screen showing a low-battery warning message. If you ever see that popup screen, it is time for you to change your batteries. Maybe now would be a good time to go buy some NiMH rechargeable AA batteries.

At the top menu level you can scroll left and right through six options: My Files, NXT Program, View, Bluetooth, Settings, and Try Me. We'll have a quick look at each of these menu options in the sub-sections below.

My Files

This section of the NXT menu system is probably where you will spend most of your time either directly or via an IDE such as BricxCC. If you drill down into My Files you will find three menu options. They are Software files, NXT files, and Sound files.

If you have any sound files on the NXT they will be listed under the Sound files menu option. You can drill down into that section and scroll through all the sound files on the NXT. Only .rso sound files will be displayed, not melody (.rmd) files. Select a sound file to run it (which actually plays the sound file), delete it, or send it to another device by drilling down another level to the file options menu.

The NXT files section lists all the on-brick or NXT programs that you have saved to the brick from within the NXT Programs main menu option. These files have an .rpg file extension. They are simple five step programs that you can create using the NXT Programs menu options. You can also create them using the very simple NPG programming language with the NBC compiler from within BricxCC or from a command prompt. Scroll through the list of NXT files to select the one you want, and run it, delete it, or send it via Bluetooth to a connected device.

All the programs you write using NBC and NXC will be listed in the Software files section of this menu. As with the other sections, you drill down into this file section and select the program by scrolling left or right through all the programs on the NXT. Once you have selected a program and drilled down to the file options level, you have the same three options as previously mentioned. Most of the time when we drill down to this level, it is to execute or run the selected program. Of course, from within BricxCC you can run a program you have open in an editor window by a single mouse click or keyboard hot key. If the program has not yet been downloaded then Ctrl+F5 will compile, download, and run the program. If the program has already been downloaded then F7 will run it on the NXT.

NXT Programs

The NXT Programs menu option is a neat way to create very simple scripts that you can run on the NXT to control motors and respond to sensor input. Drill down into this option and give it a try. The user guide contains more detailed information about how to use this section of the menu system to write NXT programs. You can find it on pages 14 and 15 of the user guide.

As previously mentioned, these programs can also be created using the simple NPG programming language as a text-based alternative to the on-brick menu-based mechanism. The NPG approach will be discussed further in chapter 7. It provides more flexibility than the menu system does and you can actually read the programs after you have written them, which you can't do with the NXT programs created using the menu system.

View

The View menu option is a handy feature. It lets you check sensor values without executing a program. You can use this section to verify light sensor readings, for example, or check sound levels.

When you select this menu option you will see a list of sensor types that you can scroll through using the left and right buttons. After selecting the right sensor type, drill down using the enter key and then scroll through a list of ports. Once you drill down into the correct port you will see the sensor value or motor rotation value shown on the LCD screen.

Bluetooth

There are several options within the Bluetooth menu. They are all described in detail on pages 21 and 34-37 in the user guide. From this menu option you can turn on or off the Bluetooth support. If you aren't using Bluetooth, turning it off will save battery power.

Once you have Bluetooth support turned on you have several other options to choose from. You can start a search for other Bluetooth devices. You can select My contacts to scroll through a list of known devices. From the Connections option you can scroll through a list of connections that are currently established with other Bluetooth devices. And within the Visibility option you can set whether other devices can find your brick.

Settings

The Settings menu provides a few basic configuration options and allows you to view version information about your NXT hardware and firmware. The options within this section are Volume, Sleep, NXT Version, and Delete files.

The Volume option lets you set the main volume level from off (0) to full volume (4). The Sleep option lets you set the sleep timer to a value from never to 60 minutes. In the NXT Version selection you will see five values listed on the screen. These show the firmware version, the AVR and BC4 versions, the build number, and your brick's unique serial number (ID). The firmware version can be changed if you install a new firmware release when an updated firmware is published. The other values are essentially part of the NXT hardware and cannot be upgraded.

The Delete files option allows you to delete from your NXT all files of four different categories: Software files (.rxe), NXT files (.rpg), Sound files (.rso), and Try me files (.rtm). If you drill down into one of these four file categories you will be asked if you really want to delete all the files of the selected type. If you really do, scroll left to the checkmark icon and select it with the enter key.

Try Me

The standard NXT firmware comes with several files pre-installed on the NXT. Each time you reload this firmware those files will be restored to the brick. To free additional flash memory space for your own programs you can delete those files. The files listed in the Try Me menu are just regular Software files except they have a .rtm file extension rather than a .rxe file extension. Internally they are exactly the same. If you want to you can compile your NBC and NXC programs with a .rtm file extension. This causes them to show up under the Try Me menu rather than My Files.

The Try Me programs that come with the NXT firmware expect that you have attached sensors and motors as described on page 7 of the user guide. Each program demonstrates some of the features provided by the sensors and motors as well as the NXT sound system and LCD screen. When you select a Try Me program and drill down to the file options level, you can only run the program or delete it. The send file option is not available for .rtm programs.

Introducing the Firmware

The NXT firmware does a good job of giving programmers the tools to take advantage of all the power that the hardware can provide. It's not perfect, but it gets the job done.

Are there any similarities between the RCX and NXT firmware? There really is no comparison. For the NXT, LEGO worked with National Instruments to develop a completely new firmware from the ground up. So the NXT firmware has little in common with the firmware used in previous generations of LEGO MINDSTORMS. That's both a good thing and a bad thing.

The downside is that there is a large body of knowledge already in the LEGO robotics community centered on the RCX firmware. Since the NXT firmware is so different, a lot of that knowledge has had to be built up from nothing all over again.

Fortunately, LEGO has done a good job of publishing software development kits (SDKs) and documentation about the way the NXT firmware works. Best of all, LEGO has actually released all the source code of the standard NXT firmware as open source. The LEGO Group's attitude of openness and cooperation with members of the LEGO robotics community has been wonderful.

 WEBSITE: *For access to the NXT firmware source code, as well as SDK documentation, visit the NXT'reme section of the LEGO MINDSTORMS website at* http://mindstorms.lego.com/Overview/NXTreme.aspx

For more information about the firmware modules and their IOMap structures, see Appendix C. Now that we've learned about the NXT firmware we will turn our attention to some building basics. The next chapter focuses on building our ability to build robots without using bricks.

Building Without Bricks

Topics in this Chapter

- TECHNIC Bricks
- Units
- Beams
- Making the Connection
- Gears
- Online Resources

Chapter 4

LEGO has gradually changed the building style in all of their TECHNIC sets. The MINDSTORMS line, including the NXT set, is no exception to this shift in building styles. When the RIS set debuted in 1998, the definition of a TECHNIC beam was very different. Before focusing on the current brick-free building boom, let's review the old way for a moment.

TECHNIC Bricks

Back when the RCX first came out, beams, like all forms of bricks in the world of LEGO, had studs on them. A one by four TECHNIC beam was the same as a one by four brick, except that it also had 3 holes through the side of the brick (see Figure 4-1). You could stack bricks and plates on beams and, as long as you followed the standard LEGO dimensional rules, everything worked.

Figure 4-1. TECHNIC bricks and TECHNIC beams

One of the key rules of "old" TECHNIC was that three plates equal one brick. In other words, the height of one LEGO brick or TECHNIC beam was the same as three plates stacked on top of each other. Another basic rule was that the height to width ratio of a one unit wide brick or TECHNIC beam was 6/5 or 1.2. The height to width ratio of a plate is 2/5 or 0.4. One more basic rule worth mentioning is that if you stacked beams together the holes would not be whole unit distances apart vertically.

The distance between the center of a hole in one TECHNIC beam and the hole in a beam stacked directly on top of it was equal to the height of the beam, whereas the horizontal distance between two holes in a TECHNIC beam was (and still is today) equal to the width of a brick. This meant that to get a whole unit distance between two holes in a vertical structure built out of TECHNIC beams you needed a mixture of beams and plates (see Figure 4-2). One beam plus two plates plus a second beam would produce the desired whole unit vertical distance $(1.2 + 0.4 + 0.4 = 2.0)$.

Figure 4-2. Vertical distances with TECHNIC bricks

Today these LEGO parts are no longer called beams. They are simply called TECHNIC bricks and that is really what they are – bricks with holes in the side. And as with other studded elements these bricks are becoming harder to find in new TECHNIC sets.

There are both good and bad aspects to the design of the TECHNIC brick. Bricks are more complicated to make than what we now call TECHNIC beams. The odd 1.2 height to width ratio can make construction difficult at times. And the way that holes are offset from the studs of a brick makes it so that you have to skip a hole when you place two TECHNIC bricks end to end.

On the other hand, stacking bricks together is about as simple as building with LEGO can get. The interface with non-TECHNIC bricks and plates was obvious and simple. Finally, the rectangular nature of bricks meshed very well with the concept of building and reinforcing frame structures. There's something about setting a brick on its end and having it stand there without any support that helps you think orthogonally.

Of course, these old rules still apply to building LEGO MINDSTORMS creations. You can still use TECHNIC bricks and regular LEGO bricks. LEGO elements work the same as they always have. You just won't find very many studs in the NXT set and very few TECHNIC bricks. If you want to build with bricks you'll have to use other sources to obtain these parts.

Units

Now that we are building without bricks, we need to learn the new rules regarding units of measure in a stud-free world. The nice thing is that the new rule is *width equals height*. A beam is as wide as it is tall. If you stack a beam horizontally on top of another beam the vertical distance between two holes is one unit.

When measuring beams, instead of trying to remember whether to count studs or holes like we had to with TECHNIC bricks, all we need to remember now is to count holes. On the LEGO website you can visit the TECHNIC design school for fantastic tutorials about building with TECHNIC parts. The TECHNIC 101 lesson is a great place to start.

WEBSITE: *Try out the tutorials in the TECHNIC Design School at* http://technic.lego.com/technicdesignschool/default.asp

As described in the TECHNIC design school tutorials, the basic unit of length is known as a module and the capital letter M is used to refer to those units. A beam that is 3M long has 3 holes. While M refers to the distance between the centers of two holes in a beam it also happens to be the width of a one unit wide brick as well as all beams. Since beams are as tall as they are wide, a beam is also 1M tall. And the studs on bricks and plates are 1M from center to center.

Of course, not all measurements need to be strictly horizontal or vertical. That's when knowing your geometry is very useful. You'll come to love the Pythagorean Theorem. It describes the relationship between the length of the hypotenuse of a right triangle and the length of the other two sides. You know: $a^2 + b^2 = c^2$. Using this little equation, you can figure out the number of holes you need in a beam that connects diagonally between two perpendicular beams. See the TECHNIC design school tutorials for some examples.

NOTE: *The length of a beam includes the distance from the start of the beam to the center of the first hole and the distance from the center of the last hole to the end of the beam. So a beam that is 4M long has 4 holes but the total length from the center of the first hole to the center of the last hole is only 3M. Remember this when you read the design school tutorial that talks about the Pythagorean theorem. A 3-4-5 right triangle actually involves beams that are 4M, 5M, and 6M long. The number of holes in a straight beam (aka the beam length) is always the center-to-center length plus one.*

Beams

TECHNIC beams come in two basic varieties: straight or bent (see Figure 4-3). The NXT set includes a variety of each type. Straight beams are great for rectilinear construction. Bent beams are great for bracing and for building robots that have a more stylish appearance rather than just a basic box look and feel.

You may notice that some of the bent beams in Figure 4-3 have two different types of holes. Most beams have round holes for inserting different pins. Sometimes a beam will start and end with a special cross-axle hole into which you can insert a standard LEGO axle.

Figure 4-3. Straight and bent beams

These cross-axle holes are terrific for creating joints that cannot rotate and for connecting the two pieces at 90-degree angle intervals. You can place axles through the usual round holes as well, but in that case the axle rotates freely while the beam remains in a fixed position. If an axle is set into the axle hole at the end of a beam, when the axle rotates the beam will rotate at the same time. The use of cross-axle holes at the end of bent beams is one reason why these types of beams are often referred to as liftarms.

Making the Connection

Connecting beams and gears together is what building a robot is all about. To make these connections work, we will use two basic types of connectors: pins and axles. Let's have a look at these two VIPs (very important pieces).

TECHNIC pins

Pins come with or without friction. If you want a tight fitting connection where rotation is restricted somewhat by friction between the pin and the beam hole, then you will want to use friction pins. The vast majority of the connections you make when building robots with the NXT set involve friction pins. Friction pins come in 2M and 3M lengths. The NXT set has several other specialized forms of the friction pin, shown in Figure 4-4.

Figure 4-4. Friction pins

The following tables show the part numbers and names for some of the most important elements that we will use to build robots. The part number is the LDRAW part number that is used by many different LEGO computer aided design (CAD) software tools. This number is not the same as the element number used by LEGO internally. The Peeron

Name is the name for the element that is used on the Peeron.com website. Peeron.com is perhaps the most useful website on the Internet for finding out where to get the parts you need to build your latest LEGO creation. The end of this chapter lists other fantastic online resources that you can use to your advantage as you experiment with LEGO.

 WEBSITE: *Visit Peeron.com to search for LEGO elements and examine part inventories from a huge selection of LEGO sets at* http://www.peeron.com/

Element	Part Number	LEGO Digital Designer Name	Peeron Name
	2780	Connector peg with friction	Technic Pin with Friction and Slots
	6558	Connector peg with friction 3M	Technic Pin Long with Friction
	50	Ball with friction snap	Technic Pin with Friction with Towball
	43093	Connector peg with friction /cross axle	Technic Axle Pin with Friction
	32054	2M friction snap with cross hole	Technic Pin Long with Stop Bush
	32136	Module bush	Technic Pin 3L Double

Table 4-1. Friction pin element names

The NXT set also includes non-friction or regular TECHNIC pins (see Figure 4-5). These are much less common than the friction variety. There are three 2M long pins and thirteen 3M long pins included in this set. It is interesting to contrast these small quantities with the large number of friction pins. The NXT set comes with a whopping 82 2M friction pins and 34 3M friction pins. Friction pins are used in building TECHNIC and MINDSTORMS creations far more often than non-friction pins.

In addition to the two basic types of non-friction pins, there are two elements in the NXT set that are extremely useful for designing and building robots. The LEGO Digital Designer software calls them the "beam 3M with 4 snaps" and the "angular beam 90 degrees with

Figure 4-5. Pins without friction

4 snaps". These special elements are extremely useful when you are building without bricks because they simplify the process of changing the orientation of parts.

You will often need to position your beams in 3D-space so that holes, pins, and axles are oriented along the x-axis, the y-axis, and the z-axis. Changing orientation can sometimes be tricky but parts like these complex pins make it a bit easier. This is because these elements have pins and holes that are oriented along multiple axes. The last element shown in Figure 4-5 is particularly useful, since it features connection points oriented along all three axes. Table 4-2 lists the frictionless pins in the NXT set.

There are several other types of connector elements included in the NXT set which help you build along all three axes in your design (see Table 4-3). You will learn a lot by trying out many different combinations of these elements. Try rotating the orientation of your beams through the x, y, and z-axes. The key to developing your building skills is to experiment with these stud-free elements in your designs.

Element	Part Number	LEGO Digital Designer Name	Peeron Name
	3673	Connector peg	Technic pin
	X202	3M Connector peg	Technic Pin Long
	3749	Connector peg/cross axle	Technic Axle Pin
	48989	beam 3M with 4 snaps	Technic Axle Joiner Perpendicular 3L with 4 Pins
	55615	angular beam 90° with 4 snaps	Technic Beam 3 x 3 Bent with Pins

Table 4-2. Non-friction pin element names

Element	Part Number	LEGO Digital Designer Name	Peeron Name
	6536	cross block 90°	Technic Axle Joiner Perpendicular
	32184	double cross block	Technic Axle Joiner Perpendicular 3L
	42003	cross block 3M	Technic Axle Joiner Perpendicular with 2 Holes
	32034	angle element 180° (2)	Technic Angle Connector #2
	924	angle element 135° (4)	Technic Angle Connector #4
	32014	angle element 90° (6)	Technic Angle Connector #6
	32013	angle element 0° (1)	Technic Angle Connector #1
	32291	TECHNIC cross block 2x1	Technic Axle Joiner Perpendicular Double
	41678	TECHNIC cross block/ fork 2x2	Technic Axle Joiner Perpendicular Double Split
	44809	HTO V beam 90 degrees	Technic Pin Joiner Perpendicular Bent
	32557	TECHNIC cross block 2x3	Technic Pin Joiner Dual Perpendicular
	6553	Catch	Technic Pole Reverser Handle
	32039	catch with cross hole	Technic Connector with Axlehole
	2905	Triangle	Technic Triangle
	45590	Damper 2M	Technic Axle Joiner Double Flexible

Table 4-3. Miscellaneous connector element names

TECHNIC axles

Axles are just plain fun. They are also fundamental to building robots that move or manipulate things. Not only are they used to transfer mechanical energy throughout your robot but they are also used as basic structural elements in your designs. As you have already seen above, several pin connector elements and bent beams have axle holes in addition to holes for pins. Using pins and axles together to form a strong frame is often the best approach to building without bricks.

In the NXT set there are a variety of different axle lengths. The selection provided by the set is by no means a complete representation of the lengths and types of axles available in different LEGO sets. Axles are generally organized into even lengths and odd lengths. In modern TECHNIC sets, odd-length axles are grey, or "Medium Stone" as the color is referred to on the Peeron website, while even-length axles are black. Older sets were not as strict about the colors used for odd- and even-length axles, often using red, white, black, and various shades of grey as the particular set design required. The NXT set includes many short axles and fewer long axles, as is usually the case in TECHNIC sets. The short axles are often used structurally, so more of them are included. The longer axles are generally involved in drive trains or power transmission, meaning you won't need as many in a typical robot.

In addition to the plain axles in lengths of 12M, 10M, 8M, 7M, 6M, 5M, 4M, 3M, and 2M there are a few other axle-based or axle-related elements included in the NXT set. Some we have seen in the previous section, since they are related to both pins and axles. The remaining specialty axle elements are shown in Figure 4-6. This figure also shows the tow ball elements, which can be used to create linkages that transfer force in tension while allowing for rotation about the tow ball as it is held within the socket.

Figure 4-6. Specialty axle and tow ball-related elements

There are probably an infinite number of ways that you can incorporate axles into your structural designs. We'll look at a couple of simple approaches in this section, but once again the best way to learn is by doing. In Figure 4-7 you can see an example of a pair of 3M axles combined with a double cross block. This structure could be used to plug an attachment into a robot frame. Just pull out the double cross block along with its two axles and the attachment can be easily swapped with a different one.

Figure 4-7. Using 3M axles to plug in an attachment

Another example combines some 3M axles, 3M connector pegs with friction, a 7M axle, a pair of full bushes, connector pegs with friction/cross axle, and both bent and straight beams. Figure 4-8 shows how all these can be combined into something that resembles a bumper. We may see this combination of parts again in a later chapter.

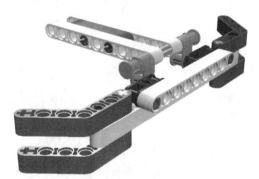

Figure 4-8. Combining axles and pins to build a hinged bumper

The last example in this section demonstrates how to combine a few simple elements together in order to help transition beams from one spatial orientation to another. Figure 4-9 combines the cross block 2x1 with the cross block fork 2x2 using a 2M axle. This device allows us to pin together a beam with holes oriented along one axis and a bent beam with holes oriented along a different axis.

Figure 4-9. Changing orientation with connectors and axles

Now grab some beams, axles, and connector elements. Join them together and see how strong you can make a structure. You'll be amazed what you can do with axles when you add in some axle joiners, full and half bush elements, pulleys, tow balls with cross axles and even Bionicle eyes. Feel free to think outside the box and experiment a little. And definitely take the time to study how the LEGO professional designers use these elements in their designs. You'll learn a lot by examining the techniques used in TECHNIC instructions.

Gears

One disappointment I have with the NXT set is the quantity and selection of gears that are included. That's not to say that what we get is not great, but you can never have too many gears. I'd especially appreciate more large gears. Fortunately, there are many ways to grow your gear collection without spending a lot of money. Figure 4-10 shows the collection of different types of gears that are included in the NXT set.

Figure 4-10. Gears included in the NXT set

Except for the turntable gear and the worm gear, all the gears in the NXT set are a full 1M wide or thick. You'll run across the very popular 12-tooth bevel gear and use it in your robots before long. Although it isn't included in the NXT set, its popularity in many other TECHNIC sets and its usefulness will lead you to it sooner or later. It is one of a few half-width gears that LEGO makes. Gears generally have a radius length that is also expressed in terms of the standard LEGO unit of measure. A quick way to calculate the radius of most LEGO gears is to divide the number of teeth by sixteen.

WEBSITE: *See the TECHNIC design school website for an excellent tutorial on using gears at* http://technic.lego.com/technicdesignschool/lesson.asp?id=1_b.

When trying to mesh two spur gears together along the same axis of rotation, the sum of the lengths of each radius must correspond to the distance between the holes you are using to position the gears in your robot design. Bevel gears are a special case (see Table 4-4). When you mesh bevel gears with perpendicular axes of rotation, you will need to account for the proper radial distances in two dimensions.

The knob wheels (see Table 4-4) mesh with the "teeth" of each wheel, fully overlapping the width or thickness of the other wheel so no special allowances are required. However, the double bevel gears included in the NXT set do not mesh across the full width of the gear. With these gears acting as bevel gears, you will need to make some adjustments when you position them in your robots. It's at times like this that the half bush and full bush elements come in very handy (Figure 4-11).

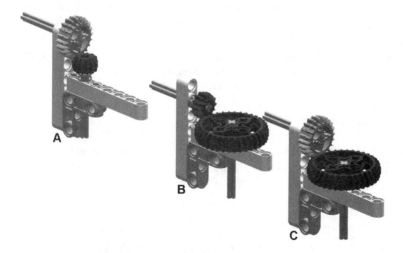

Figure 4-11. Spacing out bevel gears

Element	Part Number	LEGO Digital Designer Name	Peeron Name
	3647	8 Tooth	0.5M
	32270	12 Tooth Double Bevel	0.75M
	4019	16 Tooth	1.0M
	32072	Knob Wheel	1.0M
	32269	20 Tooth Double Bevel	1.25M
	3648	24 Tooth	1.5M
	x403	36 Tooth Double Bevel	2.25M
	3649	40 Tooth	2.5M

Table 4-4. Gear names and radius lengths

The NXT set includes two specialized gears – the worm gear and the turntable. Worm gears are great for driving a spur gear with a very high gear ratio since the worm has only one tooth. However, it suffers from a great deal of sliding friction which can slowly devour both the gear and the axle it is mounted to as well as any bracing pieces that hold the worm gear in mesh with its partner. One of the nice aspects about the worm gear is that while you can use the worm to drive the gear it is meshed with in either direction that gear cannot drive the worm at all. It is not very efficient, and high torque applications will just make the negative aspects of friction even worse. LEGO makes a number of special gearbox elements that work with the worm gear and a 24-tooth spur gear. These come in handy so you will probably want to be on the lookout for sets that include gearbox elements.

TRY ME: Try to build an assemblage using gears so that turning a crank causes a wheel to spin very fast. Conversely, you could reverse this scenario so that turning a crank causes a wheel to turn slowly but powerfully.

WEBSITE: *The Education Resource Set, which is available from the LEGO Education store website, is a great way to acquire additional elements such as the gearbox, gear racks, and differential gears.* http://www.legoeducation.com/store/detail. aspx?CategoryID=121&by=9&pl=10&ID=1277.

The turntable is the largest gear element that LEGO makes. It is a very special piece that is often used in conjunction with a worm gear. The worm can be meshed with the 56-tooth outer gear on the turntable to provide a slow turning joint. This type of joint is found in many applications, including robotic arms. In addition to driving the turntable via a worm, you can also use several of the spur gears on the outer gear. You can also drive the turntable via the inner 24-tooth gear, but fewer gears can be used in this case due to the way the teeth on the inner gear are oriented along the inner gear's circumference. Both the worm gear and the turntable can be seen in Figure 4-10 above. You should look for opportunities to use these gears in your own robotic projects. We'll put them to use later in this book.

There are a few LEGO gear types that deserve a mention, even though they are not included in the NXT set. One is the TECHNIC gear rack 1x4 element, which is like a small one by four plate except that it has gear teeth along the top rather than studs. This piece has a few rack relatives such as the 1x8, 1x10, 1x12, and 1x14 gear rack with holes (Figure 4-12). The longer versions with holes are designed to slide on tiles or the top of a smooth TECHNIC beam with a spur gear such as an 8-tooth gear driving the rack from side to side. The 1x4 gear rack is designed to build longer racks using multiple pieces placed end to end on top of bricks or plates.

Figure 4-12. Gear rack elements

One other gear that I really miss in the NXT set is the differential gear, which is actually a combination of a specialized differential housing element along with three 12-tooth bevel gears (which are also missing from the NXT set). Together with a pair of axles these pieces make up a differential gear, which is the backbone of drive-steering vehicles, not to mention an assortment of other robotic applications

Figure 4-13. Differential gear and 12-tooth bevel gears

The lack of a differential is not crushing, however. Using a handful of the gears that come in the NXT set and various connector elements, it is possible to build your own differential that works very well. The NXTasy website has a great repository of building instructions, including instructions for how to build a differential out of parts from the NXT set.

WEBSITE: *Visit* http://nxtasy.org/category/nxt-repository/projects/technic/ *for a simple differential gear design that you can build yourself. While you are there, check out some of the other instructions!*

Online Resources

There are many excellent online resources related to building robots with LEGO pieces. If you ever need inspiration or information, check out some of the websites in Table 4-5.

Webpage	Description
www.texbrick.com/	Thomas Avery's fantastic site with great information about building large structures using LEGO.
www.brickshelf.com/cgi-bin/gallery.cgi?f=8965	Gus Jansson's brickshelf gallery about LEGO geometry.
www.badlink.com/lego_mindstorms/gears_drivetrain.htm	A great site with lots of links to other sites about LEGO robotics.
http://creator.lego.com/designschool/courses.asp	Interesting tutorials about building with LEGO and combining elements from different types of sets into your own projects.

Table 4-5. Inspiration and educational websites

Buying additional LEGO parts to feed your robot creations might be something that you'll want to do before long, if you haven't already started. There are many different options on the web for making these purchases. The LEGO shop-at-home website is a great place to visit if you are looking to add to your collection of TECHNIC parts.

The items listed in Table 4-6 are cost-effective ways to quickly build up the parts you need for that complicated robot you have in mind. Another approach you might try is to check out Goodwill stores for used LEGO sets. Local retail stores such as Target or Toys R Us also have bargains on older sets. You can find inventories for a massive number of LEGO sets on the Peeron website mentioned earlier.

Webpage	Description
http://shop.lego.com/Product/?p=10072	Beams
http://shop.lego.com/product/?p=10073	Connectors
http://shop.lego.com/product/?p=10076	Gears
http://shop.lego.com/Product/?p=8287	Motors and extra parts
http://shop.lego.com/ByTheme/Leaf.aspx?cn=17&d=70	MINDSTORMS NXT

Table 4-6. Purchasing additional pieces

In the next chapter we'll take a very close look at the power that Bricx Command Center brings to NXT programming.

Bricx Command Center

Topics in this Chapter

- Connecting to the Brick
- The Programmer's Editor
- Drag and Drop Programming
- Exploring Your Code
- Managing Macros
- Compiling and Running
- Exploring the Tool Windows
- Getting Help

Chapter 5

Professionals often pay thousands of dollars for development tools that enhance the process of writing programs. In this chapter we will explore a development tool that is every bit as refined as commercial tools, yet costs nothing. For embedded systems like the ARM processor running within the NXT brick, it helps if the tool can communicate with, control, and inspect the state of the device. These tools will simplify the coding process, resulting in faster development times.

In this chapter we will examine one of the best development tools for LEGO MINDSTORMS programming. It is available for the NXT and every other LEGO programmable brick ever created. Originally conceived for the RCX, the RCX Command Center (RcxCC) has been transformed over the years into an all brick suite of tools with support for many programming languages, including MindScript, LASM, Not Quite C (NQC), brickOS C/C++, brickOS Pascal, leJOS Java, and pbForth. RcxCC was renamed along the way as Bricx Command Center. The name "Bricx" is pronounced "bricks" to reflect the expansion from just the RCX to all LEGO programmable bricks.

As previously mentioned, BricxCC also provides language support for Not eXactly C (NQC for the NXT) and NeXT Byte Codes, as well as RICScript and NPG. BricxCC also includes many useful tools for the LEGO programmable bricks, plus a number of tools designed specifically for the NXT brick. Let's start up the software and see what it can do.

Connecting to the Brick

When you launch BricxCC, by default the first thing you see is the Find Brick dialog (Figure 5-1). This is where you tell BricxCC how to connect to your brick, the type of brick, and the type of operating system or firmware.

The first time you run BricxCC, the default Port and Brick Type are not configured for the NXT. We'll see how to change these defaults, but for now just type in "usb" in the Port field and pick NXT from the Brick Type list. Whether you are using the NBC/NXC enhanced firmware or the standard NXT firmware, you will want to leave the Firmware set to

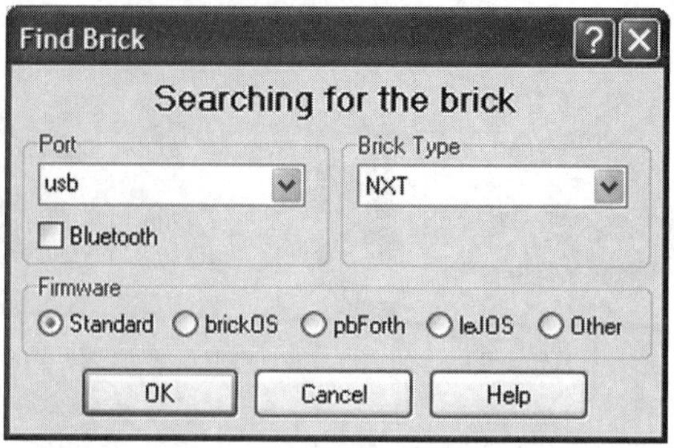

Figure 5-1. The Find Brick dialog

Standard. Connect the NXT via a USB cable, turn it on, and then click OK. BricxCC should connect to the brick and display the main window. If it fails to connect, an error message will appear indicating that certain options are unavailable.

The Port list contains a number of entries that are loaded from a file called nxt.dat. You can find this file in the same directory as BricxCC. exe. If the file doesn't exist when you start up BricxCC, it will be created automatically. BricxCC writes an NXT brick resource name and alias to this file for each NXT brick it finds via USB or Bluetooth.

The fastest way to connect to the NXT is to pick an alias or a brick resource name from the list of Ports. For Bluetooth connections, you should pick the alias or brick resource name that contains the name of your NXT. Bluetooth resource names should have BTH at the beginning, which signifies that it is a Bluetooth resource.

```
BTH::NXT=BTH::NXT::00:16:53:01:E2:CB
BTH::JCH2=BTH::JCH2::00:16:53:FF:01:56::5
NXT=USB0::0X0694::0X0002::00165301E2CB::RAW
JCHNXT=USB0::0X0694::0X0002::001653FF0030::RAW
JCH2=USB0::0X0694::0X0002::001653FF0156::RAW
```

The *alias* is the text to the left of the equal sign as shown in the code sample above. You can edit nxt.dat using notepad and change the aliases to whatever you prefer. The *brick resource name* is the string to the right of the equal sign. These values must not be modified or it will no longer connect to the brick. If you know your brick's serial number and Bluetooth name, then you can construct the Bluetooth or USB resource name manually by following the pattern above.

The main BricxCC window is shown in Figure 5-2. The Templates window is displayed by default in a separate window. Older versions of BricxCC used a very large window that could fill the entire screen but now the templates are organized into a tree structure by categories. This makes it easier to access the individual templates. We'll learn more about the template window later in this chapter.

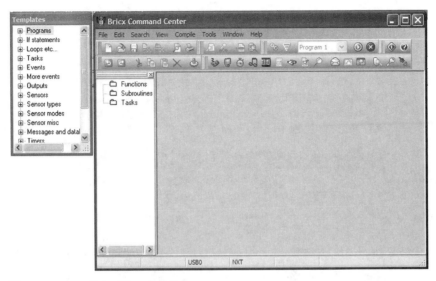

Figure 5-2. The BricxCC main window

The template tree window can dock in the same area as the code explorer window, within the main BricxCC form (Figure 5-3). I recommend docking it below the code explorer window. BricxCC remembers the state and position of all its windows, so it will start up in the same physical configuration the next time you run the application. You can also show or hide the template window at any time by pressing the F9 key or by selecting the View|Templates menu option from the main BricxCC menu.

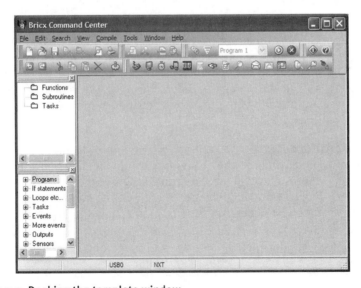

Figure 5-3. Docking the template window

MDI vs Tabbed windows

Initially BricxCC uses the Multiple Document Interface (MDI) form of window management, which used to be common in windows applications. This is the default mode but it is not the only mode available for you to use. If you prefer to use a tabbed multiple window mode instead, you can change this option via the BricxCC Preferences dialog (Figure 5-4). Select Edit|Preferences to open the dialog. On the General tab you will see the "Use MDI mode" check box option.

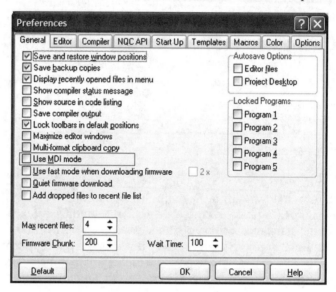

Figure 5-4. Selecting the multiple window mode

To switch to the tabbed multi-document mode, just remove the check from this option as shown in Figure 5-4. Unlike other settings in the Preferences dialog, this change will not take effect immediately. BricxCC will switch to the new mode after you exit the application completely and start it up again. Before we do that, though, let's make a few more modifications to the default settings while we have the Preferences dialog open.

Default startup options

When you launch BricxCC the behavior of the application is controlled by the settings on the Start Up tab. You can configure whether to show the Find Brick dialog or not and whether or not to try to connect to the brick as the application launches. These settings are shown in Figure 5-5.

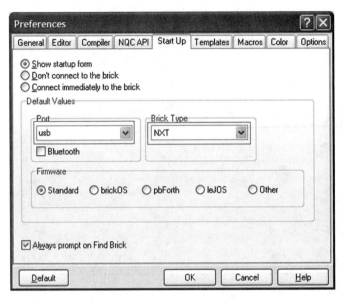

Figure 5-5. BricxCC start up options

BricxCC uses RCX as the default brick type. It initially has no value for the Port. To make starting up BricxCC even simpler, change these values to NXT as the brick type and "usb" as the port value. The changes to the default settings are emphasized in the screen shot above by surrounding the controls with a dark rectangle. Just pick NXT from the Brick Type drop-down list and type "usb" into the Port edit field.

Compiler settings

Let's go ahead and tweak the default values for the compiler while we have the Preferences dialog open. Since BricxCC supports many different brick types and many different programming languages there are a number of options on the Compiler tab as you can see in Figure 5-6. Let's focus on just a couple of the options that will further simplify writing code for the NXT using the NXC programming language.

Since we will program the NXT using the standard firmware and since NXC will be our primary programming language, we should tell BricxCC to use NXC as the preferred language. This option is shown with a surrounding rectangle in the above screenshot. By selecting this option, BricxCC will assume that every new editor window contains NXC code even if the file has not yet been saved with a file extension. Normally the file extension tells BricxCC which programming language is contained in the file but in the absence of an extension it uses this configuration setting to help it decide which compiler to use and which syntax highlighter to use, as well as which help to display.

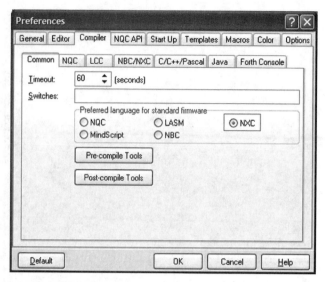

Figure 5-6. Preferred language

Switching to the NBC/NXC tab on the Compiler page, we can make a few more changes to further enhance our BricxCC experience (Figure 5-7). These changes are optional but they are highly recommended. The options to change are surrounded by rectangles in the screenshot.

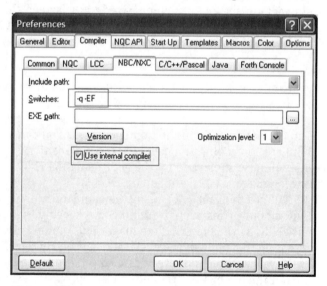

Figure 5-7. NXC and NBC settings

We want to tell BricxCC to use its built-in copy of the NBC/ NXC compiler by checking the "Use internal compiler" option. This dramatically speeds up the process of downloading compiled programs to the NXT using Bluetooth. If you are connected via USB the difference is much less noticeable but it prevents you from accidentally using an old version of the compiler if you happen to have a copy somewhere on your computer.

Since we will be taking advantage of the bug fixes and improvements in the enhanced NBC/NXC firmware, we need to tell the compiler to use these features as well. This information is passed to the compiler by using the Switches edit field as shown in the screenshot above. The -EF switch tells the compiler that we are using the enhanced firmware. It will then be able to generate code that uses the features that are not provided in the standard NXT firmware. Normally the NXT will beep after a program is downloaded to it. It makes it easier to tell that the download has completed successfully. If you prefer a quiet brick, however, you can optionally add the -q switch to the Switches field. This turns off the confirming beep at the completion of the download operation.

Now we are ready to close the Preferences dialog with our default settings properly configured. Exit BricxCC and start it up again so that the changes to the multiple document mode come into effect as described earlier. After we confirm the Find Brick operation, the BricxCC window will be displayed as shown in Figure 5-8.

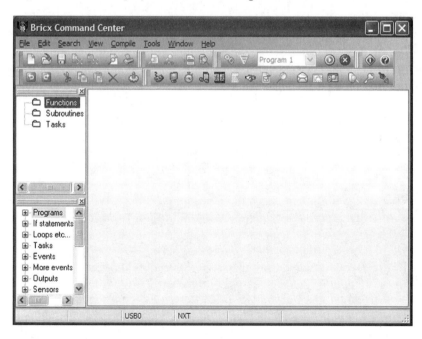

Figure 5-8. BricxCC in tabbed multiple document mode

Notice that the status bar at the bottom of the window shows that our brick type is NXT and that we are connected via USB. The status bar can be hidden or shown by selecting the View|Status bar menu option. You can also show or hide each of the six BricxCC toolbars using the View|Toolbars sub-menu as shown in Figure 5-9. If you prefer to use keystrokes, you can use Alt+V, b to display the sub-menu and then press any of the underlined letters to toggle the toolbar's visibility.

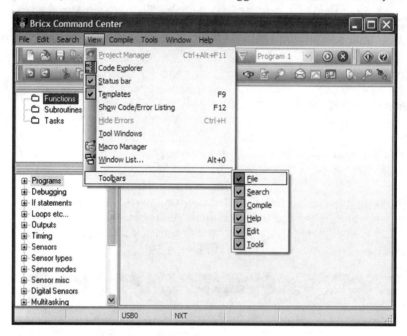

Figure 5-9. Showing or hiding toolbars

Next we will take a look at the many powerful features provided by the BricxCC editor.

The Programmer's Editor

Since we will be spending a lot of time in an editor within BricxCC, it is important that it provides us with features and flexibility to fit our preferred programming style. Fortunately, the BricxCC editor is very flexible, allowing you to control its many capabilities. We'll examine the features it provides and the options we can configure in this section.

When you have files open in BricxCC, each editor window has its own tab containing the name of the file (Figure 5-10). If you open too many files to fit on screen, a pair of left and right arrow buttons will appear at the top right corner of the editor windows. Use these buttons to scroll through the editor tabs. You can also quickly switch from one tab to another in sequence by pressing Ctrl+Tab.

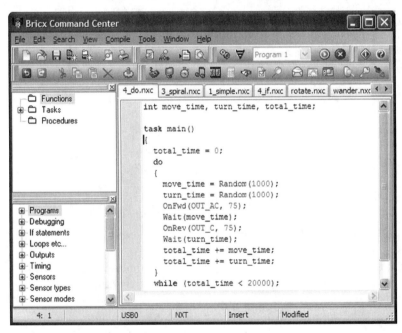

Figure 5-10. Editing files in BricxCC

If you look down at the status bar you will see the word Modified when the current editor window contains unsaved changes. The Save button will also become enabled if a file has been modified. The status bar also shows whether the editor is in Insert or Overwrite mode, which you can toggle via the Insert button on your keyboard. As with other editors, when you are in Insert mode, characters that you type will be inserted at the cursor position. While in Overwrite mode, typed characters will replace existing text. Finally, the status bar displays the current cursor position in the editor window. As you move the cursor around using the arrow keys or by clicking with the mouse, these values are automatically updated.

General Editor Preferences

The BricxCC editor provides the standard editing features that you would expect from any text editor. You can select text, cut, copy, paste, undo, redo, and so forth. To select each of these features, use the buttons on the Edit toolbar, Edit menu items, and standard keyboard shortcuts. In addition to the basic editing commands, this menu is also used to access the Preferences dialog. Let's have a look at the options for the editor that you can configure to your own liking via the Preferences dialog (Figure 5-11).

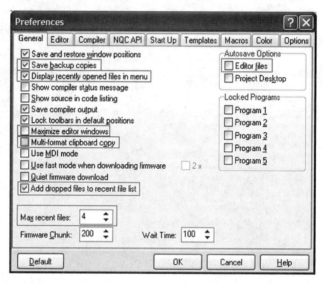

Figure 5-11. Editor options on the General tab

On the General tab you can configure several options that control the BricxCC editor, as shown in Figure 5-11 above. By default BricxCC saves a backup copy of the current file whenever you save the editor contents. The backup version will have a .bak file extension. BricxCC also adds recently opened files to the File menu. Use the "Max recent files" field near the bottom of the page to change the number of recent files. If you want files that you open via drag and drop to be added to the recent files list, check that option here. Also available is the option to automatically save editor files whenever you compile, the option to maximize editor windows when running in MDI mode, and the option to copy selections to the clipboard using multiple formats, including plain text, HTML, and RTF.

The Editor Page

Most of the options for configuring your BricxCC editor are found on the Editor tab, as shown in Figures 5-12 and 5-13. The default settings are displayed below. For details about any of the options in the Preferences dialog you can consult the BricxCC help system. As with all the BricxCC dialogs, you can use the "What's This" help button in the title bar and click the option to view a hint about the item you select. Just left click the question mark button in the title bar and then left click on the button, edit field, or check box you want to see a hint about. Use the Help button to view the full help text for the Preferences dialog.

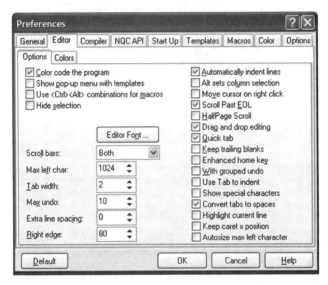

Figure 5-12. Options for the BricxCC editor

You can adjust the tab width in the editor. The maximum number of undo steps can also be configured. If you want to turn off the syntax highlighting in the editor you can remove the check in the "Color code the program" option. You can also disable the drag and drop editing feature if that is your preference. If you want to select a column of text easily you can enable the "Alt sets column selection mode" option. If it is enabled, whenever you make a selection using your mouse with the Alt key pressed it will select a column of text rather than the normal selection mode.

The "Scroll Past EOL" option determines the editor's behavior when you arrow to the end of a line or click in the blank space to the right of the last character in a line. If you turn off this option, regardless of where you click the cursor will be placed just after the last character and using the right arrow key will automatically wrap around to the beginning of the next line. With this option enabled, as you press the right arrow key the cursor will just continue to move toward the right

66

edge of the editor window without wrapping. You may want to make sure the "Keep trailing blanks" option is disabled while in this mode or you could wind up adding unintentional spaces at the end of lines in your code.

If you favor strict tab levels you may want to disable the Quick tab option. When this option is enabled it overrides the normal tab settings with a tab method that sets the amount of spacing based on the text in the line above the current line. The Quick tab mode makes it easy to quickly tab to a level of indentation set in the previous line, but that may not match your configured tab stop settings. This option is turned on by default.

The enhanced home key option configures the editor so that when you press Home on your keyboard it first moves the cursor to the current level of indentation. Exactly where the cursor will be positioned in this mode is dependant upon your code in the editor window. Pressing Home a second time will position the cursor at the beginning of the line.

Another underappreciated feature in BricxCC is the editor's ability to draw structure lines, which you can enable on this page (see Figure 5-13). These are vertical lines that are drawn in the editor window to show the nested structure of your program in a manner similar to what you see in the Visual Studio or Delphi editors. To turn on structure lines, simply pick a structure line color other than clNone. You can also tell BricxCC to highlight the active line by setting the Active line color to a value that is not the same as the Background color.

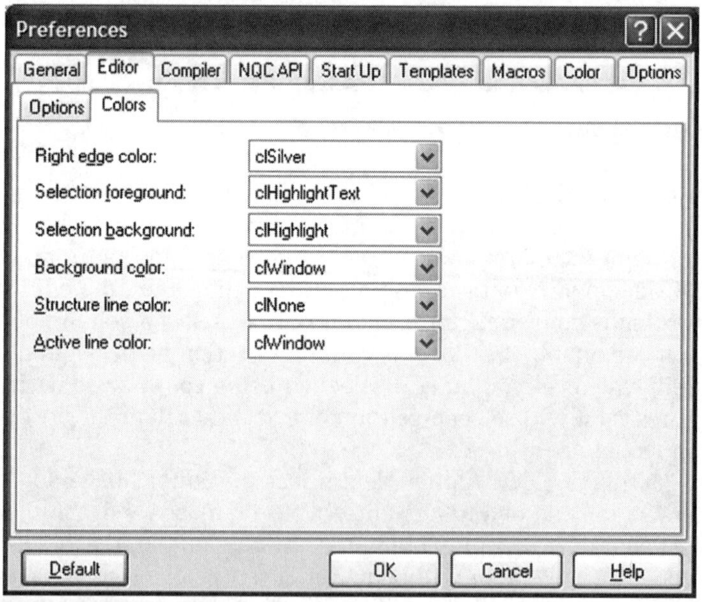

Figure 5-13. Editor color options

The Macros Page

Within the Preferences dialog you can also configure basic editor macros. These <Ctrl> <Alt> macros provide an easy way to enter often-repeated code snippets into your program. The Macros page of the Preferences dialog is shown in Figure 5-14 below. This editor feature can be enabled or disabled via an option on the Editor page.

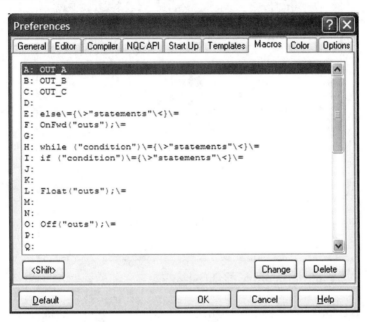

Figure 5-14. Editor ⟨···Ctrl···⟩⟨···Alt···⟩ macros

If the <Ctrl> <Alt> macros option is enabled then whenever you press Ctrl+Alt and any single letter from A-Z or number from 0-9 the editor will insert the macro associated with that key. You can also define Shift versions, which will be executed if you press Shift+Ctrl+Alt and the letter or number of the macro you want to execute. The syntax for writing these macros is the same as syntax used for BricxCC templates, which we will examine in the next section.

The Color Page

The ability to show the structure of your program using what is called syntax highlighting is an important part of making programming easy in BricxCC. You have complete control over this feature via the settings on the Color page of the Preferences dialog (Figure 5-15).

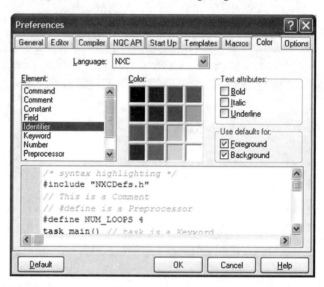

Figure 5-15. Configuring the syntax highlighting

Make sure when you are using this page that you pick the programming language that you want to configure via the Language drop-down list. Once you have NXC or any other programming language selected, you can choose an element in the list-box and then reconfigure how it should be displayed in the editor by left or right clicking on the color grid. A left click will pick a new foreground color, while a right click will pick a new background color. You can also select the bold, italic, or underline text attributes if desired. As you make changes you will see the sample code window update to display how a program will be drawn in the editor window if you save your revised settings.

The Options Page

The last page in the Preferences dialog is a catchall location for several
different editor settings that you should be familiar with. The Options
page is shown in Figure 5-16 below. Let's take a quick look at how these
settings can impact your BricxCC experience.

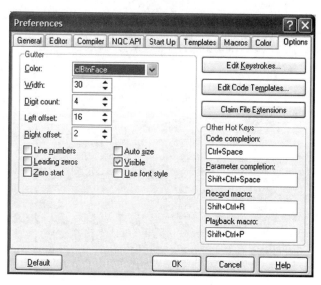

Figure 5-16. Configuring settings on the Options page

The gutter is the area to the left of the editor window that at first
glance may not seem to be very useful. In its default state it doesn't do
a lot. If you like seeing line numbers then the gutter will be very useful
to you. You can enable the line numbers option on this page. Once you
make that change, whenever you edit a program in BricxCC the gutter
will contain line numbers. This page lets you configure several different
options for the gutter, including its width, background color, number of
digits to use for each line number, whether to show leading zeros, and
whether to start numbering with zero or one. We'll see another feature
of the gutter shortly.

Other options on this page include setting hot keys for code
completion, parameter completion, as well as macro recording and
playback. These features will be explored shortly. For now just make
note of the default hot keys and remember that this is where you need
to go if you decide to change them. From this page we can also edit the
editor keystrokes, edit code templates, and configure BricxCC to claim a
number of different file extensions for its own use.

Configuring Editor Keystrokes

Figure 5-17 shows the dialog that is displayed when you click the Edit Keystrokes button. Using this dialog you have complete control over the behavior of the BricxCC editor. Everything you do in BrixCC, beyond simply typing letters and numbers, involves an editor command of some sort.

Figure 5-17. The BricxCC keystroke editor

When you press the left arrow key, for example, the editor responds to the ecLeft command. You could, if you wished, change this to do something completely different. From simple text selection and basic cursor movement to indenting blocks of text and matching open and close braces, the editor commands provide you with many powerful features that make writing programs easier. If you are not fond of the default set of keystrokes you are able to completely change them out with your own preferences. You can even configure multiple keystrokes for the same command if you want. When you add or edit a keystroke you will see the dialog shown in Figure 5-18 below.

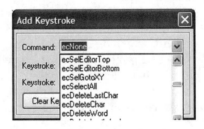

Figure 5-18. Adding a keystroke command

Using the Add Keystroke dialog, pick the editor command from the drop-down list and then press the keystroke or pair of keystrokes required to execute this command. While the predefined set of keystrokes covers most of the available editor commands, there are a number of them which do not have keystrokes defined initially. You can browse through the list of editor commands and see if you want to add keystrokes for some of the ones that are not currently defined.

BricxCC Code Templates

When you click the Edit Code Templates button on the Options page you will see the Code Templates dialog (Figure 5-19). Code templates are executed in the BricxCC editor using the ecAutoCompletion editor command. By default this command is assigned to the Ctrl+J keystroke. A code template is similar to the Templates and Macros, but it operates somewhat differently. This BricxCC feature is exactly the same as the code templates feature originally found in the Delphi IDE.

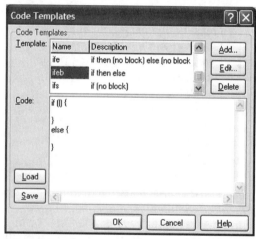

Figure 5-19. BricxCC Code Templates

Using the Code Templates dialog, you can define a short template name such as "ife" or "forb". You can add a short description that tells what this template does. In the Code field you type in the definition of the template. The vertical bar (|) is used to set the location of the editor cursor after the template has been expanded into the editor window.

To use the template, type its name into the editor and then press the keystroke that you have defined for the ecAutoCompletion command. When you press the keystroke it will automatically complete the named template if it finds a match using the name you typed. Unlike the <Ctrl><Alt> macros, you don't need to remember the keystroke you happen to have defined for each macro. You just need to know the name you assigned to your template and Ctrl+J or the keystroke of your choice.

Claiming File Extensions

The last configuration feature in the Preferences dialog is the ability
to configure BricxCC as the registered application for file extensions
associated with several different programming languages. When you
run BricxCC it automatically claims .nqc, .nqh, .nbc, .nxc, .rs, and
.npg extensions if they are not already registered. If you want to register
BricxCC as the application to use when opening other source code
files, click on the Claim File Extension button on the Options page
(Figure 5-20).

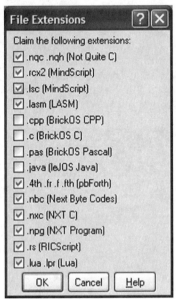

Figure 5-20. Registering BricxCC file extensions

From the File Extensions dialog you can register or un-register
any of the listed file extensions for BricxCC. Once you confirm your
selections with the OK button, BricxCC will automatically launch
whenever you double click on a file with one of the extensions you
chose. When a file having one of your chosen file extensions is selected,
the Explorer context menu, which you can display via a right mouse
click, will contain a New and Print menu option in addition to the
default Open option. All three of these menu items will launch BricxCC
and perform the specified operation.

The File Menu and Toolbar

The BricxCC File menu and toolbar provide easy access to many standard editor features (Figure 5-12). You can open multiple files in one operation via the File|Open option or Ctrl+O (see Figure 5-21). Another way to open files in the editor is to drag and drop selected files from an explorer window to the BricxCC window. You can create new files in BricxCC with Ctrl+N or File|New. These options are also available via the standard toolbar buttons on the File toolbar. On this menu you'll also find options for closing files, saving files, printing, and print previewing. You'll also find a list of recently opened files on the File menu.

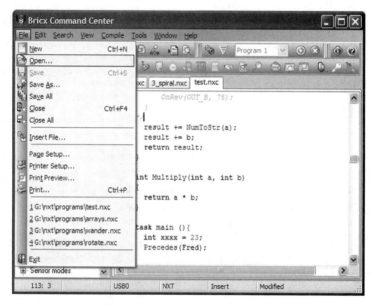

Figure 5-21. File menu options

The Page Setup dialog contains many options for configuring the page layout for printing. You can configure margins, turn on line numbering, and set header and footer options however you like. You can also use the handy Print Preview dialog to see how your program will look before you print it based on the configuration options you select in the Page Setup dialog. These two options are often lacking in other development tools. You can also use the Insert File option on this menu to insert the contents of another file into the edit window at the current cursor position.

The Edit Menu and Toolbar

The BricxCC Edit menu and toolbar expose a full suite of standard editing features (Figure 5-22). In addition to providing access to these basic editing features and to the Preferences dialog, there are a few additional options.

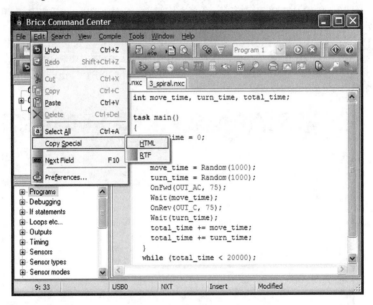

Figure 5-22. Edit menu options

The Edit menu includes the Copy Special sub menu. These options let you copy the current selection or the entire editor contents to the clipboard in either HTML or Rich Text Format (RTF). The Next Field (F10) menu option is a special feature that you can use to quickly jump forward through the current program from one quoted string to another. And as we saw earlier, the number of levels of undo is a configuration setting in the Preferences dialog.

The Search Menu and Toolbar

The BricxCC Search menu and toolbar give you easy access to standard find and replace features (Figure 5-23). There are also a few additional options for you to use that go beyond simple searching.

The find and replace features are similar to these same features in Microsoft Word or Notepad. The Replace dialog is shown in Figure 5-24. As you can see it lets you choose a number of options such as case sensitive searches, search for whole words only, search from the current position or caret, or to restrict the search to the current text selection. And you can tell BricxCC to search either forward or backward.

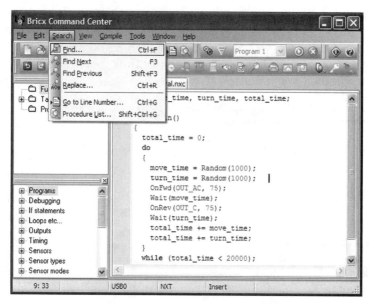

Figure 5-23. Search menu options

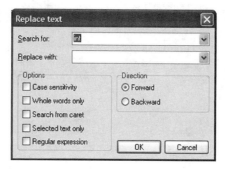

Figure 5-24. Replacing text

A nice addition to the basic search functionality is the ability to use regular expressions. If you aren't familiar with regular expression searching and replacing then you should definitely check out some of the many tutorials and quick start guides that can be found on the web. When you enable the regular expression (regex) option you can use them in both the search text and in the replace text. The regex engine in BricxCC is not a complete implementation of every feature available in regular expressions. It does, however, do a pretty good job of providing for most of your regular expression needs.

WEBSITE: A nice regular expression cheat sheet can be downloaded from http://opencompany.org/download/regex-cheatsheet.pdf. *Also be sure to check out the information about regular expressions at* http://en.wikipedia.org/wiki/Regular_expression.

Another option on the Search menu is "Go to Line Number". When you select this option, usually by pressing Ctrl+G, you will see the dialog shown in Figure 5-25. Simply type a number here and press Enter or click OK and the editor window will instantly move the edit position to the specified line number.

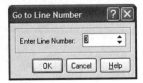

Figure 5-25. Jump to a line number

The final option on the Search menu is the Procedure List (Figure 5-26). This is a very useful feature. When you display the Procedure List using the menu item, the toolbar button, or the Shift+Ctrl+G hot key, BricxCC quickly parses your program code and displays a list of all the tasks, subroutines, and functions that you have defined in your program.

Procedure	Type	Line
f(n) FooBar	Function	70
f(n) Fred	Task	42
f(n) George	Subroutine	305
f(n) main	Task	124
f(n) Multiply	Function	119
f(n) MyTestSub	Subroutine	13

string FooBar (int a, string b)   1/6

Figure 5-26. The Procedure List dialog

After the dialog appears, start typing characters in the name of the function or task that you want to locate in the program. The characters you type will incrementally filter the list. Using the toolbar buttons at the top of this window you can set whether it should match only the start of the name or anywhere within the name. Use either the mouse or keyboard to arrow to the desired routine and press Enter or double click. BricxCC will position the editor right at the task or function you were looking for.

Bookmarks

Another nice feature of the BricxCC editor is the ability to define up to ten unique positions in the editor window and quickly jump to any of them using a convenient hot key. The saved positions are called bookmarks. There are ten editor commands to set or clear a numbered bookmark. There are also ten editor commands that will jump to a specific bookmark number.

You can see how bookmarks are displayed in Figure 5-27. I've also turned on line numbers in the gutter. The bookmark numbered 1 is shown to the left of line number 452. The default bookmark hot keys are Shift+Ctrl+0 through Shift+Ctrl+9 to set or clear a bookmark and Ctrl+0 through Ctrl+9 to jump to an existing bookmark. Of course, you can change these hot keys if you want to via the Options page of the Preferences dialog.

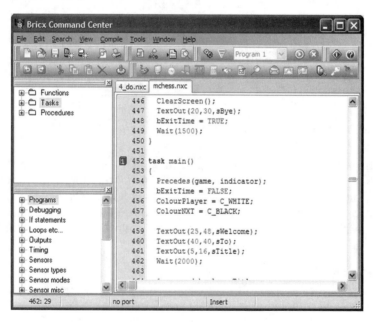

Figure 5-27. Setting bookmarks

To clear this bookmark you can simply press Shift+Ctrl+1 on the same line as the existing bookmark. Another way to quickly clear a bookmark is to left click on the bookmark image in the gutter. If you want to move the bookmark just place your cursor at the new location and press the same keystroke to set the mark at the current position. The bookmark stores not only the line number but the horizontal position of the edit cursor as well.

The edit window also has a context menu, which provides easy access to both setting and jumping to bookmarks. This menu is shown in Figure 5-28.

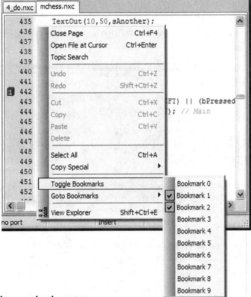

Figure 5-28. Editor context menu

In addition to the basic editing features that you see on this menu, the context menu also has the Open File at Cursor option with the Ctrl+Enter hot key. This lets you quickly open a file referenced in an *#include* statement using a single keystroke. The Close Page option, along with its Ctrl+F4 hot key, lets you easily close files without taking your hands off the keyboard.

Code completion

One of the best features of the BricxCC editor is called Code Completion. Activate code completion using the Ctrl+Space hot key or whatever hot key you configured for this feature on the Options page of the Preferences dialog. An example of what happens when you press the code completion hot key is shown in Figure 5-29.

As you type, items disappear from the list when they no longer match the pattern already in the editor. If you backspace the list updates to reflect the revised pattern. You can use the mouse or arrow keys to select an item from the popup window and press Enter or double click to replace the partial code with the completed code. The list filtering is case sensitive so it helps you learn the proper case to use for NXC keywords, API commands, and pre-defined constants. This feature is available for NBC, NXC, RICScript, and NPG programs. Using code completion as you write your NXT programs is a great way to learn programming.

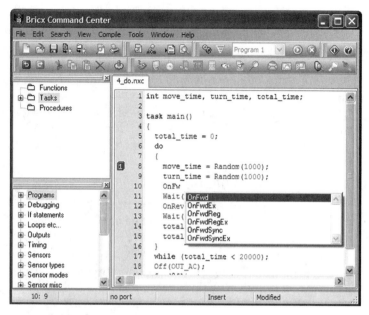

Figure 5-29. The code completion popup window

Parameter completion

Another helpful tool in your power programming toolkit is BricxCC's Parameter Completion feature (Figure 5-30). As you saw earlier, the default hot key for invoking parameter completion is Shift + Ctrl + Space. You can use this hot key or pick one that suits you better.

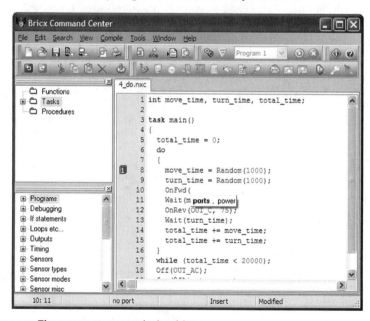

Figure 5-30. The parameter completion hint

The hint that pops up when you use the parameter completion hot key will show you what parameters the API function expects. As you type parameters into the editor the hint text will update to show the current argument in bold. These hints help teach you the right number and type of arguments that are needed to call API routines.

TRY ME: *Give parameter completion a try. Open a new document and type "TextOut(" then press Shift-Ctrl-Space. You can also wait a second or two after typing the bracket and BricxCC will volunteer the parameters for you.*

Drag and Drop Programming

One of the defining features with the NXT-G programming language is the drag and drop functionality. You can easily drag a block of code into your program and connect it in sequence with other blocks of code. Each time you drag a block to your program, such as the Move block or the Display block, you need to edit several parameters using a separate area of the editor window. Once you select another block, you can't see all those values any more just by looking at the block wired into the program. You also often need to pull down a drawer of inputs and outputs from a block and use your mouse to draw wires connecting variables or other blocks' inputs and outputs together. It sounds simple but it can become awkward and inefficient as your code grows. But the idea of being able to program via drag and drop is appealing.

That's where the BricxCC templates come into play. Earlier we docked the template tree window to the left of our editing area for easy access. By default, any time you click on a template item in the tree it is inserted into the editor window at the cursor position. You can also configure the template tree to require a double click to insert a template. Simply right click within the area and a context menu appears (see Figure 5-31).

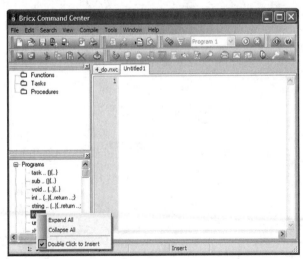

Figure 5-31. Setting the template window to use double clicks

In double-click mode the template window allows you to drag and drop templates from the tree into your program. Just like in NXT-G, you can drag blocks like the RotateMotor API command and drop it wherever you want to "wire" it into your program. Unlike NXT-G, however, all the parameters to the block are plainly visible in the text of your program and not hidden away in a separate part of the editor.

BricxCC comes with a huge assortment of NBC and NXC templates already defined and ready to use. But you are not limited to using just these. If you don't like the order of templates or you want to reduce the number of template choices, you can make these type of changes within the Preferences dialog. The Templates page of the Preferences dialog is shown in Figure 5-32 below.

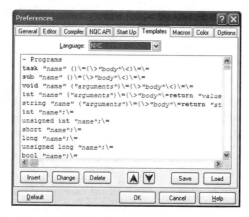

Figure 5-32. Editing the default templates

Select the programming language you want to edit using the drop-down list at the top of the page. The editor on this page syntax-highlights the template code using the settings you have chosen on the Color page. You can type directly into the editor window or you can use the Insert, Change, and Delete buttons along with the up and down arrow buttons to configure the templates exactly how you want them. This page also lets you save the contents of the template editor window to a file or load a previously saved list of templates from a file.

The syntax used when editing templates deserves a brief explanation. Since each template is defined on a single line in the template editor, you need to have some way to indicate carriage returns and indentation levels in the template text. A carriage return without additional indentation is defined in a template by using the \= token. A carriage return followed by indentation to the next tab stop is defined using the \> token. To specify a carriage return followed by a reduction in indentation to the previous tab stop use the \< token.

Quoted strings in templates are called fields. Use the F10 key to jump from one field to the next after you have inserted your template into the editor. When you insert a template, the first field is selected so that you can start typing to replace it with the desired code.

The tree structure that you see when the templates are loaded into the template tree window is defined by using the - and the | characters at the start of a line. The text following the dash and vertical bar is the caption shown for that node in the tree.

You can also select templates in the editor using a right-click context menu, but I find that entering templates using the template tree window to be much more useful. If you wish to use the right-click context menu, bring up the Editor page of the Preferences dialog and select the "Show pop-up menu with templates" option. Once enabled, this menu will replace the normal context menu that we saw earlier. The vertical bar and dash characters mentioned above play a more important role in the popup menu template feature. A vertical bar will generate a new column of menu items whereas the dash simply creates a menu separator line.

Exploring Your Code

The BricxCC Code Explorer allows you to effortlessly navigate through your source code. As you edit your code, BricxCC is continually checking for new tasks, functions, or subroutine declarations. As it finds new items, it adds them to the code explorer tree under the appropriate category. If you need to quickly find a particular function, all you need to do is expand the code explorer tree and double click on the function you are looking for. BricxCC will take you right where you want to go in the editor.

The code explorer tool has its own options dialog. Access it via the right click context menu and the Properties menu option as shown in Figure 5-33 below. The Close option on the context menu will hide the code explorer window. Select the View Editor option on this menu or

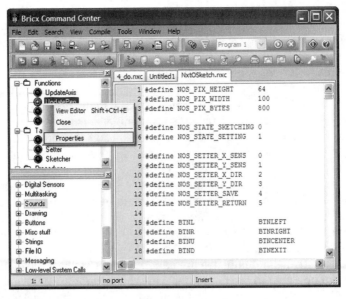

Figure 5-33. The code explorer context menu

press Shift+Ctrl+E to switch focus from the code explorer to the active editor window.

The properties dialog lets you configure several options. The dialog is shown in Figure 5-34. Here you can chose whether functions are listed in the tree alphabetically or by their order in the source code file. You can also disable the option to automatically show the code explorer window.

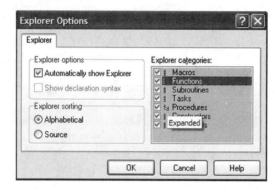

Figure 5-34. Code explorer options

The explorer categories list lets you control a few more options. If you uncheck a category it will no longer display in the code explorer tree until you select it again. To the right of the checkbox is a small icon that you can click to switch it back and forth between the collapsed state and the expanded state. This lets you decide whether categories in the tree are normally collapsed or expanded. I usually leave the categories collapsed but as with so many other features in BricxCC you get to decide what works best for you.

Managing Macros

By now you are probably thinking that we've run out of BricxCC features to explore in this chapter. The good news is there are several more features in BricxCC that will make you a powerful programmer. The next hidden gem is BricxCC's keyboard macro manager. This is a feature that is completely distinct from the <Ctrl><Alt> macros that we saw earlier. Keyboard macros are a feature built into the editor in BricxCC and they are closely tied to the editor's commands.

Earlier we saw how to configure the hot keys for recording and playing back macros in the Preferences dialog. By default, use the Shift+Ctrl+R hot key to start and stop the macro recorder. When you want to play back a macro you can simply press Shift+Ctrl+P.

If that was all there was to keyboard macros in BricxCC it would only rate as a good feature. However, to make it great, access the Macro Manager option on the View menu in BricxCC (Figure 5-35).

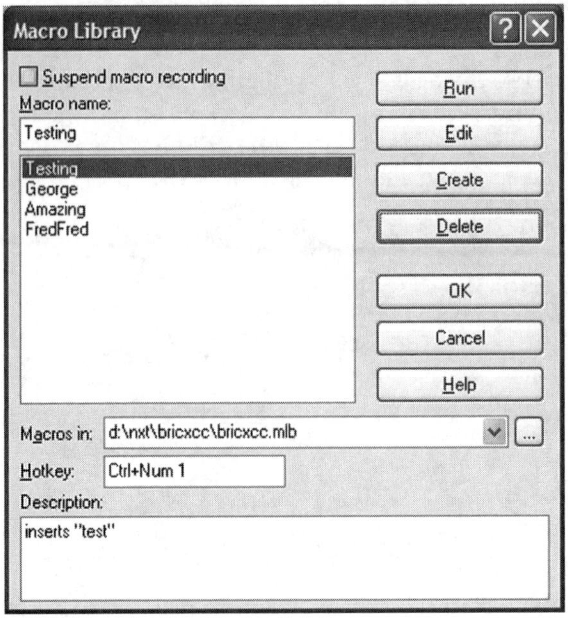

Figure 5-35. The macro library

This dialog is where the real power of keyboard macros is exposed. Every time you record a macro within an editor window the macro manager adds a copy to your library of keyboard macros. You can suspend this feature by checking the "Suspend macro recording" option at the top of the window. Even with this option selected you can still record and playback macros. They just won't be added to the library.

Macros start off with a generic MacroN name where N is an increasing integer value that depends on how many macros you have in the library already. You can select a generically named macro from the list and type in a more appropriate name in the Macro name field. Enter a short sentence or two in the Description field at the bottom of this window to help you remember what the macro does when you run it. BricxCC even lets you assign a hot key to each of the macros in your macro library. You can also run the selected macro from the manager dialog by clicking the Run button.

In addition to managing and configuring a library of keyboard macros, this dialog also gives you the ability to edit the macros you have previously recorded. You can even create new macros without using the recorder at all. When you click either the Edit button or the Create button, you will see the Macro Editor dialog, which is shown in Figure 5-36.

Figure 5-36. The macro editor window

This dialog contains a mini programmer's editor that operates very similarly to the main editor in BricxCC. It even uses syntax highlighting when displaying the macro source code. If you edit an existing macro, you will see a text-based representation of the keyboard macro that lists each of the editor commands you used when you recorded the macro. When creating a new macro using this dialog, the body of the macro is initially empty (aside from the macro begin-end structure as shown in the code sample below).

```
macro unnamed
begin

end
```

The macro editor also has built-in code completion for the editor commands that you can use when writing a keyboard macro. The macro language itself is very simple. There are no looping constructs or conditional statements like you find in NXC. However, you can define a sequential list of editor commands that should be executed whenever the macro is played back. Figure 5-37 shows the code completion function in action while editing a keyboard macro.

Figure 5-37. Code completion for keyboard macros

If you haven't already done so, I highly recommend that you spend some time recording and playing back macros in the editor. When put to good use, the keyboard macro feature in BricxCC will enable you to automate the most arduous programming tasks.

Compiling and Running

There isn't a lot to cover when it comes to compiling, downloading, and running your programs on the NXT. It just works. These options are exposed via hot keys, menu options on the Compile menu, and toolbar buttons. Simply press F5 to compile your code, F6 to compile and download, and Ctrl+F5 to compile, download, and run the program on the NXT. You can also run the current program if it is already downloaded using the F7 key. Press F8 to stop any program that is currently running on the NXT.

But there is more to compiling than just pressing a button. Sometimes your program does not compile and the compiler returns error messages to help you figure out the problem. BricxCC displays these error messages at the bottom of the editor window as shown in Figure 5-38.

You'll often see more than one error message for a single problem in your code. That's because the compiler tries to keep going after the first error, but it will often be out of sync with the source code so it may think there are problems when there really aren't any. If you fix the first error message, you may see that all the others go away. As you can see

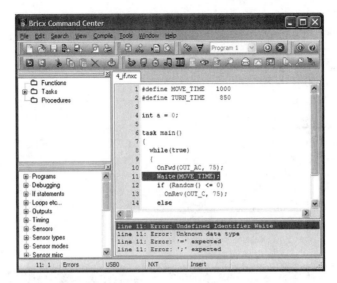

Figure 5-38. Compiler error messages

in Figure 5-38 above, BricxCC has automatically highlighted the line containing the first error. If you click on other lines in the error window, BricxCC will jump to the specified line number in the editor. This makes it extremely easy to navigate through your code to fix any errors found by the compiler.

Another often overlooked feature in BricxCC is the F12 hot key which is associated with the View|Show Code/Error Listing menu option. If your program compiles successfully, this window will contain a complete listing of the assembly language generated by the compiler. If there are errors when you compile, the full error text output appears in this window. I use it a lot to look for additional information about errors or to examine the low level code produced by my NXC program. Figure 5-39 shows this dialog with the error details from the example in the previous screenshot.

```
Full errors in 4_if.nxc
# Error: Undefined Identifier Waite
File "C:\DOCUME~1\Owner\LOCALS~1\Temp\temp.nxc" ; line
#      Waite(
#-----------------------------------------------------
# Error: Unknown data type
File "C:\DOCUME~1\Owner\LOCALS~1\Temp\temp.nxc" ; line
#      Waite(
#-----------------------------------------------------
# Error: '=' expected
File "C:\DOCUME~1\Owner\LOCALS~1\Temp\temp.nxc" ; line
#      Waite(1
#-----------------------------------------------------
# Error: Unknown data type
File "C:\DOCUME~1\Owner\LOCALS~1\Temp\temp.nxc" ; line
#      Waite(1000);
#-----------------------------------------------------
# Error: ';' expected
File "C:\DOCUME~1\Owner\LOCALS~1\Temp\temp.nxc" ; line
```

Figure 5-39. F12 code/error listing window

Each time you compile your program, the error message window at the bottom of the editor will hide. If errors occur it will become visible again as we saw earlier. But you can also hide the error window manually if you desire using the Ctrl+H hot key or the View|Hide Errors menu option.

Exploring the Tool Windows

Now we've come to a fun feature in BricxCC, the Tool menu. It is filled with a diverse set of tools to interact directly with your NXT (Figure 5-40). A few of these items have stand-alone utility siblings, which we will examine in Chapter 6. We will skip over these for now in favor of the tools that have no other equal elsewhere. Just remember when you get to the next chapter that a lot of the utilities you learn about there can also execute from within BricxCC as well.

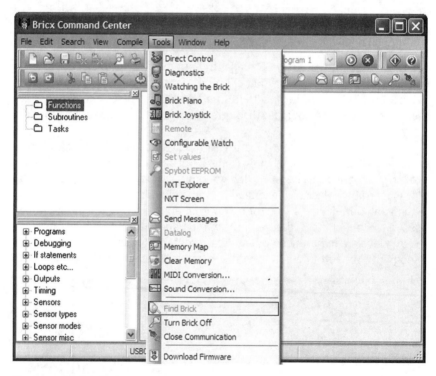

Figure 5-40. The Tool menu options

Nearly all of the tools in BricxCC are non-modal, which means you can have different tool windows open at the same time. All the tools that we examine in this chapter can be used simultaneously. Many of the tools are accessible via both a menu option and a toolbar button for your convenience. Selecting either the menu option or the toolbar button will switch between showing the tool and hiding the tool. If a tool window is obscured by the main BricxCC window you can also use the View|Tool Windows menu option to bring all visible tool windows to the front.

Direct Controller

Starting at the top of the menu we see the Direct Controller tool (Figure 5-41). This tool is also the first button on the Tools toolbar.

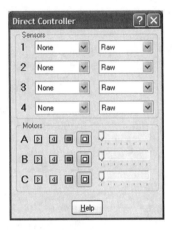

Figure 5-41. The Direct Controller tool

With the Direct Controller you can configure any of the four input ports to a specific sensor type and mode. If you want to interact with a sensor from within BricxCC, it is important to set these to the right values. If you plug an NXT light sensor into port 1, for example, it will not operate properly until you select the Light Active or Light Inactive sensor type in the drop-down list. If you select the active version then you'll notice that the LED turns on. You can also try plugging in a motor to one of the three output ports and control it using the buttons and power level slider. The power level slider will adjust the speed of your motor in increments of 14 so there are a total of 7 levels that you can select with the slider. The red button stops the motor with braking applied while the yellow button allows the motor to coast to a stop.

Diagnostics

The next tool on the menu is the Diagnostics tool (Figure 5-42).
This tool is the second button on the Tools toolbar.

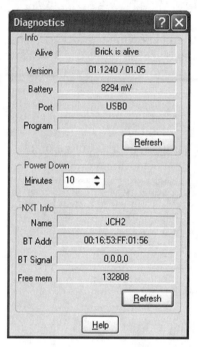

Figure 5-42. The Diagnostics tool

You can monitor the battery level using this tool and set the power
down time in minutes. The Diagnostics tool displays the firmware and
protocol versions as well. If a program is running on the NXT, its name
is shown. You can also retrieve the Bluetooth name and address as well
as the Bluetooth signal level for all four connections. Finally, this tool
also shows the number of bytes of free flash memory left on the brick.

Watching the Brick

Perhaps the most frequently used BricxCC tool window is the Watching
the Brick tool (Figure 5-43). This tool is the third button on the Tools
toolbar. You will probably use this tool together with other tool windows
such as Direct Controller or Brick Joystick. You can also use it while
your program is running on the NXT so that you can monitor and
analyze sensor and motor data.

Each watch item can be enabled or disabled using the checkbox
to the left of the item name. For motors it is a combination of the A, B,
or C checkbox and the watch name that determines whether a watch
is active or not. The All and None buttons can be used to either select
all the watches or to select none of the watches. The Poll Now button
will retrieve all the watch data a single time, while Poll Regular will

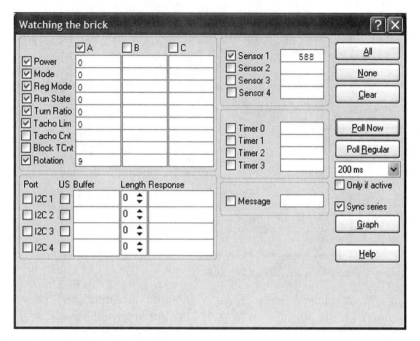

Figure 5-43. The Watching the Brick tool

retrieve the data at the interval you select. You can also check the "Only if active" option to tell the tool that it should only poll for data while it is active.

If you want to poll an I²C device you need to use the special section at the bottom left corner of this tool rather than the analog sensor watches labeled Sensor 1 through Sensor 4. You will also need to configure the port for use with a low-speed device using the Direct Controller tool. If you check the "US" checkbox then BricxCC will assume you mean to poll using the Ultrasonic Sensor's standard 0x02 0x42 read command with a single byte response. If you need to send a fancier command or read more than one byte, leave US unchecked and type in the data you want to write to the sensor in two digit hexadecimal pairs such as 02 41.

Graphing Watches

When you poll for data, such as a light sensor value or the rotation count of motor A, you have the option to collect each data point and graph it. Use the Graph button to display the BricxCC graph window and start collecting data points (Figure 5-44). Data for this graph came from the Ultrasonic sensor attached to port 1 with regular polling every 200 milliseconds.

If you have multiple watches selected in the watch window the graph will contain multiple series. By default the chart is a 3D line

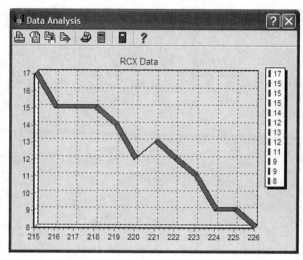

Figure 5-44. Graphing a watch in BricxCC

graph. The legend will show recent data points if there is only one series but when multiple series exist it will show the name of each series along with the color associated with it. If you don't like the default chart settings, you can bring up the Chart Configuration dialog and reconfigure it however you wish. The General tab of the configuration dialog is shown in Figure 5-45 below.

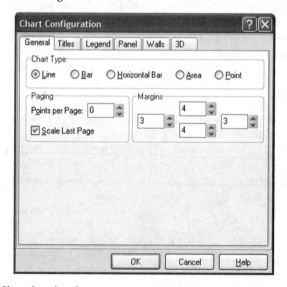

Figure 5-45. Choosing the chart type

On the General tab you can choose between five different basic chart types: line, bar, horizontal bar, area, and point. You can configure the chart margins and the paging settings as well. Each of the pages in this dialog presents different chart options that you are able to tweak and modify however you see fit. BricxCC will redraw the chart to reflect the new chart options once you confirm your changes. For best results you should not switch chart types after you have already started collecting data. The graph doesn't show the existing data points correctly since it has to delete the existing series and create new ones for the newly chosen chart type. Instead, choose your desired chart type and then start polling for data via the watch window.

On the Titles page, as shown in Figure 5-46, you can change the chart title and add text to the footer if you desire. You can turn on a frame for the title or footer and configure the line style, width, and color. And you can specify whether the title and footer should be centered, left aligned, or right aligned.

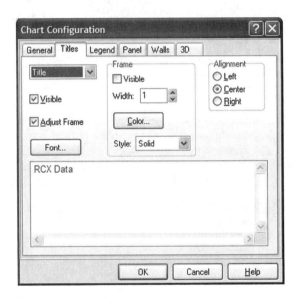

Figure 5-46. Configuring chart titles

Use the options on the Legends page, shown in Figure 5-47, to specify exactly how you want the legend to look on your chart. You can turn it off if you prefer or you can tweak just about any property you can imagine. Consult the BricxCC help system for more information about these options or try them out.

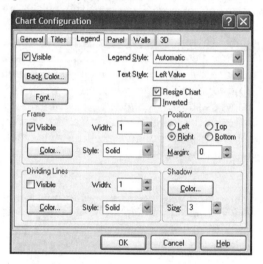

Figure 5-47. Configuring the legend

The chart panel is the entire background of the chart from the top left corner of the graph window, below the toolbar, to the bottom right corner of the form. You can configure the color of this background on the Panel page of the chart configuration dialog, as shown in Figure 5-48. If you prefer you can turn on a gradient and configure its colors and direction.

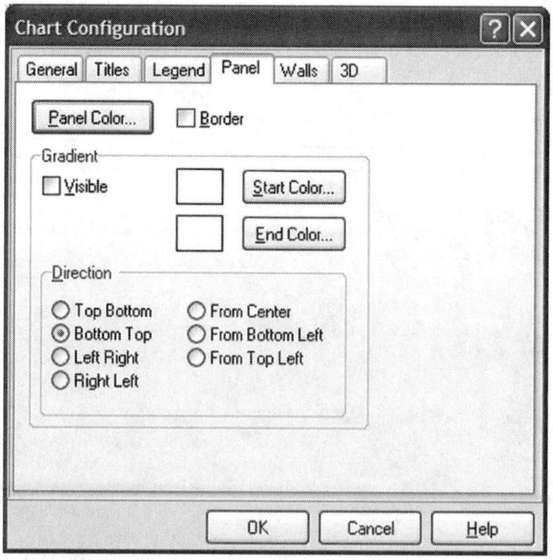

Figure 5-48. Configuring the chart panel

You can also configure a few settings for the three chart walls on the Walls page of the chart configuration dialog (Figure 5-49). Set the background color, the border visibility and style, and the size and transparency of each wall. The size value lets you specify how thick to draw the chart wall.

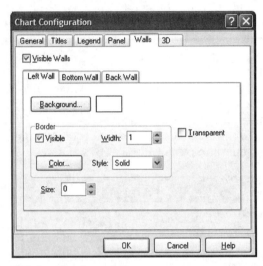

Figure 5-49. Configuring the chart walls

Figure 5-50 shows the 3D page of the chart configuration dialog, where you can control the overall 3D appearance of the chart. If you don't want a 3D look you can disable that option here. You can set the depth of the 3D effect and whether to draw the graph with a fixed orthogonal orientation. If you disable the Orthogonal option you can

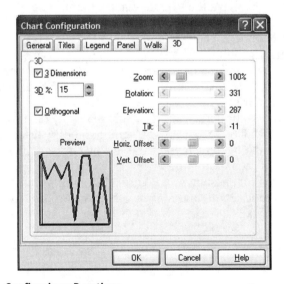

Figure 5-50. Configuring 3D options

drag the Rotation, Elevation, and Tilt sliders to rotate the graph in three dimensions. The Zoom slider lets you zoom in and out of the graph. You can shift the graph horizontally and vertically as well using the horizontal offset and vertical offset sliders.

The graph toolbar gives you a number of options. Use Ctrl+P or click the print button to print the graph. The chart configuration dialog is available via Ctrl+G or through the second toolbar button. The third button is the copy to clipboard option. This option copies all the data collected by the graph tool to the clipboard. Use the Ctrl+C hot key for easy access to this feature.

The next option is the Export chart feature. Figure 5-51 shows the export chart dialog where you can choose the bitmap, metafile, or enhanced metafile format for the chart image export. The image can either be copied to the clipboard or saved directly to a file. The hot key for exporting the chart image is Ctrl+E.

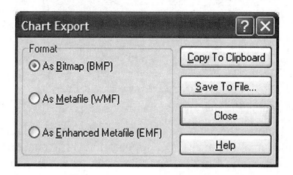

Figure 5-51. Exporting the chart as an image

Not only do you have the overall chart configuration options shown above but each series in the chart has its own set of options as well. The exact set of options depends on the selected chart type. Here we will look at just the options for the default line chart type. Experiment on your own with configuring series options for the other chart types. Use the "What's This" help feature for helpful hints for each option and consult the main BricxCC help system for additional details.

You can access the series configuration dialog by holding the control key and left clicking on the series you want to modify. Alternatively, use the select series dialog, which you can view by clicking the select series toolbar button or by pressing the Ctrl+S hot key. The series type page of the configuration tool is shown in Figure 5-52. For line charts you will see three tabs: Line, Point, and Marks (other chart types have their own unique set of pages).

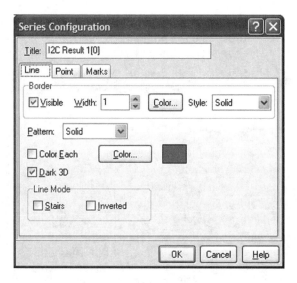

Figure 5-52. Configuring a series style

For a line series you can select whether or not to show the line border and configure its width, color, and line style. The 3D version of the line will have a fill pattern, which you can choose here. You can set the color for this series or pick the "Color Each" option, in which case the color of each section of the line will be automatically chosen for you.

On the Point page of the series configuration dialog you can turn on the drawing of each data point using the options shown in Figure 5-53 below. The points are Square by default but you can choose Circle, Triangle, and Down Triangle options instead. The width and height of each point is also configurable.

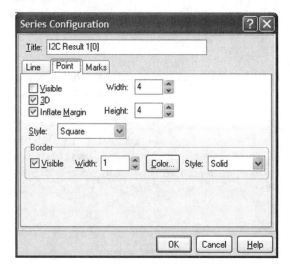

Figure 5-53. Configuring series point settings

The last page lets you turn on data markers that display the actual data value on the chart itself. See Figure 5-54 for a screenshot of the options available for the series data markers.

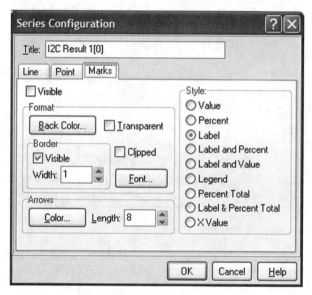

Figure 5-54. Configuring series mark values

That's about it for graphing data in BricxCC. There are a lot of options to choose from but you can always just use the default settings if you don't want to customize. All of these capabilities are available for the NXT when using the BricxCC watch windows. The possibilities are practically endless. And you can use the graphing feature with all the previous generations of LEGO programmable bricks such as the Spybot or RCX brick. With BricxCC you can graph not only the RCX watch data but also the RCX datalog.

Brick Joystick

The next tool we'll examine in this chapter is the Brick Joystick tool (Figure 5-55). It is the fifth button on the Tools toolbar.

This tool is a lot of fun. You can configure which motors are used for driving and steering and the style of mobile robot you have designed. You can also set the desired motor speed using the slider at the bottom of the window. Then, using your mouse or keyboard, you can press the various arrow keys and watch your robot move around. The great part is that you can even use a Joystick connected to your computer to control the motors!

Figure 5-55. The Brick Joystick tool

Configurable Watch

Now let's look at the Configurable Watch tool (Figure 5-56). This tool is the seventh button on the Tools toolbar. It is basically a slightly modified version of the main BricxCC watch window that we examined above.

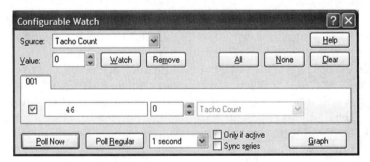

Figure 5-56. The Configurable Watch tool

The main difference between this tool and the Watching the Brick tool is that you pick watch sources and values at the top of the window and then add a watch for that combination using the Watch button. Whenever you click the Watch button, a new page is added to the page control in the middle of the window. Conversely, whenever you click the Remove button, the active page is removed from the list of watches. The remainder of this tool, including the ability to graph multiple watches and automatically poll for data at specified intervals, functions exactly like the main BricxCC watch window.

Send Messages

The Send Messages tool window is shown in Figure 5-57. This tool is the tenth button on the Tools toolbar. You can use it to send messages to the NXT either via Bluetooth or USB.

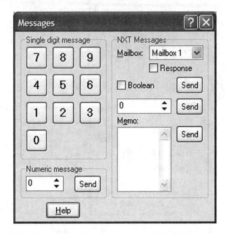

Figure 5-57. The Send Messages tool

All the messages that are sent using this tool adhere to the NXT's standard Direct Command protocol. The direct command packet contains a MessageWrite command (0x09). You need to specify not only the data to send but also the mailbox or queue on the NXT where you want to store the data.

In NXT-G the block for reading messages can handle three different message body formats: boolean, numeric, and string. NBC and NXC have API commands that perform the same function. A boolean message will always have two bytes in the message body. The first byte is either 1 or 0 for the true or false boolean value. The second byte is always a null terminator, zero. A numeric message always contains five bytes. The first four bytes are a 32 bit value (little-endian) containing the number you want to send. The fifth byte is, once again, a null terminator. A string message simply contains all the bytes in the string message plus a zero at the end. The maximum number of bytes that can be sent, including the terminator, is 59.

Memory Map

The next tool is the Memory Map tool. You can see what it looks like in Figure 5-58. It is the twelfth button on the Tools toolbar. If you have used this tool in BricxCC for bricks other than the NXT, you may notice that the data for the NXT is very different from what you have seen before.

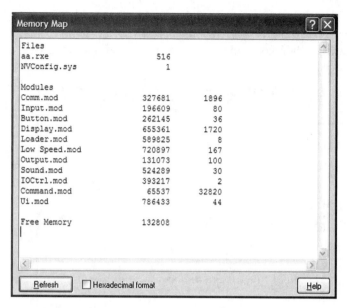

Figure 5-58. The Memory Map tool

With the NXT the tool contains a list of all the files on the NXT along with their size. Following this list is a list of all the firmware modules and their module identifier and module IOMap size.

Configure Tools

The last standard item on the Tools menu is the Configure Tools option. This option gives you the ability to add other tools to the Tools menu so that you can easily launch them from within BricxCC. When you choose this option you will see the window in Figure 5-59 below. Any executable that you add using this dialog is added on the Tools menu below the Configure Tools option.

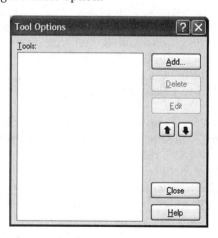

Figure 5-59. Configuring new tools

Use this window to add, delete, and edit your own tool items. The up arrow and down arrow buttons let you order the tools to your liking. When you add or edit a tool the dialog in Figure 5-60 is displayed.

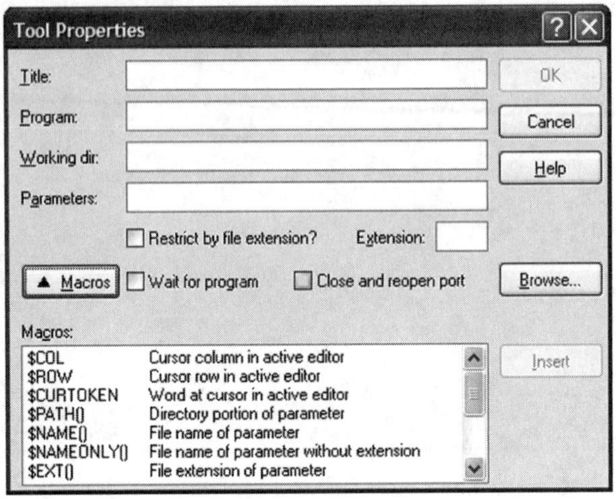

Figure 5-60. The tool properties dialog

Here you can configure the title to show on the Tools menu for your tool. You can use the Browse button to select the executable you want to launch or you can just type in the executable name in the Program field. It is important that the working directory match the path you enter for the program. When you use the Browse button both of these fields are automatically configured correctly.

You can also specify parameters to pass to the executable when it is launched. The tool parameters can use special macros in order to send the program information from BricxCC, such as the editor filename or the word at the cursor location in the active editor window. The BricxCC help system describes each of the supported macro tokens in detail. Other settings here let you choose whether or not to wait for the program to complete, whether or not to close the connection to the brick and reopen it after running the program, and whether or not to hide this tool unless the active editor contains a file having a specific file extension.

The two dialogs shown above for configuring additional Tools are shared with another area of BricxCC that involves compiling your programs. In the Preferences dialog you can define a set of pre-compile and post-compile steps that BricxCC will automatically execute in sequence before and after each compile operation. The buttons used to configure these steps are found on the Common tab on the Compiler page as shown in Figure 5-61.

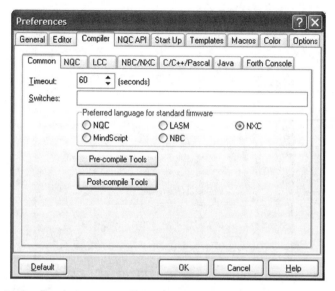

Figure 5-61. Configuring pre-compile and post-compile steps

Since you can restrict tools to only execute for certain file extensions, you can define tools that are only designed to run before compiling NBC code, for example, and that tool will not launch when you are editing an NXC program. Normally these additional compiler steps wait for the program to complete before going on to the next step.

Getting Help

It is worth mentioning BricxCC's robust help system. This system will help you learn all the features that the IDE provides, and guide you along the path of learning the supported programming languages. Especially important is context sensitive help for language keywords and API commands.

If you are editing a program and you are not sure what parameters to use for the RotateMotorExPID command, you have the definition of the command at your fingertips. The help system lets you search using an index or simply press the help hot key to display the relevant page right away.

Figure 5-62 shows the main BricxCC help contents page. You can drill down into the topic of your choice using this tree structure, or you can click the Index tab and type in a keyword. The BricxCC help system opens the page that you select either via the Contents page or via the Index.

Figure 5-62. The BricxCC help contents

When you press F1 (or the key you assign to the ecContextHelp editor command) the editor will pass the current token in the editor to the help system so that it can show the appropriate help page. As you see in Figure 5-63, the resulting window contains a description of the selected function, along with all the pertinent information about how to use it correctly.

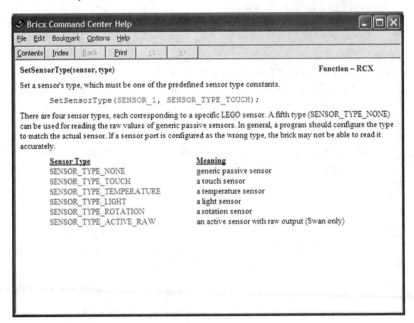

Figure 5-63. Context sensitive language help

As we have seen previously, the BricxCC help system also provides the "What's This" help hints that explain various tool options throughout the IDE. These hints are often all the information you'll need to learn how to best use the BricxCC tools and the features provided by the programmer's editor to strengthen your abilities as an NXT programmer. In the next chapter we'll take a look at a number of additional tools you can use to augment the power of BricxCC.

NXT Utilities

Topics in this Chapter

- Creating Melodies
- Creating Sounds
- Creating Pictures
- Decompiling NXT executables
- Exploring the NXT
- Virtual NXT

Chapter 6

I n addition to using Bricx Command Center on the Windows platform for your NXT programming needs, you can find several freely available NXT utilities on the web. These utilities are extraordinarily useful for all kinds of tasks related to programming the NXT. Knowing where to find and how to use these utilities can dramatically increase your programming power. Many of them are also available for non-Windows platforms, so those of you who are running Mac OSX or Linux are not left out.

In this chapter we will explore some of the best offerings available on the web and see how they can be used to help us create and control our custom NXT robots. Most of the utilities are hosted as part of the BricxCC project on SourceForge. The NXT Utilities page on the project website is the one stop shop for downloading these tools.

 WEBSITE: *Download the utilities in this chapter from the NXT Utilities page on the BricxCC website at* http://bricxcc.sourceforge.net/utilities.html.

Creating Melodies

A popular feature of the NXT and its firmware is the ability to play files containing PCM sound wave samples wrapped up in the NXT sound file format (with a .rso extension). The NXT software comes with many sounds that can be used in an NXC, NBC, or NXT-G program to make your robot speak or play realistic sound effects. Using an NXT sound file conversion utility, like the one we will discuss in the next section, you can easily create additional sounds for use in your programs.

Unfortunately, even though NXT sound files are fairly low quality, using just single-channel, 8-bit samples with an 8-kilobyte sample rate, they can still take up large amounts of your brick's flash memory when they are used in a program. The First LEGO League (FLL) published a document entitled, "Writing Efficient NXT-G Programs", which recommends that you "minimize use of Sound and Display blocks" in order to more optimally use the available flash memory. With flash memory limited to around 128 kilobytes, it is important to pay close attention to how much space is taken up by sound files.

WEBSITE: *Read the FLL recommendations for writing efficient NXT-G programs at* www.firstlegoleague.org/sitemod/upload/Root/ WritingEfficientNXTGPrograms2.pdf.

The size problem associated with sound files brings us to a little-known feature of the NXT firmware called melody files. What is a melody file? You might think of a melody file as being something like a MIDI file. Where NXT sound files and WAVE files contain sound samples (8 thousand of them per second in sound files) melody and MIDI files contain notes or tones that are played for a specific duration. MIDI files are much more complex than melody files but the analogy helps explain the difference between NXT sound and melody files. NXT melody files have the .rmd file extension.

If a sound lasts for more than a split second, the number of samples that are stored in the file causes the file size to grow rapidly. With a melody file, playing a single tone for ten seconds takes only four bytes, two bytes for the frequency or tone and two bytes for the duration. An uncompressed NXT sound file lasting ten seconds would take something like 80,000 bytes. A compressed NXT sound file of the same length will be 40,000 bytes long; ten-thousand times larger than the melody file.

Of course, melody files can't reproduce speech or incredible sound effects. The music that is played using a melody file won't have the quality that is usually found in a sound file. But the size of even very long melody files is much smaller than the size of short sound files.

File creation

You can create melody files in a few different ways. Each has its pluses and minuses. In addition to the many tools we saw in the previous chapter, Bricx Command Center also has a Brick Piano tool that is used to manually construct a song using a series of tones and rests (Figure 6-1). You can save the resulting song in many different formats, one of which is the NXT melody file format. For simple melodies this is an easy way to create a melody file. The Brick Piano is also available as a standalone utility that supports Windows, Mac OSX, and Linux platforms.

When creating a melody file, be sure to set the Language option at the bottom of the window to NXT Melody. You can use this same tool to generate either NBC or NXC source code that will play back the song you create using the virtual keyboard. If you are creating NBC or NXC code, you can copy the code to the clipboard via the Copy button. When generating an NXT Melody file, the Copy button copies frequency and duration values in text format to the clipboard but the NXT Melody is not generated, unlike with NBC and NXC code. The following code sample shows you what the contents of the clipboard look like when you have NXC as your selected language.

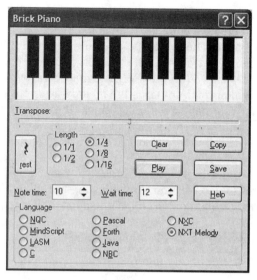

Figure 6-1. Brick Piano utility

```
PlayTone(659,400);
Wait(480);
PlayTone(698,400);
Wait(480);
PlayTone(784,400);
Wait(480);
```

You can use your mouse to transpose the mini virtual keyboard up and down through a six-octave range. Select whether to record whole notes, half notes, quarter notes, eighth notes, or sixteenth notes using the Length radio button options. Press the rest button to record a rest with a particular length. You can also adjust the basic note times and rest times using the numeric edits shown in Figure 6-1. These values are in hundredths of a second. When you are ready to record a note, click on the appropriate virtual piano key.

On platforms with NXT communication support built-in, when you click on a piano key the NXT will play the note you have selected. On supported platforms, clicking Play will playback the entire song on the NXT brick. The rest of the functionality in this utility works the same regardless of the operating system.

If you want to clear all your recorded notes and start again, press the Clear button. However, once you have completed your masterpiece, click the Save button and type in a filename when prompted. Your composition is stored as an NXT Melody or as NBC or NXC source code if you selected one of those options.

MIDI conversion

An easier way to create a complex tune is to convert an existing MIDI file into a melody file. The difficulty with this approach is that many MIDI files use multiple tracks, each of which only play some of the notes that comprise the song. Without a MIDI editor or track viewer, it may be hard to tell which track contains the melody as opposed to the bass guitar or drums. You may have the best luck finding ring tone MIDI files on the Internet, which often use just a single track.

The MIDI Conversion tool in BricxCC is used to convert MIDI files into the NXT melody file format as well as several other formats (Figure 6-2). Using this tool, you pick which track you want to convert and the format to use for the conversion. As with the Brick Piano, this tool is also available in a standalone flavor with support for multiple platforms.

Figure 6-2. MIDI conversion utility

This utility is quite simple to use. You pick your output language from the options available. For our purposes, pick NXT Melody as we did with the Brick Piano tool. Notice that the Clipboard destination option is disabled when you select NXT Melody as the desired language.

The parameters group has several options that let you control how the notes in the MIDI file are converted into frequencies and durations in the NXT Melody file. In most cases you can leave these values unchanged and you'll be pleased with the resulting melody file. The primary parameters are the track selection and the tempo value. If the tempo specified in the MIDI file is very different from the tempo you select here then the resulting melody will play either faster or slower. If you pick track zero then the first non-empty track in the MIDI file is used for conversion.

Use the transpose option to shift the notes up or down octaves. The gap value is used during the conversion to determine whether a new note should be played or whether to continue with the existing note. Larger gap values mean that there must be a larger note gap in the MIDI file before a new note will be generated during the conversion process.

If you want MIDI pitch bend commands to play a role in the conversion, enable this option. With this option selected you can adjust the sensitivity level, which controls how far a note must bend before it will be recorded as a different note in the resulting output. You may want to experiment with all of these settings to see their impact on the resulting melody file.

TRY ME: *Try downloading a MIDI file of your favorite song from the Internet and converting it to an NXC tune using the conversion utility.*

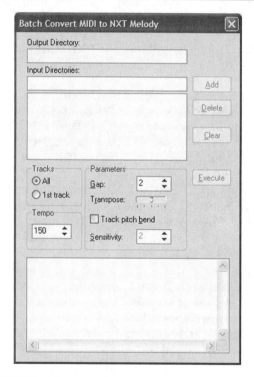

Figure 6-3. NXT MIDI batch conversion

Batch conversion

The quickest way to create multiple complex melody files is to use the standalone NXT MIDI batch conversion utility, which makes it possible to convert many MIDI files all at once (see Figure 6-3). You can choose to either convert just the first track or all the tracks in the MIDI file.

The batch conversion mechanism does not support combining multiple MIDI tracks into a single melody file. However, you can process the MIDI file using a standard MIDI editor to combine tracks before performing the conversion to melody files.

WEBSITE: *Download Audacity, a popular open-source MIDI editor: http://audacity.sourceforge.net*

The settings in the batch convert utility have the same semantics and operational behavior as the MIDI conversion utility above. In the batch utility you type in a directory for all the resulting melody files. You also enter one or more source locations where the MIDI files can be found. Once you have configured the parameters and selected the input and output locations, you can click execute to begin the batch conversion process. As the utility converts the specified track(s) from each MIDI file, it writes the destination filename to the memo field at the bottom of the window.

Like the other standalone utilities, the NXT MIDI batch conversion utility is available for multiple platforms, including Mac OSX, and Linux.

Creating Sounds

If you have room for large files on your brick and you require higher quality sounds than that provided by NXT melody files, then you will have to fall back to the far larger NXT sound files. Often you will want to create your own sound files. This is, fortunately, very easy to do using the two utilities in this section.

As we have previously discussed, NXT sound files contain PCM samples. And as of version 1.05 of the NXT firmware, the NXT sound files can also contain IAM ADPCM samples. It is easier to piggyback on existing software tools that record these types of sound files and then convert them to the NXT sound file format.

There are many freely available utilities that can record sounds in various formats. Likewise there are many different utilities available on the Internet that can convert back and forth between different sound file formats. All we really need is a utility which can take one of those many formats, such as the WAVE format, and convert back and forth between that format and the NXT sound file format.

Wav2Rso

This is where the wav2rso utility comes into the picture (Figure 6-4). As the name indicates, this simple utility converts .wav files into .rso files. It can also convert .rso files into .wav files. You can download wav2rso for Windows, Mac OSX, and Linux platforms.

The wav2rso utility lets you pick any number of input files using the Select Files button. The Directory button is used to select the destination directory for the converted files. If you are converting WAVE files into

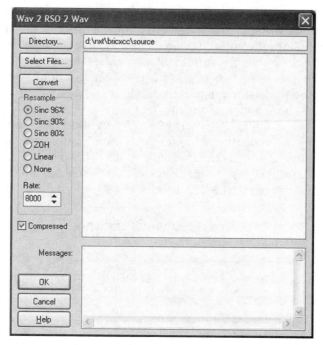

Figure 6-4. Wav2Rso conversion

NXT sound files then you also need to select the desired output sample rate, which can range from 2k to 16k samples per second. If the input sample rate is different from the desired output sample rate then a sample rate conversion must occur.

Sample rate conversion is also known as resampling. If you convert from a lower sample rate to a higher sample rate it is called upsampling. Decreasing the sample rate is called downsampling. Whenever you change the sample rate of a digital audio waveform, you should pay attention to the Nyquist frequency. This frequency is half the sampling frequency of the signal processing system. If the Nyquist frequency is greater than the maximum bandwidth or frequency of the waveform being sampled then no aliasing will occur. This means if you convert a WAVE file to an NXT sound file with a 2k-sample rate, then the maximum frequency that can be represented without errors is somewhat less than 1kHz.

A waveform containing frequencies above the Nyquist frequency, as well as frequencies slightly less than the Nyquist frequency, will result in the sampled data containing aliasing errors. This is when frequencies in the source end up being represented in the sampled waveform as lower frequency aliases or duplicates. Frequencies above the Nyquist frequency are indistinguishable from various lower frequencies and are therefore introduced as flaws or artifacts in the sampled data.

In order to avoid errors in the output, a low-pass filter should be applied to the original data to remove frequencies above the Nyquist frequency. Since the cut-off frequency of a low-pass filter is not a hard limit (a brickwall filter) you actually have to set the cut-off below the Nyquist frequency so that as the low-pass filter cuts out frequencies, by the time the Nyquist frequency is reached there are no more frequencies left in the source with any amplitude or energy.

The wav2rso utility supports five different resampling methods, which vary in the quality of the resulting samples from very good to poor. The best quality is the sinc 96% option. The name "sinc" comes from sine cardinal. The normalized sinc function is an important part of digital signal processing. Its definition is shown in Figure 6-5.

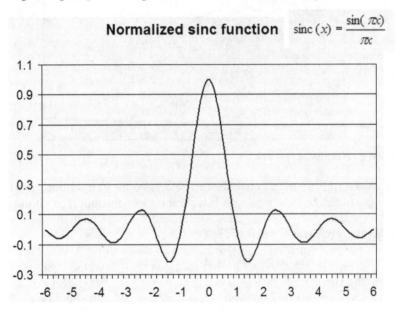

Figure 6-5. The normalized sinc function

The first three resampling methods use sinc function interpolation with limited bandwidths ranging from 96% to 80%. Other options represent decreasing levels of quality as you make selections further down the list of resampling methods. You can also choose whether or not to compress the resulting NXT sound file. Once you've selected the input files, the output directory, and the options for the conversion you simply click the Convert button (see Figure 6-4 above).

The utility will process all the input files and write a log message to the memo at the bottom of the window for each file it processes. The message will indicate whether it was able to perform the conversion or not. If the conversion completes successfully, the utility will log the name of the resulting file. In each case, all WAVE files are converted into NXT sound files while all NXT sound files are converted into WAVE files. Input WAVE files must be uncompressed PCM formatted files in order for the utility to process them successfully.

WavRsoCvt

Another option for converting back and forth between WAVE and NXT sound files is the command line utility called wavrsocvt (Figure 6-6). It runs on the same platforms as wav2rso. Since it is a command line utility you can easily incorporate it into scripts that chain together other conversion tools in order to convert other sound formats. Similarly you could call this utility from a script that takes the output sound file and downloads it to the NXT.

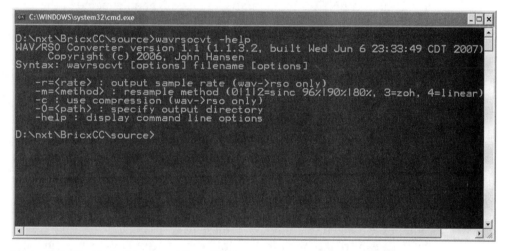

Figure 6-6. WavRsoCvt command line utility

You can specify the conversion options as command line switches as shown in Figure 6-6 above. If you need to compress the sound file, use –c on the command line. Specify the sample rate with –r=N where N is a number from 2000 to 16000. Finally, you can specify the resample method using –m=N where N is 0 through 4, as you can see detailed in the figure above. Method 0 is the highest quality band-limited sinc interpolation method. Method 4 is the lowest quality linear method.

Creating Pictures

NXT pictures are often used to spice up a robotics program. The Alpha Rex robot uses two heart pictures to display something like a beating heart on the LCD while the robot moves around. Ross Crawford's mini chess NBC program uses several graphics to draw the chess pieces on the screen. Arno van der Vegt's incredible Space Invaders NBC program also uses NXT pictures with amazing results.

The LEGO MINDSTORMS NXT software ships with nearly one hundred little pictures that you can use in NXT-G programs or programs written in NBC or NXC. They are a great starting point for introducing images in your own programs. They won't, however, fill every need when it comes to graphics. To create your own picture files, there is no better utility than the excellent nxtRICedit written by Andreas Dreier. We'll see more of this utility later in this chapter.

WEBSITE: *Download the nxtRICedit utility for creating picture files at* http://nxtasy.org/2006/10/01/nxtricedit-pc-ric-editor/

With the LEGO MINDSTORMS NXT software you can browse through the list of available pictures whenever you wire a Display block into your NXT-G program. As you highlight a file name in the list, the preview window shows you the image. This may give you the impression that an NXT picture is kind of like a bitmap file - that the RIC file format is some form of an image file format akin to .bmp or .gif or maybe .jpg. But that could not be further from the truth. These picture files are actually very different from your typical image file.

An NXT picture actually contains a script or simple code that consists of a list of drawing commands. These commands, or opcodes, are executed whenever you use the Display block in NXT. They call the GraphicOut API function in NBC and NXC, or directly execute the low-level syscall opcode in NBC with the DrawGraphic function along with a TDrawGraphic argument structure. The opcodes in the NXT picture are executed in the order that they occur within the file as it is read by the firmware in response to the DrawGraphic system call. Like other executable files on the NXT, these pictures must be continuous, which means that they cannot be fragmented across multiple sectors in the NXT filesystem. Other types of files, such as data files or sound and melody files, do not have this restriction.

Another important issue you should be aware of, regarding not only NXT pictures but also any other type of file on the NXT, is that every file on the NXT takes up one or more chunks of file system (flash) memory and each chunk is 256 bytes in size. You may have an NXT picture that is quite small, perhaps only 40 bytes or so, but it will still use up one 256-byte chunk of flash memory. This means that when you create an RIC file you may want to optimize it so that its size doesn't go slightly over another 256-byte boundary. Keep this in mind when you are creating sound files and melody files as well. For a detailed look at the NXT picture format, see Appendix D.

Parameterized Pictures

You can parameterize NXT pictures using nxtRICedit. This para-meterized value is an index into the Variables array that you can pass into the GraphicOutEx function in both NBC and NXC.

By using parameterized values in your NXT picture, you can use a single NXT picture over and over again to draw very different images on the NXT screen by passing different values into GraphicOutEx via the Variables array. The small program below demonstrates a simple approach to drawing a parameterized NXT picture.

```
int Values[] = {0};
void Display(int n) {
  Values[0] = n*10;
  GraphicOutEx(Values[0], 0,
               "letters.ric", Values, true);
  Wait(200);
}

task main() {
  while(true) {
    for(int i=0; i<9; i++ )
      Display(i);
  }
}
```

The letters.ric file used in this example is very simple. It is a single row of nine 10x8 pixel images stitched together in a tiled fashion one after the other. Each of the small images contains a hand-drawn letter. You'll find a screenshot in the next section that shows how this image looks in the nxtRICedit utility. The total image width is 96 pixels since the last letter is only six pixels wide. The image can be written in RICScript using the code you see below.

```
// letters.ric
desc(0, 96, 8);
sprite(1,
  0xCC3F0FC3E0783E0F8208781E,
  0xCC3F0FC3F0FC3F0FC318FC3F,
  0xCC0C030330CC330CC318CC30,
  0xFC0C030330CC330CC1B0CC30,
  0xFC0C0303F0CC330FC1B0FC30,
  0xCC0C0303E0CC330F80E0FC30,
  0xCC3F030300FC3F0DC0E0CC3F,
  0xCC3F030300783E0CC040CC1E);
copybits(0, 1, arg(0), 0, 10, 8, 0, 0);
```

Each argument in the sprite command following the data address represents a row of bits (or pixels) in the sprite image. The 24 hexadecimal digits represent a total of 96 bits, since there are 8 bits per 2-digit hexadecimal number. RICScript requires an even number of hexadecimal digits in each row bits argument. Since there are eight row bits arguments following the data address there are eight rows of pixels in the sprite image. The only parameterized value in this picture happens to be in the copybits command. You can see the arg(0) parameter which indicates that this particular value, the copy bits source rectangle X value, is parameterized and it will use the first element in the Variables array passed into the GraphicOutEx API function.

A fancier example of how you can take advantage of parameterized NXT pictures is demonstrated in the code samples and screen captures below. The NXT picture and the sample NXC code were both originally created by Andreas Dreier, the author of nxtRICedit. First let's start with a single NXT picture containing tiled images of black and white chess pieces on black and white squares. The total sprite size is 104 pixels wide by 14 pixels tall. In other words, the Sprite command's RowBytes value is 13 and its Rows value is 14. The goal is to use these small images as part of a chess program written for the NXT.

To draw the entire board state in a single GraphicOutEx call the NXT picture will also need to have a total of 64 CopyBits commands. Each of them is parameterized so that they draw a different portion of the Sprite data at the appropriate chessboard position. The source rectangle has a parameterized X value. All of the remaining values in the source rectangle and both the destination point values are hard coded in the picture file.

The following code sample shows how this picture could be defined using RICScript. Notice the 14 row bits arguments to the sprite command with each value containing a total of 26 hexadecimal digits. For the sake of brevity, the remaining 60 copybits commands are not shown. The total size of the resulting RIC file is only 1472 bytes.

```
// Chess1.ric
sprite(1,
  0xFEFEFEFEFEFEFEFEFEFEFEFEFE,
  0xFEFEBEBEEEFEFEFE82BA86C686,
  0x86C2BEBAC6EEFED6EEB6BABABA,
  0xBABA86B6EEC6FEEEEE8E86AA86,
  0xBEC2BA8EEEEEEFED6EEB6BAB6B6,
  0xBEFA86B6F6FEFEFEEEBA86CABA,
  0xFEFEFEFEFEFEFEFEFEFEFEFEFE,
  0x00000000000000000000000000,
  0x00004040100000007C44783878,
  0x783C4044381000281048444444,
  0x44447848103800101070785478,
  0x403C44701010000281048444848,
  0x40047848080000001044783444,
```

```
          0x00000000000000000000000000);
copybits(0, 1, arg(0), 7, 7, 7, 0, 0);
copybits(0, 1, arg(1), 0, 7, 7, 7, 0);
copybits(0, 1, arg(2), 7, 7, 7, 14, 0);
copybits(0, 1, arg(3), 0, 7, 7, 21, 0);
// (the remaining 60 commands are not shown)
```

Once we have this NXT picture defined then we are ready to write our NXC program to draw the chessboard on the NXT screen. Each time a chess piece is moved we modify the appropriate entries in our array of 64 variables and call GraphicOutEx again. The initial board state and a few moves can be drawn with the very simple NXC code shown below.

```
int b[] = {
  64, 72, 80, 88, 96, 80, 72, 64, // 1
  56, 56, 56, 56, 56, 56, 56, 56, // 2
  48, 48, 48, 48, 48, 48, 48, 48, // 3
  48, 48, 48, 48, 48, 48, 48, 48, // 4
  48, 48, 48, 48, 48, 48, 48, 48, // 5
  48, 48, 48, 48, 48, 48, 48, 48, // 6
  40, 40, 40, 40, 40, 40, 40, 40, // 7
  32, 24, 16,  8,  0, 16, 24, 32  // 8
};
// A    B    C    D    E    F    G    H

#define Vacant 48

#define Move(_from, _to) \
  b[_to] = b[_from]; \
  b[_from] = Vacant; \
  GraphicOutEx( 8,8,"Chess1.ric" , b, true); \
  Wait(2000);

task main() {
  // setup board
  GraphicOutEx( 8,8,"Chess1.ric" , b, true);
  Wait(2000);
  Move(A2, A3); // white pawn from A2 to A3
  Move(B7, B5); // black pawn from B7 to B5
  Move(A3, A4); // white pawn from A3 to A4
  while( true );
}
```

Figure 6-7 shows how this code will look as it draws on the NXT screen. The chess file and rank references such as A3 and B6 in the above code are just simple preprocessor macros that make indexing into the board state array very easy. A1 is equal to zero, A2 equals 8, and so forth through all 64 elements in the array.

Figure 6-7. Chess using parameterized NXT pictures.

The chess example just shown requires that you have the enhanced NBC/NXC firmware installed on your NXT. That's because it uses more than 16 parameters in the NXT picture, which is the maximum number of parameters supported by the standard firmware. Being able to take advantage of the full 256 parameters in a picture file is a good reason to use a version of the NXT firmware that has this bug fixed.

nxtRICedit

Shortly after Andreas Dreier got his hands on the NXT, he completed the first version of his NXT picture file utility called nxtRICedit. It is the best available standalone tool for creating NXT pictures. You'll find that it makes creating your own pictures simple. Figure 6-8 shows the letters. ric picture loaded in nxtRICedit.

The tools on the left side of the window provide many different ways to create and manipulate your NXT picture. You can set pixels as well as draw lines, rectangles, filled rectangles, circles, and text. The current image can be positioned within the editor pane using the arrow buttons. You can flip the image about the X or Y-axis. And you can invert the black and white colors in the image. In addition to drawing sprites for use in your pictures, you can also add any of the supported opcodes to your picture with full support for value parameterization.

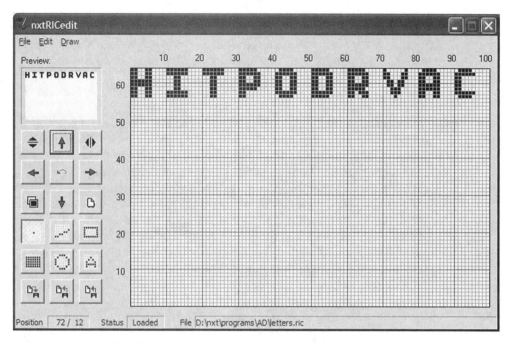

Figure 6-8. nxtRICedit utility

With this tool you can even import images in other formats and convert them into sprite commands which you can incorporate into your pictures quickly. The conversion from a multi-color image into the monochrome form required for RIC files is controlled via a contrast slider. It even lets you preview the resulting sprite image as you move the slider left and right to adjust the contrast setting.

TRY ME: *Using nxtRICedit, draw a robot face or other image, save the RIC file and upload it to your NXT for display.*

RICScript

Another option for creating NXT picture files is to use either BricxCC or the NBC command line compiler along with the RICScript language. From the utilities page on the BricxCC SourceForge project website you can also download a simple utility, called RIC Loader, which is available for multiple platforms. It provides a simple user interface for creating a new NXT picture and adding picture commands to it. We'll see more detail regarding the RICScript language in chapter 7.

Decompiling Executables

It may not be every day that you need to reconstruct the source code for an NXT program from the executable itself but there is a good chance that it will come in very handy at some point. Maybe you accidentally lose your program's source code. Or maybe you just want to find out how a program works and all you have is the .rxe file.

The NXT Program Dumper utility is a tool you can use if you ever need to decompile .rxe files. Of course, it also works with Try Me .rtm files and with the RPGReader.sys program. The decompiled output of this utility is NBC source code. Since variable names cannot be reconstructed in their original form when you decompile an executable, the resulting NBC code can be a bit daunting. However, if you are patient and take your time, working in stages, it is not terribly difficult to figure out the code and to modify it with comments and variable name replacements so that it is easy to follow. Figure 6-9 contains a screenshot of the NXT Program Dumper utility after it has decompiled a small executable.

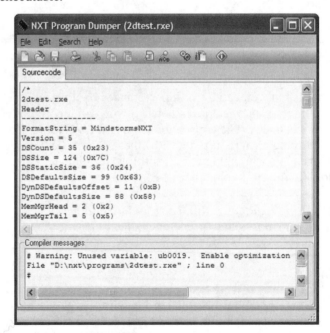

Figure 6-9. NXT Program decompiler

When you select the Decompile option from the File menu or via the toolbar, you can browse for .rxe, .rtm, or .sys files and open one with the utility. The Sourcecode memo will contain the decompiled NBC code for the file that you select. This utility can also compile the code in the Sourcecode memo and create an executable file just like you can do from within BricxCC or using the NBC command line compiler.

Another way to decompile programs is using the NBC command line compiler. Just pass the .rxe filename and the -x command line switch on the command line. The compiler will output the decompiled NBC code to the terminal window. You can also redirect the output to a file using standard shell commands.

```
nbc -x 2dtest.rxe
nbc -x RICTest.rxe > tmp.nbc
```

Exploring the NXT

As you write programs for the NXT, you will probably eventually grow frustrated by the method provided by the NXT user interface for managing the files on the brick. It works fine but it isn't the most efficient way to find a file and execute it. And when you want to delete a few files it can be tedious to move back and forth through the menu system to select and confirm all the required menu options.

The LEGO MINDSTORMS NXT software doesn't help a lot here either. The Memory tab on the NXT window dialog provides basic support for listing files of different types, uploading files, downloading files, and deleting files off the NXT. But it still isn't as easy as it should be. That's where the NXT Explorer utility comes in handy (Figure 6-10).

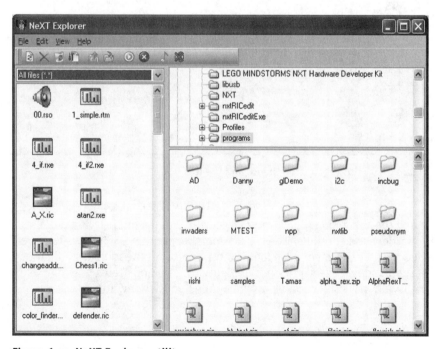

Figure 6-10. NeXT Explorer utility

On the left side is a list view showing a filtered list of the files on the NXT. The filter setting is above the list. To the right is a view of the files on your computer. At the top is a tree view of the directory structure and below that is a list view of the files found in the directory you have selected in the tree. Both the files on the computer and the files on the NXT are filtered by the selection you make in the filter drop-down list. Normally all files are shown but you can select a filter which will restrict both lists of files to just certain NXT file types if you want. Figure 6-11 shows some of the file type filter options.

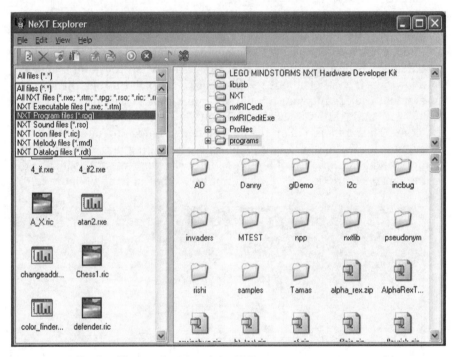

Figure 6-11. Filtering files on the PC and the NXT

You can select a different view for the NXT list as well as the PC list. Using the View menu, pick from large icon, small icon, list, and details as your view style options. These correspond to standard Windows Explorer views. The View menu also lets you turn on or off the toolbar at the top of the window. The screenshot in Figure 6-12 demonstrates how to make your view style selection using the options on the View menu.

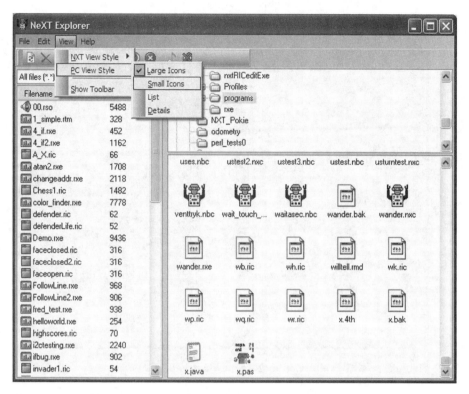

Figure 6-12. Setting the PC and NXT view style.

Using the NXT Explorer you can select files on your computer and drag them to the NXT list to copy them onto the NXT. You can do the same operation in the other direction by selecting one or more files in the NXT list and dragging them to the PC list view. You can also select the Upload or Download option on the toolbar or the File menu to upload or download the file(s) you have selected in either view.

Deleting files is as simple as selecting files on the NXT and selecting the Delete option on the toolbar or via the File menu. Select the files using standard Explorer mechanisms with your mouse or keyboard. You can refresh the list of files in both views using the Refresh option. The Defragment option actually copies all the files to your computer, deletes the files off the NXT, and then copies the files back onto the NXT in a sorted fashion so that the files that must be contiguous come first in order from largest to smallest. Figure 6-13 shows all the options available to you to manage your NXT file system using this marvelous utility.

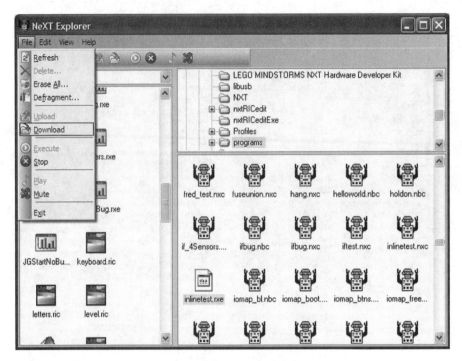

Figure 6-13. NXT Explorer options on the File menu.

Virtual NXT

Wouldn't it be nice if your own computer displayed the NXT LCD? The NXT LCD is far superior to the LCD screen on the RCX brick but it isn't perfect. It can be hard to see at times if lighting conditions aren't ideal. It would also be great if you could interact with the NXT user interface without having to physically press the buttons on the brick.

The NXT Screen utility gives you the ability to see a virtual NXT on your computer screen and interact with its user interface (Figure 6-14). The utility displays an image of the NXT with its LCD screen and four buttons. Use your mouse and keyboard to press the buttons and watch the NXT respond to your control without ever touching the brick.

When the utility begins it does not automatically start polling the brick for images. You can manually poll for a screen image using Ctrl+N or via the right click popup menu Poll Now option. Or you can start continuous polling via the Poll option on the popup menu or via the Ctrl+P hot key. You can set how frequently you want to refresh the image via 10 different preset refresh rates ranging from 50 milliseconds to 1 minute by using the hot keys Ctrl+1 through Ctrl+0. You can also control this setting via the popup Refresh Rate submenu options as shown in Figure 6-15.

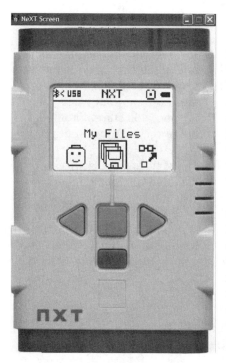

When you click on the buttons on the NXT image while a program is not running on the NXT, the brick will respond as if you pressed that button on the brick itself. The PC and the brick will make the click sound and the NXT menus will respond to the click as expected. You can drill down and scroll through the various user interface options on the NXT using the virtual buttons.

Whether your PC clicks or not can be enabled or disabled via the Play Clicks menu option on the context menu. You can also control what sound is played by the utility for the clicks by putting a WAVE file of your choice in the same directory as the utility executable with the name click.wav. Instead of the default click sound you will hear this wave file each time you click a virtual NXT button.

Figure 6-14. The NXT Screen virtual NXT utility

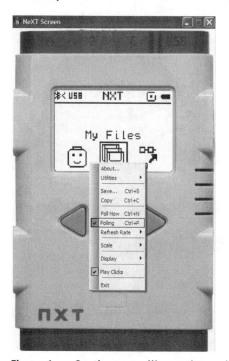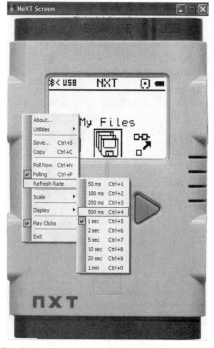

Figure 6-15. Continuous polling option and refresh rate

Alternatively, the up, down, left, and right arrow keys on your keyboard can substitute for the NXT buttons. The left and right arrows correspond to the NXT's left and right arrow buttons. Pressing the up arrow key will press the orange enter button on the NXT and the down arrow key on your computer presses the dark grey escape button. The only limitation to the virtual buttons is that they only work when you are interacting with the NXT user interface. Once you start a program, the buttons are completely under the control of the running program and remote access is not available.

One of the cool features that this utility provides is the ability to take screen captures of the NXT screen much like you would take a screen capture of your computer screen. Instead of Alt+PrtSc you just use Ctrl+C to copy the screen image to the clipboard or you can use Ctrl+S to save the screen image to a file in JPG, BMP, or PNG format. You can also select a different resolution for the utility if you prefer. The Scale submenu lets you pick several different scale values from 1x to 4x in 0.25 increments. The menu options for adjust the NXT Screen scale factor are shown in Figure 6-16 below.

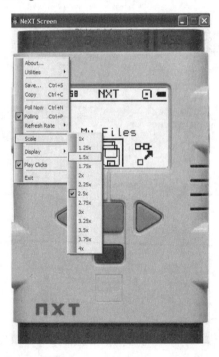

Figure 6-16. NXT screen scale factors

The NXT screen utility also lets you execute three special functions via the popup menu Utilities submenu. From this menu you can change your NXT brick name. You can also reset the brick's Bluetooth system back to its factory default settings. And you can reboot your NXT into SAMBA mode. Normally this would never be required but it might come in handy in rare situations. If you ever need to download a new firmware version to the NXT such as if you want to install the enhanced NBC/NXC firmware so that you can take advantage of all its bug fixes, speedups, and enhanced features then you would be better off just using the BricxCC firmware download option. These utility menu options are shown in the screenshot in Figure 6-17 below.

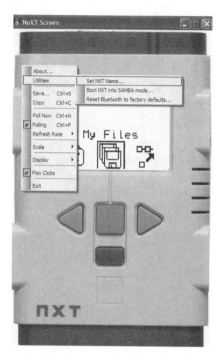

Figure 6-17. NXT Screen utility menu options

All of the utilities we have seen in this chapter can help enhance your NXT programming experience. Learn how to use them to your advantage. Develop your skills and abilities with them through experimentation. They will help you put the power into your NXT programming.

Programming the NXT

Topics in this Chapter

- Introduction to C
- NXC Overview
- NQC Compatibility
- RICScript Overview
- NXT Programs

Chapter 7

T his chapter briefly introduces the C programming language. After that, we'll shift our focus to Not eXactly C. We'll look at the most frequently used features in NXC as well as some of its less well-known capabilities. A complete description of the NXC API can be found in the *NXC Programmer's Guide*, plus a quick reference to the NXC API is contained in Appendix A. If you followed through Chapter One, you already have NXC installed on your operating system. If not, follow the steps outlined in the Getting Started section in Chapter One.

 WEBSITE: *Download a copy of the* NXC Programmer's Guide *from the NXC documentation section of the NBC website. This document is the complete specification for the NXC programming language.* http://bricxcc.sourceforge.net/nbc/

This chapter also provides an overview of the RICScript programming language for creating RIC picture files. We will end the chapter with a quick look at the very simple NPG programming language for writing tiny NXT Programs that run using the RPGReader. sys program on the brick.

Introduction to C

The C programming language has been around since engineers at AT&T's Bell Laboratories first created it back in 1972. C descends from a programming language called B, which descends from BCPL.

C continues to be widely used for such things as writing firmware and virtual machines for programmable bricks, as well as the software running on the computers in your car and on space shuttles. You'll find it in all types of computer programs, from 3D games with intense graphics to simple serial port interface routines. Its popularity remains high in spite of more modern programming languages such as C#, Java, C++, Ruby, and Lua having been created and popularized in the years since C was first created.

In 1983 the American National Standards Institute (ANSI) began the process of developing the first official standard for the C programming language. Prior to this, Brian Kernighan and Dennis Ritchie released their famous book, *The C Programming Language*, which was first published in 1978. The ANSI standard was finally completed in 1989 and accepted by the International Standards Organization (ISO) as IOS/IEC 9899-1990.

Not Quite C

When the RCX brick was first introduced by LEGO back in 1998, a number of adults became interested in expanding the capabilities of the standard software. One of the earliest innovators was Dave Baum. He developed a new programming language for the RCX that used a syntax that was very similar to the C programming language. Due to the limitations imposed by the standard RCX firmware, the language was not quite an implementation of C for the RCX. Dave's efforts resulted in the birth of Not Quite C.

NXC Overview

As we saw earlier, the firmware running on the NXT is very different from the standard RCX firmware. This made extending NQC to support the NXT complicated. Rather than waste a lot of time trying to mold NQC to fit the new firmware, it was decided to develop a new compiler from scratch, but with the goal of making the new programming language feel a lot like its parent so that users of NQC could easily make the transition. The new language is a lot like C, more so even than NQC, and it is a lot like NQC. Much like NQC, it targets the standard NXT firmware. As a result of all these factors the new language was named Not eXactly C or NXC.

The NXC Language

This section describes the NXC language itself. This includes the lexical rules used by the compiler, the structure of programs, statements and expressions, and the operation of the preprocessor.

NXC is a case-sensitive language, just like C and C++. That means that the identifier "xYz" is not the same identifier as "Xyz". Similarly, the "if" statement begins with the keyword "if" but "iF", "If", or "IF" are all just valid identifiers – not keywords.

The lexical rules describe how NXC breaks a source file into individual tokens. This includes the way comments are written, the handling of whitespace, and valid characters for identifiers.

Comments

Two forms of comments are supported in NXC. The first are traditional C comments. They begin with /* and end with */. These comments are allowed to span multiple lines, but they cannot be nested.

```
/* this is a comment */
/* this is a two
```

```
      line comment */
/* another comment...
   /* trying to nest...
      ending the inner comment...*/
   this text is no longer a comment! */
```

The second form of comments supported in NXC begins with // and continues to the end of the current line. These comments are sometimes known as C++ style comments. As you might guess, the compiler ignores comments. Their only purpose is to allow the programmer to document the source code.

```
// a single line comment
```

Whitespace

Whitespace consists of all spaces, tabs, and newlines. It is used to separate tokens and to make a program more readable. As long as the tokens are distinguishable, adding or subtracting whitespace has no effect on the meaning of a program. For example, the following lines of code both have the same meaning:

```
x=2;
x   =  2  ;
```

Some of the C++ operators consist of multiple characters. In order to preserve these tokens, whitespace cannot appear within them. In the example below, the first line uses a right shift operator ('>>'), but in the second line the added space causes the '>' symbols to be interpreted as two separate tokens and thus generate an error.

```
x = 1 >> 4; // set x to 1 right shifted by 4 bits
x = 1 > > 4; // error
```

Numerical Constants

Numerical constants may be written in either decimal or hexadecimal form. Decimal constants consist of one or more decimal digits. Hexadecimal constants start with 0x or 0X followed by one or more hexadecimal digits.

```
x = 10;  // set x to 10
x = 0x10; // set x to 16 (10 hex)
```

String Constants

String constants in NXC, just like in C, are delimited with double quote characters. NXC has a string data type that makes strings easier to use than they are in C. Behind the scenes, a string is automatically converted into an array of bytes, with the last byte in the array being a zero. The final zero byte is generally referred to as the null terminator for a C-style string.

```
TextOut(0, LCD_LINE1, "testing");
```

Identifiers and Keywords

Identifiers are used for variable, task, function, and subroutine names. The first character of an identifier must be an upper or lower case letter or the underscore ('_'). Remaining characters may be letters, numbers, and an underscore.

A number of potential identifiers are reserved for use in the NXC language itself. These reserved words are call keywords and may not be used as identifiers. A complete list of keywords appears in Table 7-1 below.

`__RETURN__`	`Char`	`Long`	`sub`
`__RETVAL__`	`const`	`Mutex`	`switch`
`__STRRETVAL__`	`continue`	`priority`	`task`
`__TMPBYTE__`	`default`	`repeat`	`true`
`__TMPWORD__`	`do`	`return`	`typedef`
`__TMPLONG__`	`else`	`safecall`	`unsigned`
`Abs`	`false`	`Short`	`until`
`Asm`	`for`	`Sign`	`void`
`Bool`	`goto`	`Start`	`while`
`break`	`if`	`Stop`	
`Byte`	`inline`	`string`	
`Case`	`int`	`struct`	

Table 7-1. NXC Keywords

Variables

All variables in NXC are defined using one of the types listed in Table 7-2.

Type Name	Information
Bool	8 bit unsigned
byte, unsigned char	8 bit unsigned
Char	8 bit signed
unsigned int	16 bit unsigned
short, int	16 bit signed
unsigned long	32 bit unsigned
long	32 bit signed
mutex	Special type used for exclusive code access
string	Array of byte with a null terminator
Structure	User-defined structure types
Array	Arrays of any type

Table 7-2. Variable Types

Variables are declared using the keyword for the desired type, followed by a comma-separated list of variable names and terminated by a semicolon (';'). Optionally, an initial value for each variable may be specified using an equals sign ('=') after the variable name. Several examples appear below:

```
int x;      // declare x
bool y,z;   // declare y and z
long a=1,b; // declare a and b, initialize a to 1
```

Global variables are declared at the program scope (outside of any code block). Once declared, they may be used within all tasks, functions, and subroutines. Their scope begins at declaration and ends at the end of the program.

Local variables may be declared within tasks and functions. Such variables are only accessible within the code block in which they are defined. Specifically, their scope begins with their declaration and ends at the end of their code block. In the case of local variables, a compound statement (a group of statements bracketed by '{' and '}') is considered a block:

```
int x;   // x is global

task main()
{
    int y;  // y is local to task main
    x = y; // ok
    {   // begin compound statement
       int z;   // local z declared
       y = z; // ok
    }
    y = z; // error - z no longer in scope
}

task foo()
{
    x = 1;  // ok
    y = 2;  // error - y is not global
}
```

Structures

NXC supports user-defined aggregate types known as structs. These are declared very much like you declare structs in a C program.

```
struct car

{
  string car_type;
  int manu_year;
};

struct person
```

```
{
  string name;
  int age;
  car vehicle;
};

myType fred = 23;
person myPerson;
```

After you have defined the structure type you can use the new type to declare a variable or nested within another structure type declaration. Members (or fields) within the struct are accessed using a dot notation.

```
myPerson.age = 40;
anotherPerson = myPerson;
fooBar.car_type = "honda";
fooBar.manu_year = anotherPerson.age;
```

You can assign structs of the same type but the compiler will complain if the types do not match.

 TRY ME: *Create your own structure called alien that has properties for the number of eyes, number of arms, number of legs, color and species. Try declaring several different alien species and accessing their members.*

Arrays

NXC also supports arrays. Arrays are declared the same way as ordinary variables, but with an open and close bracket following the variable name.

```
int my_array[]; // declare an array with 0 elements
```

To declare arrays with more than one dimension simply add more pairs of square brackets. The maximum number of dimensions supported in NXC is 4.

```
bool my_array[][]; // declare a 2-dimensional array
```

Global arrays with one dimension are initialized at the point of declaration using the following syntax:

```
int X[] = {1, 2, 3, 4}, Y[]={10, 10}; // 2 arrays
```

The elements of an array are identified by their position within the array (called an index). The first element has an index of 0, the second has index 1, and so on. For example:

```
my_array[0] = 123; // set first element to 123
my_array[1] = my_array[2]; // copy third into second
```

To initialize local arrays or arrays with multiple dimensions, use the ArrayInit function. The following example shows how to initialize a two-dimensional array using ArrayInit. It also demonstrates some of the supported array API functions and expressions.

```
task main()
{
  int myArray[][];
  int myVector[];
  byte fooArray[][][];

  ArrayInit(myVector, 0, 10); // 10 zeros
  ArrayInit(myArray, myVector, 10); // 10 vectors
  ArrayInit(fooArray, myArray, 2); // 2 myArrays

  myVector = myArray[1];
  fooArray[1] = myArray;
  myVector[4] = 34;
  myArray[1] = myVector;

  int ax[], ay[];
  ArrayBuild(ax, 5, 6);
  ArrayBuild(ay, 2, 10, 6, 43);
  int axlen = ArrayLen(ax);
  ArraySubset(ax, ay, 1, 2); // ax = {10, 6}
  if (ax == ay) {
    // compare two arrays
  }
}
```

NXC also supports specifying an initial size for both global and local arrays. The compiler automatically generates the required code to correctly initialize the array to zeros. If a global array declaration includes both a size and a set of initial values, the size is ignored in favor of the specified values.

```
task main()
{
  int myArray[10][10];
  int myVector[10];
//  ArrayInit(myVector, 0, 10);
//  ArrayInit(myArray, myVector, 10);
}
```

The calls to ArrayInit are not required since we specified the equivalent initial sizes in the array declarations above. In fact, the myVector array is not needed unless we have a use for it other than to initialize myArray.

Statements

The body of a code block (task or function) is composed of statements. Statements are terminated with a semi-colon (';'), as shown in the example code above.

Variable Declaration

Variable declaration, as described in the previous section, is one type of statement. It declares a local variable (with optional initialization) for use within the code block. The syntax for a variable declaration is shown below.

```
arg_type variables;
```

Here arg_type must be one of the supported types in NXC. Following the type are variables, which must be a comma-separated list of names with optional initial value as shown in the code fragment below.

```
name[=expression]
```

Arrays of variables may also be declared.

```
int array[n][=initializer];
```

You can also define variables using user-defined aggregate structure types.

```
Struct TPerson {
  int age;
  string name;
};
TPerson bob; // cannot be initialized at declaration
```

Assignment

Once declared, variables may be assigned the value of an expression using the syntax shown in the code sample below.

```
variable assign_operator expression;
```

There are nine different assignment operators. The most basic operator, '=', simply assigns the value of the expression to the variable. The other operators modify the variable's value in some other way as shown in Table 7-3.

The code sample below shows a few of the different types of operators that you can use in NXC expressions.

```
x = 2; // set x to 2
y = 7; // set y to 7
x += y;    // x is 9, y is still 7
```

Control Structures

The simplest control structure is a compound statement. This is a list of statements enclosed within curly braces ('{' and '}'):

```
{
    x = 1;
    y = 2;
}
```

Operator	Action
=	Set variable to expression
+ =	Add expression to variable
− =	Subtract expression from variable
* =	Multiple variable by expression
/ =	Divide variable by expression
% =	Set variable to remainder after dividing by expression
& =	Bitwise AND expression into variable
| =	Bitwise OR expression into variable
^| =	Bitwise exclusive OR into variable
|| =	Set variable to absolute value of expression
+ − =	Set variable to sign (-1,+1,0) of expression
> > =	Right shift variable by expression
< < =	Left shift variable by expression

Table 7-3. Operators

Although this may not seem very significant, it plays a crucial role in building more complicated control structures. Many control structures expect a single statement as their body. By using a compound statement, the same control structure can be used to control multiple statements.

The if-statement evaluates a condition. If the condition is true, it executes one statement (the consequence). An optional second statement (the alternative) is executed if the condition is false. The two syntaxes for an if-statement are shown below.

```
if (condition) consequence
if (condition) consequence else alternative
```

The condition of an if-statement must be enclosed in parentheses, as shown in the code sample below. Notice how a compound statement is used in the last example to allow two statements to execute as the consequence of the condition.

```
if (x==1) y = 2;
if (x==1) y = 3; else y = 4;
if (x==1) { y = 1; z = 2; }
```

The while-statement is used to construct a conditional loop. The condition is evaluated, and if true the body of the loop is executed, then the condition is tested again. This process continues until the condition becomes false (or a break statement is executed). The syntax for a while loop appears in the code fragment below.

```
while (condition) body
```

It is very common to use a compound statement as the body of a loop. In the sample below the single statement body has been replaced by a compound statement in order to execute more than one statement in the loop.

```
while(x < 10)
{
    x = x+1;
    y = y*2;
}
```

A variant of the while-loop is the do-while loop. The syntax for this control structure is shown below.

```
do body while (condition)
```

The difference between a while loop and a do-while loop is that the do-while loop always executes the body at least once, whereas the while loop may not execute it at all.

Another type of loop is the for-loop. It uses the syntax shown below.

```
for(statement1 ; condition ; statement2) body
```

A for loop always executes statement1, and then it repeatedly checks the condition. While it remains true, it executes the body followed by statement2. The for-loop is equivalent to the code shown below, which uses a while-loop instead.

```
statement1;
while(condition)
{
    body;
    statement2;
}
```

The repeat-statement executes a loop a specified number of times. This control statement is not a standard C looping construct. NXC inherited it from NQC.

```
repeat (expression) body
```

The expression determines how many times the body will be executed.

NOTE: *It is only evaluated a single time and then the body is repeated that number of times. This is different from both the while and do-while loops, which evaluate their condition each time through the loop.*

A switch-statement executes one of several different blocks of code depending on the value of an expression. One or more case labels precede each block of code. Each case must be a constant and unique within the switch statement. The switch statement evaluates the expression, then looks for a matching case label. It will then execute any statements following the matching case until either a break statement or the end of the switch is reached. A single default label may also be used - it will match any value not already appearing in a case label. Technically, a switch statement has the following syntax:

```
switch (expression) body
```

The case and default labels of a switch statement are not statements in themselves - they are labels that precede statements. Multiple labels can precede the same statement. These labels have the syntax shown below.

```
case constant_expression :
default :
```

A typical switch statement might look like this:

```
switch(x)
{
    case 1:
        // do something when X is 1
        break;
    case 2:
    case 3:
        // do something else when x is 2 or 3
        break;
    default:
        // do this when x is not 1, 2, or 3
        break;
}
```

NXC also supports using string types in the switch expression and constant strings in case labels.

The goto statement forces a program to jump to the specified location. Statements in a program can be labeled by preceding them with an identifier and a colon. A goto statement then specifies the label that the program should jump to. For example, this is how an infinite loop that increments a variable is implemented using goto:

```
my_loop:
    x++;
    goto my_loop;
```

The goto statement should be used sparingly and cautiously. In almost every case, control structures such as if, while, and switch make a program much more readable and maintainable than using goto.

NXC also defines the until-macro for compatibility with NQC. This construct provides a convenient alternative to the while loop. The actual definition of until is shown below.

```
#define until(c)        while(!(c))
```

In other words, until will continue looping until the condition becomes true. It is most often used in conjunction with an empty body statement:

```
until(SENSOR_1 == 1);// wait for sensor press
```

The asm Statement

The asm statement is used to define many of the NXC API calls. The syntax of the statement is:

```
asm {
 one or more lines of assembly language
}
```

The statement simply emits the body of the statement as NeXT Byte Codes (NBC) code and passes it directly to the NBC compiler backend. The asm statement is often used to optimize code so that it executes as fast as possible on the NXT firmware. The following example shows an asm block containing variable declarations, labels, and basic NBC statements as well as comments.

```
asm {
//      jmp __lbl00D5
      dseg segment
        sl0000 slong
        sl0005 slong
        bGTTrue byte
      dseg ends
      mov sl0000, 0x0
      mov sl0005, sl0000
      mov sl0000, 0x1
      cmp GT, bGTTrue, sl0005, sl0000
      set bGTTrue, FALSE
      brtst  EQ, __lbl00D5, bGTTrue
   __lbl00D5:
}
```

A few NXC keywords have meaning only within an asm statement. These keywords provide a means of returning string or scalar values from asm statements and for using temporary integer variables of byte, word, and long sizes.

ASM Keyword	Meaning
RETURN	Used to return a value other than _RETVAL__ or __STRRETVAL__
RETVAL	Writing to this 4-byte value returns it to the calling program
STRRETVAL	Writing to this string value returns it to the calling program
TMPBYTE	Use this temporary variable to write and return single byte values
TMPWORD	Use this temporary variable to write and return 2-byte values
TMPLONG	Use this temporary variable to write and return 4-byte values

Table 7-4. ASM Keywords

The asm block statement and these special ASM keywords are used throughout the NXC API. You can have a look at the NXCDefs.h header file for several examples of how they are used. To keep the main NXC code as "C-like" as possible, and for the sake of better readability, NXC asm block statements can be wrapped in preprocessor macros and placed in custom header files which are included using #include. The following example demonstrates using macro wrappers around asm block statements.

```
#define SetMotorSpeed(port, cc, thresh, fast, slow) \
  asm { \
  set theSpeed, fast \
  brcmp cc, EndIfOut__I__, SV, thresh \
  set theSpeed, slow \
EndIfOut__I__: \
  OnFwd(port, theSpeed) \
  __IncI__ \
}
```

Other NXC Statements

A function call is a statement of the form:

```
name(arguments);
```

The arguments list is a comma-separated list of expressions. The number and type of arguments supplied must match the definition of the function itself. Optionally, the return value of the function may be assigned to a variable.

You can start or stop tasks with the following statements. The stop statement is only supported if you are running the enhanced NBC/NXC firmware on your NXT.

```
start task_name;
stop task_name;
```

You can adjust the priority of a task using the priority statement. Setting task priorities also requires the enhanced NBC/NXC firmware.

```
priority task_name, new_priority;
```

Within loops (such as a while loop) you can use the break statement to exit the loop or exit a switch statement (as shown above). You can also use the continue statement to skip to the top of the next iteration of the loop.

```
break;
continue;
```

It is possible to cause a function to return before it reaches the end of its code by using the return statement with an optional return value.

```
return [expression];
```

Many expressions are not legal statements. A notable exception is the expression using increment (++) or decrement (--) operators.

```
x++;
```

The empty statement (just a bare semicolon) is also a legal statement.

Expressions

Values are the most primitive type of expressions. More complicated expressions are formed from values using various operators. The NXC language only has two kinds of built in values: numerical constants and variables.

Numerical constants in the NXT are represented as integers. The type depends on the value of the constant. NXC internally uses 32 bit signed math for constant expression evaluation. Numeric constants are written as either decimal (e.g. 123) or hexadecimal (e.g. 0xABC). Presently, there is very little range checking on constants, so using a value larger than expected may have unusual effects.

Two special values are predefined: true and false. The value of false is zero (0), while the value of true is one (1). The same values hold for relational operators (e.g. <): when the relation is false the value is 0, otherwise the value is 1.

Values are combined using operators. Several of the operators are only used in evaluating constant expressions, which means that their operands must either be constants, or expressions involving nothing but constants. The operators are listed in Table 7-5 by order of precedence from highest to lowest.

Operator	Description	Associativity	Restriction	Example
abs()	Absolute value	n/a		abs(x)
sign()	Sign of operand	n/a		sign(x)
++, ——	Post increment, Post decrement	Left	Variables only	x++
—	Unary minus	Right		-x
~	Bitwise negation	Right		~123
!	(unary) Logical negation	Right		!x
*, /, %	Multiplication, division, modulo	Left		x * y
+, —	Addition, subtraction	Left		x + y
< <, > >	Left and right shift	Left		x << 4
<, >, <=, >=	Relational operators	Left		x < y
==, !=	Equal to, not equal to	Left		x == 1
&	Bitwise AND	Left		x & y
^	Bitwise XOR	Left		x ^ y
|	Bitwise OR	Left		x | y
&&	Logical AND	Left		x && y
||	Logical OR	Left		x || y
?:	Conditional value	n/a		x==1 ? y : z

Table 7-5. Expression Operators

Where needed, parentheses are used to change the order of evaluation:

```
x = 2 + 3 * 4;       // set x to 14
y = (2 + 3) * 4;  // set y to 20
```

Conditions

Comparing two expressions forms a condition. There are also two constant conditions - true and false - that always evaluate to true or false respectively. A condition is negated with the negation operator, or two conditions combined with the AND and OR operators. Like most compilers, NXC supports something called "short-circuit" evaluation of conditions. This means that if the entire value of the conditional can be logically determined by only evaluating the left hand side of the condition, then the right hand side will not be evaluated.

Table 7-6 below summarizes the different types of conditions.

Condition	Meaning
true	always true
false	always false
Expr	true if expr is not equal to 0
Expr1 == expr2	true if expr1 equals expr2
Expr1 != expr2	true if expr1 is not equal to expr2
Expr1 < expr2	true if expr1 is greater than expr2
Expr1 >= expr2	true if expr1 is greater than or equal to expr2
! condition	logical negation of a condition – true if condition is false
Cond1 && cond2	logical AND of two conditions (true if and only if both conditions are true)
Cond1 || cond2	logical OR of two conditions (true if and only if at least one of the conditions are true)

Table 7-6. Conditions

You can use conditions in NXC control structures, such as the if-statement and the while or until statements to determine exactly how you want your robot to behave.

An NXC program is composed of code blocks and variables. There are two distinct types of code blocks: tasks and functions. Each type of code block has its own unique features, but they share a common structure. The maximum number of code blocks of either type is 256.

Tasks

Since the NXT supports multi-threading, a task in NXC directly corresponds to an NXT thread. Tasks are defined using the task keyword using the syntax shown in the code sample below.

```
task name()
{
  // the task's code is placed here
}
```

The name of the task may be any legal identifier. A program must always have at least one task - named "main" - which is started whenever the program is run.

The body of a task consists of a list of statements. Scheduling dependant tasks using the Precedes or Follows API function is the primary mechanism supported by the NXT for starting other tasks concurrently. You can start and stop tasks by using the start and stop statements. There is also an NXC API command, StopAllTasks, which stops all currently running tasks. You can also stop all tasks using the Stop function. A task can stop itself via the ExitTo API function or by task execution simply reaching the end of the task. In the code sample below, the main task starts a music task and a movement task, waits ten seconds before stopping the music task, and then waits another five seconds before stopping all tasks to end the program.

```
task music() {
  while (true) {
    PlayTone(440, 500);
    Wait(600);
  }
}

task movement() {
  while (true) {
    OnFwd(OUT_A, Random(100));
    Wait(Random(1000));
  }
}

task main() {
  start music;
  start movement;
  Wait(10000);
  stop music;
  Wait(5000);
  StopAllTasks();
}
```

Functions

It is often helpful to group a set of statements together into a single function, which your code can then call as needed. NXC supports functions with arguments and return values. Functions are defined using the syntax below.

```
[safecall] [inline] return_type name(argument_list)
{
    // body of the function
}
```

The return type is the type of data returned. In the C programming language, functions must specify the type of data they return. Functions that do not return data simply return void.

The argument list may be empty, or may contain one or more argument definitions. An argument is defined by a type followed by a name. Commas separate multiple arguments. All values are represented as bool, char, byte, int, short, long, unsigned int, unsigned long, strings, struct types, or arrays of any type. NXC also supports passing argument types by value, by constant value, by reference, and by constant reference.

When arguments are passed by value from the calling function to the called function, the compiler must allocate a temporary variable to hold the argument. There are no restrictions on the type of value that may be used. However, since the function is working with a copy of the actual argument, the caller will not see any changes it makes to the value. In the example below, the function foo attempts to set the value of its argument to 2. This is perfectly legal, but since foo is working on a copy of the original argument, the variable y from the main task remains unchanged.

```
void foo(int x)
{
    x = 2;
}

task main()
{
    int y = 1; // y is now equal to 1
    foo(y); // y is still equal to 1!
}
```

The second type of argument, const *arg_type*, is also passed by value, but with the restriction that only constant values may be used. This is rather important since there are a few NXT functions that only work with constant arguments.

```
void foo(const int x)
{
    PlayTone(x, 500); // ok
    x = 1; // error - cannot modify argument
```

```
}

task main()
{
    foo(2); // ok
    foo(4*5);   // ok - expression is still constant
    foo(x); // error - x is not a constant
}
```

The third type, *arg_type* &, passes arguments by reference rather than by value. This allows the called function to modify the value and have those changes occur in the caller. However, only variables may be used when calling a function using *arg_type* & arguments:

```
void foo(int &x)
{
    x = 2;
}

task main()
{
    int y = 1; // y is equal to 1

    foo(y); // y is now equal to 2
    foo(2); // error - only variables allowed
}
```

The fourth type, const *arg_type* &, is rather unusual. It is also passed by reference, but with the restriction that the called function is not allowed to modify the value. Because of this restriction, the compiler is able to pass anything, not just variables, to functions using this type of argument. Due to certain firmware restrictions, however, passing an argument by reference is best used only if you need to modify the value within the function.

Functions must be invoked with the correct number and type of arguments. The code example below shows several different legal and illegal calls to function foo.

```
void foo(int bar, const int baz)
{
    // do something here...
}

task main()
{
    int x; // declare variable x
    foo(1, 2); // ok
    foo(x, 2); // ok
    foo(2, x); // error - 2nd argument not constant!
    foo(2); // error - wrong number of arguments!
}
```

You can optionally mark NXC functions as inline functions. This means that each call to a function will create another copy of the function's code. Unless used judiciously, inline functions can lead to excessive code size.

If a function is not marked as inline then an actual NXT subroutine is created and the call to the function in NXC code will result in a subroutine call to the NXT subroutine. The total number of non-inline functions (aka subroutines) and tasks must not exceed 256.

Another optional keyword that can be specified prior to the return type of a function is the safecall keyword. If a function is marked as safecall then the compiler will synchronize the execution of this function across multiple threads by wrapping each call to the function in Acquire and Release calls. If a second thread tries to call a safecall function while another thread is executing it, the second thread will have to wait until the function returns to the first thread.

The Preprocessor

NXC also includes a preprocessor that is modeled after the standard C preprocessor. The C preprocessor processes a source code file before the compiler does. It handles such tasks as including code from other files, conditionally including or excluding blocks of code, stripping comments, defining simple and parameterized macros, and expanding macros wherever they are encountered in the source code.

The NXC preprocessor implements the following standard preprocessor directives: #include, #define, #ifdef, #ifndef, #endif, #if, #elif, #undef, ##, #line, and #pragma. Its implementation is close to a standard C preprocessor, so most things that work in a generic C preprocessor should have the expected effect in NXC. Any significant deviations are explained below.

#include

The #include command works as expected, with the caveat that the filename must be enclosed in double quotes. There is no notion of a system include path, so enclosing a filename in angle brackets is forbidden.

```
#include "foo.h"  // ok
#include <foo.h> // error!
```

NXC programs can begin with #include "NXCDefs.h" but they don't need to. This standard header file includes many important constants and macros, which form the core NXC API. NXC no longer requires that you manually include the NXCDefs.h header file. Unless you specifically tell the compiler to ignore the standard system files, this header file is included automatically.

#define

The #define command is used for simple macro substitution. Redefinition of a macro is an error. The end of the line normally terminates macros, but the newline may be escaped with the backslash ('\') to allow multi-line macros. The backslash character must be the very last character in the line or it will not extend the macro definition to the next line. The code sample below shows how to write a multi-line preprocessor macro.

```
#define foo(x)  do { bar(x); \
                baz(x); } while(false)
```

The #undef directive may be used to remove a macro's definition.

(Concatenation)

The ## directive works similar to the C preprocessor. It is replaced by nothing, which causes tokens on either side to be concatenated together. Because it acts as a separator initially, it can be used within macro functions to produce identifiers via combination with parameter values.

Conditional Compilation

Conditional compilation works similar to the C preprocessor. Any of the preprocessor directives shown in Table 7-7 may be used in your NXC programs.

Directive	Meaning
#ifdef symbol	If symbol is defined then compile the following code
#ifndef symbol	If symbol is not defined then compile the following code
#else	Switch from compiling to not compiling and vice versa
#endif	Return to previous compiling state
#if condition	If the condition evaluates to true then compile the following code
#elif	Same as #else but used with #if rather than with #ifdef or #ifndef

Table 7-7. Conditional compilation directives

NQC Compatibility

NXC is not 100% compatible with NQC. In many cases the NXT firmware required changes to the basic API functions for controlling motors and reading sensor values. The RCX firmware also provided support for resource acquisition and event handling that is not provided by the NXT firmware, so some NQC constructs could not be implemented in NXC. But there is a great deal in common between the two programming languages.

Controlling Outputs

The basic outputs in NXC are named the same as in NQC. They are OUT_A, OUT_B, and OUT_C. These are constants that tell the firmware which motor you want to control. In NQC you could combine two outputs by adding them together. This is an area where NXC differs from NQC.

NXC defines constants that represent sets of motors: OUT_AB, OUT_AC, OUT_BC, and OUT_ABC. You should almost always use these constants whenever you want to control a motor in NXC. In NQC you could set the power, the direction, and the on/off state of a motor in separate API calls. In NXC these settings have all been combined into a single API call. There are other low-level API routines, which let you control each of these values individually, but the details are hidden in the NQC compatibility routines.

In NXC the basic motor control functions are OnFwd(ports, power), OnRev(ports, power), Float(ports), and Off(ports). Float and Off have the same semantics as their identically named NQC cousins. OnFwd and OnRev, however, both require that you also specify a power level along with the ports argument. Power levels for NXT motors can be anywhere from 0 to 100 rather than just 0 to 7 with the RCX. Reversing the motor's direction is accomplished by negating the power level.

```
OnFwd(OUT_AB, 75); // run A and B fwd at 75% power
OnRev(OUT_A, 50); // run A in reverse at 50% power
```

Beyond simply turning on a motor or set of motors at a specific power level there are other commands that let you control the motors in different ways. The NXT firmware has the ability, for example, to synchronize the rotation of a pair of motors. To take advantage of this feature you should use the OnFwdSync(ports, power, turnratio) and OnRevSync(ports, power, turnratio) functions, which also allow you to choose a non-zero turn ratio for adjusting the relative power levels between two motors. The OnFwdReg(ports, power, regmode) and OnRevReg(ports, power, regmode) methods let you specify either the synchronization mode or regulated power mode, which enables the firmware to automatically adjust the power level if the specified level is not being achieved due to friction or some other external influence.

```
OnFwdSync(OUT_AB, 75, 0); // no turning
OnFwdReg(OUT_A, 75, OUT_REGMODE_SPEED); //reg speed
```

NXC also provides a number of API functions that let you use tachometer-limited motion. In other words, you tell the motors how far you want them to turn and the firmware will automatically stop the motors when they reach their target. Use the `RotateMotor(ports, power, tacholimit)` command, or one of its several variants, to specify a tachometer limit for one or more motors. This function call will not return until the motors have reached the specified limit or until some other task sets the power level of the first motor to zero.

```
RotateMotor(OUT_A, 75, 720); // 2 full turns
```

We'll look at controlling the outputs in a lot more detail in Chapter 10. You can find more information about the NXC motor API functions in Appendix A.

Using Sensors

Basic sensor control in NXC is extremely similar to what you may be familiar with from having used NQC. The NXT has four sensor ports rather than just three so in NXC you will find sensor port constants `S1`, `S2`, `S3`, and `S4`. NXC also defines macros for reading analog sensor values: `SENSOR_1`, `SENSOR_2`, `SENSOR_3`, and `SENSOR_4`. In NQC you could use these macros both to read a sensor value and to configure a sensor port but in NXC you have to use the `S1-S4` constants for port configuration, plus you can only read a sensor's value from the `SENSOR_N` constants. As with NQC you can also read a sensor value using the `SensorValue(port)` API function.

```
x = SensorValue(S1);
y = SENSOR_2; // same as SensorValue(S2);
```

The NXT supports all the same sensor types that you could use in NQC along with a few new sensor types that are specific to the NXT. Just like in NQC you can either use the `SetSensor(port, config)` API function along with a sensor configuration constant or you can individually set the sensor type and mode using calls to `SetSensorType(port, type)` and `SetSensorMode(port, mode)`. The sensor configurations available in NXC are listed in Table 7-8 below.

Configuration	Type	Mode
SENSOR_TOUCH	Touch	Boolean
SENSOR_LIGHT	RCX Light	Percentage
SENSOR_ROTATION	RCX Rotation	Rotation
SENSOR_CELSIUS	Temperature	Celsius
SENSOR_FAHRENHEIT	Temperature	Fahrenheit
SENSOR_PULSE	Touch	Pulse Count
SENSOR_EDGE	Touch	Edge Count

Table 7-8. Sensor configurations

The sensor type constants are all listed in Table 7-9. The sensor mode constants are listed in Table 7-10.

Sensor Type	NBC Sensor Type	Meaning
SENSOR_TYPE_NONE	IN_TYPE_NO_SENSOR	no sensor configured
SENSOR_TYPE_TOUCH	IN_TYPE_SWITCH	NXT or RCX touch sensor
SENSOR_TYPE_TEMPERATURE	IN_TYPE_TEMPERATURE	RCX temperature sensor
SENSOR_TYPE_LIGHT	IN_TYPE_REFLECTION	RCX light sensor
SENSOR_TYPE_ROTATION	IN_TYPE_ANGLE	RCX rotation sensor
SENSOR_TYPE_LIGHT_ACTIVE	IN_TYPE_LIGHT_ACTIVE	NXT light sensor with light
SENSOR_TYPE_LIGHT_INACTIVE	IN_TYPE_LIGHT_INACTIVE	NXT light sensor without light
SENSOR_TYPE_SOUND_DB	IN_TYPE_SOUND_DB	NXT sound sensor with dB scaling
SENSOR_TYPE_SOUND_DBA	IN_TYPE_SOUND_DBA	NXT sound sensor with dBA scaling
SENSOR_TYPE_CUSTOM	IN_TYPE_CUSTOM	Custom sensor (unused)
SENSOR_TYPE_LOWSPEED	IN_TYPE_LOWSPEED	I2C digital sensor
SENSOR_TYPE_LOWSPEED_9V	IN_TYPE_LOWSPEED_9V	I2C digital sensor (9V power)
SENSOR_TYPE_HIGHSPEED	IN_TYPE_HISPEED	Highspeed sensor (unused)

Table 7-9. SensorType constants

Sensor Mode	NBC Sensor Mode	Meaning
SENSOR_MODE_RAW	IN_MODE_RAW	raw value from 0 to 1023
SENSOR_MODE_BOOL	IN_MODE_BOOLEAN	boolean value (0 or 1)
SENSOR_MODE_EDGE	IN_MODE_TRANSITIONCNT	counts number of boolean transitions
SENSOR_MODE_PULSE	IN_MODE_PERIODCOUNTER	counts number of boolean periods
SENSOR_MODE_PERCENT	IN_MODE_PCTFULLSCALE	value from 0 to 100
SENSOR_MODE_FAHRENHEIT	IN_MODE_FAHRENHEIT	degrees F
SENSOR_MODE_CELSIUS	IN_MODE_CELSIUS	degrees C
SENSOR_MODE_ROTATION	IN_MODE_ANGLESTEP	rotation (16 ticks per revolution)

Table 7-10. SensorMode constants

NXC also provides some higher-level functions for configuring and controlling the NXT sensor ports. You can configure any port to support any type of NXT or RCX sensor. Each of the API functions shown in the code sample below performs three separate tasks: set the sensor type, set the sensor mode, and wait for the sensor to become ready.

```
SetSensorTouch(S1); // NXT or RCX touch sensor
SetSensorLight(S2); // NXT light sensor (active)
SetSensorSound(S3); // NXT sound sensor (DB)
SetSensorLowspeed(S4); // NXT I2C sensor (9 volt)
```

If you call `SetSensorType` and `SetSensorMode` or `SetSensor` directly, you should follow the sensor port configuration with a call to the `ResetSensor(port)` API function.

```
ResetSensor(S1); // wait for sensor to be ready
```

You can also set a sensor value to zero, if you wish, using the `ClearSensor(port)` function. And the NXT motors all provide three separate tachometer values that you can read and reset independently. The RotationCount value is the sensor value that you should use if you want to treat the motor's built-in rotation sensor like the RCX rotation sensor, of course with far greater precision and accuracy. These functions are demonstrated in the code sample below.

```
ClearSensor(S1); // set port 1 value to zero
ResetTachoCount(OUT_A);
ResetBlockTachoCount(OUT_A);
ResetRotationCount(OUT_A);
ResetAllTachoCounts(OUT_A);
```

To read the motor rotation sensor value you can use the `MotorRotationCount(port)` function as shown in the code sample below. The other two counters are primarily for use with tachometer limited motor control.

```
if (MotorRotationCount(OUT_A) > 180) {
  Off(OUT_A);
}
```

In addition to reading a sensor value you can also read its current configuration, its raw value, and its boolean value. These commands are listed in the following code sample.

```
x = SensorRaw(S1);
y = SensorType(S2);
i = SensorMode(S3);
z = SensorBoolean(S1);
```

The I²C or Lowspeed sensors are a special case since they are not analog sensors. You cannot read an I²C device using the standard sensor routines. Each type of I²C device has its own specific set of commands that you need to send it in order to read data from it. The Ultrasonic sensor has a special NXC function called `SensorUS(port)`, which you use to read its value.

```
dist = SensorUS(S4); // read distance
```

For generic high-level communication with an I²C device you can use the I2CBytes(*port, writebuf, nbytes, readbuf*) function. You have to pass in the port, a buffer containing the data to write to the device, a variable set to the number of bytes you want to read from the device, and an empty byte array for storing the output data. The sample program below demonstrates how to use the I2CBytes function.

```
byte cmdUSReadVersion[] = {0x02, 0x00};

string ussReadVersion() {
  byte returnLength = 8;
  string bufin;
  I2CBytes(S1,
          cmdUSReadVersion,
          returnLength,
          bufin);
  return bufin;
}

task main() {
  SetSensorLowspeed(S1);
  TextOut(0, LCD_LINE1, ussReadVersion());
  Wait(10000);
}
```

There are three low level API routines for communicating with I²C devices. The I2CBytes function simply wraps these three separate calls into a single routine for the sake of simplicity. Each time you want to read from an I²C device you have to start a transaction using I2CWrite(*port, bytes, buffer*), passing in the port, the number of bytes you want to read, and a buffer containing the bytes of data you are sending to the device. Then you monitor the state of the transaction, waiting for it to complete successfully, using I2CStatus(*port, bytes*), passing in the port and a variable that indicates the number of bytes available to read. Finally, to read the response you use the I2CRead(*port, buflen, buffer*) function, passing in the port, the read buffer length, and the read buffer itself – already allocated with a sufficient size to store the data. You'll find more information about these API functions in Appendix A.

Miscellaneous Commands

The NXC API is extremely rich with functions that provide you with the power to manipulate and control the NXT in nearly every way imaginable. See Appendix A for additional information about all the NXC API functions. Some of the most frequently used functions are mentioned below.

Timing Functions

The Wait(*time*) function makes a task sleep for a specified amount of time (in 1000ths of a second). The time argument may be an expression or a constant.

```
Wait(2000); // pause for 2 seconds
Wait(Random(1000)); // wait up to 1 sec
```

The CurrentTick() function returns an unsigned 32-bit value which is the current system timing value (called a "tick") in milliseconds.

```
x = CurrentTick();
```

The FirstTick() function returns an unsigned 32-bit value which is the system timing value (called a "tick") in milliseconds at the time that the program began running.

```
x = FirstTick();
```

The SleepTime() function returns the number of minutes that the NXT remains on before it automatically shuts down.

```
x = SleepTimeout();
```

The SetSleepTime(*minutes*) function will set the NXT sleep timeout value to the specified number of minutes. To disable the automatic shutdown, set the sleep time to 0.

```
SetSleepTime(8);
```

Program Control Functions

The Stop(*bvalue*) function stops the running program if bvalue is true. This will halt the program completely, so any code following this command is ignored.

```
Stop(x == 24); // stop the program if x==24
```

The StopAllTasks() function stops all currently running tasks. This halts the program completely, so any code following this command is ignored. It is equivalent to calling Stop(true).

```
StopAllTasks(); // stop the program
```

The Acquire(*mutex*) function attempts to acquire the specified mutex variable. If another task already has acquired the mutex then the current task will be suspended until the mutex is released by the other task. This function is used to ensure that the current task has exclusive access to a shared resource, such as the display or a motor. After the current task has finished using the shared resource, the program should call Release to allow other tasks to acquire the mutex.

```
Acquire(motorMutex); // obtain exclusive access
// use the motors
Release(motorMutex);
```

The `Release(mutex)` function releases the specified mutex variable. Use this to relinquish a mutex so that it can be acquired by another task. Release should always be called after a matching call to Acquire and as soon as possible after a shared resource is no longer needed.

```
Acquire(motorMutex); // obtain exclusive access
// use the motors
Release(motorMutex); // release to other tasks
```

The `Precedes(task1, task2, ..., taskN)` function schedules the specified tasks for execution once the current task has completed executing. The tasks all execute simultaneously unless other dependencies prevent them from doing so. This function should only be called once within a task – preferably at the start of the task definition.

```
Precedes(moving, drawing, playing);
```

The `Follows(task1, task2, ..., taskN)` function schedules this task to follow the specified tasks so that it executes when any of the specified tasks have completed executing. This function should only be called once within a task – preferably at the start of the task definition. If multiple tasks declare that they follow the same task then they will all execute simultaneously unless other dependencies prevent them from doing so.

```
Follows(main);
```

The `ExitTo(task)` function immediately exits the current task and starts executing the specified task.

```
ExitTo(nextTask);
```

Math Functions

The `Random(value)` function returns an unsigned 16-bit random number between 0 and the specified value. The value can be a constant or a variable.

```
x = Random(10); // return a value of 0..9
```

The `Random()` function, when called without an argument, returns a signed 16-bit random number.

```
x = Random();
```

The `Sqrt(x)` function returns the square root of the specified value.

```
x = Sqrt(x);
```

The `Sin(degrees)` function returns the sine of the specified degrees value. The result is 100 times the sine value (-100..100).

```
x = Sin(theta);
```

The `Cos(degrees)` function returns the cosine of the specified degrees value. The result is 100 times the cosine value (-100..100).

```
x = Cos(y);
```

The Asin(*value*) function returns the inverse sine of the specified value (-100..100). The result is degrees (-90..90).

```
deg = Asin(80);
```

The Acos(*value*) function returns the inverse cosine of the specified value (-100..100). The result is degrees (0..180).

```
deg = Acos(0);
```

String Functions

The NumToStr(*value*) function returns the string representation of the specified numeric value.

```
msg = NumToStr(-2); // returns "-2" in a string
```

The FormatNum(*fmtstr, value*) function returns the formatted string using the format and value. Use standard numeric sprintf format specifiers within the format string.

```
msg = FormatNum("value = %d", x);
```

 TRY ME: *This problem can take some effort to figure out if you are new to C programming. Output ten random numbers between 50 and 100 to the LCD screen.*

Sound Functions

The PlayTone(*frequency, duration*) function plays a single tone of the specified frequency and duration. The frequency is in hertz and the duration is in 1000ths of a second. All parameters may be any valid expression.

```
PlayTone(440, 500);   // Play 'A' for 1/2 second
```

The PlayFile(*filename*) function plays the specified sound file (.rso) or a melody file (.rmd). The filename may be any valid string expression.

```
PlayFile("startup.rso");
```

The StopSound() function stops playback of the current tone or file.

```
StopSound();
```

Drawing Functions

Each of the drawing functions, except ResetScreen and ClearScreen, take an optional boolean argument that determines whether or not to clear the screen before drawing. If you leave out this argument it will default to false. For drawing numbers and text on the screen it is important to use the LCD_LINE1 through LCD_LINE8 constants. Values other than these are automatically adjusted to one of the eight fixed LCD lines. Other drawing routines do not have this restriction.

The `NumOut(x, line, value, clear = false)` function draws a numeric value on the screen at the specified x and line location.

```
NumOut(0, LCD_LINE1, x);
```

The `TextOut(x, line, msg, clear = false)` function draws a text value on the screen at the specified x and line location.

```
TextOut(0, LCD_LINE3, "Hello World!");
```

The `GraphicOut(x, y, filename, clear = false)` function draws the specified NXT picture file on the screen at the specified x and y location. If the file cannot be found then nothing is drawn and no errors are reported.

```
GraphicOut(40, 40, "image.ric");
```

The `GraphicOutEx(x, y, filename, vars, clear = false)` function draws the specified NXT picture file on the screen at the specified x and y location. It uses the values contained in the vars array to transform the parameterized drawing commands contained within the file.

```
GraphicOutEx(40, 40, "image.ric", variables);
```

The `CircleOut(x, y, radius, clear = false)` function draws a circle on the screen with its center at the specified x and y location, using the specified radius.

```
CircleOut(40, 40, 10);
```

The `LineOut(x1, y1, x2, y2, clear = false)` function draws a line on the screen from x1, y1 to x2, y2.

```
LineOut(40, 40, 10, 10);
```

The `PointOut(x, y, clear = false)` function draws a point on the screen at x, y.

```
PointOut(40, 40);
```

The `RectOut(x, y, width, height, clear = false)` function draws a rectangle on the screen at x, y with the specified width and height.

```
RectOut(40, 40, 30, 10);
```

The `ResetScreen()` function restores the standard NXT running program screen.

```
ResetScreen();
```

The `ClearScreen()` function clears the NXT LCD to a blank screen.

```
ClearScreen();
```

RICScript Overview

RICScript is not a true programming language given that you can't define variables or do any form of iteration or conditional execution. All you can do is define a list of drawing operations that draw a picture on the LCD. And you can use parameterization to make the picture program more flexible so that each time you draw it using a different set of parameter values the resulting image is different.

Whitespace in an RICScript program is ignored. Each statement ends with a semicolon. You can use block comments or single line comments exactly like in NXC. All the values in an RICScript are numeric constants. Each command can be included any number of times in a script.

In this section we will run through all the drawing commands available to you in RICScript and document how to use them to create your own NXT pictures. As we saw in Chapter 6, there are nine commands supported by the NXT firmware. Let's look at the RICScript statement associated with each opcode in the same order that we used to examine the opcode structure in the previous chapter.

Description

Add a description command to your picture file with the `desc(`*options,* *width, height*`)` statement. This command will have no effect when you draw the picture on the LCD.

```
desc(0, 20, 20);
```

Sprite

Add a sprite command by using the `sprite(`*address, data*`)` statement. Any number of data arguments can be specified with each additional argument representing an additional row of pixels in the sprite image. The data argument is either a series of ones and zeros or a string of hexadecimal values. You can also define a sprite by including an external image in the sprite command, along with an optional threshold value for converting a multi-colored image into a monochrome image. Supported image types for the import function are bmp, jpg, and png. The sprite address must be between 1 and 10.

```
sprite(1,
        100010001110111,
        011101110101010,
        100010001110111);
sprite(2, 0xffdcffdc, 0xdcffdcff, 0xa8dca8dc);
sprite(3, import("image.bmp"));
```

VarMap

If you want to perform image transformation using parameterized values then you can use the `varmap(`*address, function1, function2*`)` statement to add a varmap command to your picture. The varmap address must be between 1 and 10. Each varmap statement must have at least two function arguments but you can specify any number of transformation function points.

```
varmap(4, f(0)=0, f(64)=64);
varmap(5, f(0)=0, f(1)= 1, f(2)= 4,
          f(3)=9, f(4)=16, f(5)=25);
```

CopyBits

To draw sprite data to the NXT LCD screen you will need to add at least one copybits command to your RICScript using the copybits(*options*, *address*, *src_x*, *src_y*, *src_w*, *src_h*, *dest_x*, *dest_y*) statement. The options value is not currently used so it should be zero. The address must be the address of a sprite that you have already defined in the RICScript prior to the copybits statement. All of the arguments to this statement except for the options argument can be parameterized using the map(value) and maparg(addr, value) functions.

```
copybits(0, 1, arg(0), 7, 7, 0, 0, 0);
copybits(0, 1, arg(1), 0, 7, 7, 7, 0);
copybits(0, 1, arg(7), 0, 7, 7, 49, 0);
```

Pixel

Draw a single pixel on the LCD screen in your NXT picture by adding a pixel(options, x, y, value) statement to your RICScript. The x, y, and value arguments can all be parameterized. The value argument is not currently used. Just set it to zero along with the unused options argument.

```
pixel(0, 6, maparg(1, 6), arg(0));
```

Line

Draw a line on the LCD screen in your NXT picture by adding a line(options, x1, y1, x2, y2) statement to your RICScript. The x1, y1, x2, and y2 arguments can all be parameterized. The options argument is currently unused.

```
line(0, 0, 0, 20, 20);// diagonal line
```

Rectangle

Draw a rectangle on the LCD screen in your NXT picture by adding a rect(options, x, y, width, height) statement to your RICScript. The x, y, width, and height arguments can all be parameterized. The options argument is currently unused.

```
rect(0, 10, 10, 20, 10);// 20x10 rectangle
```

Circle

Draw a circle on the LCD screen in your NXT picture by adding a
`circle(options, x, y, radius)` statement to your RICScript. The x, y, and radius arguments can all be parameterized. The options argument is currently unused.

`circle(0, 30, 30, 10);// circle at (30, 30)`

NumBox

Draw a number on the LCD screen in your NXT picture by adding a
`numbox(options, x, y, value)` statement to your RICScript. The x, y, and value arguments can all be parameterized. The options argument is currently unused.

`numbox(0, 2, 5, 15); // draw 15 at (2, 5)`

NXT Programs

Even simpler than RICScript programs are the NPG scripts that you can write for the NXT. These compile into very small NXT Programs or On Brick Programs. They have a .rpg file extension after you compile them. Each NXT program is a simple list of 5 program steps.

Using the NXT user interface menus, you can write these types of executable programs on the brick itself by selecting the five steps via the NXT buttons. The menu system, though, limits the order that these steps can occur. With an NPG program you can pick any five of the commands shown in Table 7-11 in any order. You just need to make sure you have either the standard RPGReader.sys on your brick or the streamlined smaller version of this file that you can download from the BricxCC utilities website.

The NXT programs all expect that the right motor is OUT_B and the left motor is OUT_C. A touch sensor should be attached to port 1, the light sensor to port 2, the sound sensor to port 3 and the ultrasonic sensor to port 4.

NPG Command	Usage
EmptyMove	This command is a no-op (i.e., nothing happens)
ForwardFive	Move forward for five full rotations
BackLeftTwo	Move backward and turn to the left for two rotations
TurnLeft	Start turning to the left
TurnLeftTwo	Turn left for two rotations
BackRight	Start backing and turning to the right
TurnRight	Start turning to the right
TurnRightTwo	Turn right for two rotations
BackLeft	Start backing and turning to the left
PlayToneOne	Play A4 for two seconds
PlayToneTwo	Play A4 for five seconds
Backward	Start moving backward without any tachometer limitation
BackwardFive	Move backward for five rotations
BackRightTwo	Move backward and turn right for two rotations
EmptyWait	This command is a no-op
WaitForLight	Wait until the light level is less than 50
WaitForObject	Wait until an object is less than 25 centimeters away
WaitForSound	Wait until the sound sensor reads less than 400.
WaitForTouch	Wait until the touch sensor is pressed
WaitTwoSeconds	Wait for two seconds
WaitFiveSeconds	Wait for five seconds
WaitTenSeconds	Wait for ten seconds
WaitForDark	Wait until the light level is greater than 30
EndLoop	Loop back to the beginning of the program and keep running
EndStop	Stop the program

Table 7-11. NPG Commands

The code sample below demonstrates how you can combine these commands into a simple NPG program.

```
Forward
WaitForLight
BackLeftTwo
WaitTwoSeconds
EndLoop // start back at the beginning
```

In the next chapter we will dig into the NBC programming language. Using this low-level language, we'll see how to produce the fastest and smallest programs possible with the NXT firmware.

Advanced Programming

Topics in this Chapter

- Introduction to Assembly
- NBC Overview

Chapter 8

It has long been believed that assembly language programming is difficult to learn and use. As you will learn in this chapter, that notion is largely false for modern assembly languages and for NeXT Byte Codes (NBC) in particular. To get started, we'll take a quick look at assembly languages in general. Then we'll focus on the powerful capabilities provided by NBC.

Introduction to Assembly

Assembly language (or assembly) allows you to write code "closer to the metal", since you are interacting directly with the chip. It is essentially a text representation of a binary machine language. Every computer chip family has its own unique assembly language due to the architectural differences between processors. Machine language is the pattern of bits used to encode a processor's operations. Assembly replaces these bits with more readable symbols or opcodes. These symbols are also called mnemonics since a name like mov is a lot easier to remember than 0x1B or b00011011.

With a high-level language, a single line might represent many processor operations. Each line of assembly, however, is a single operation for the processor. An assembler, such as NBC, is responsible for translating from assembly to the machine language understood by the CPU. Sometimes, rather than running on a CPU, the machine language runs on a virtual machine.

Modern assembly languages have support for preprocessor macros, which let you define blocks of code with substitution strings. These are often called Macro Assemblers. Using macros, you can write code that looks a lot like C with function calls containing arguments and so forth. So you might simply write OnFwd(OUT_A, 75) in assembly and the preprocessor will expand this statement into one or more pure assembly language statements.

NBC Overview

NeXT Byte Codes (NBC) is a powerful assembly language with macro support that is designed specifically for programming the NXT brick. The NXT firmware has a byte-code interpreter, or virtual machine (VM), which is used to execute programs. The NBC compiler translates a source program into LEGO NXT byte-codes, which are then executed by the VM on the NXT itself. Although the preprocessor and format of NBC programs are similar to assembly, NBC is not a general-purpose assembly language – there are many restrictions that stem from limitations of the NXT virtual machine.

There are two separate parts to NBC: the language and the application programming interface (API). The NBC language describes the syntax used to write programs. The NBC API describes the system functions, constants, and macros that are used by programs. This API is defined in the header file called "NXTDefs.h". In keeping with NQC and NXC, you do not need to explicitly include this file at the beginning of your NBC programs because it is automatically included by the compiler.

This chapter describes the NBC language and portions of the NBC API. It will provide you with the information you need to write your own NBC programs. A quick reference guide to the NBC API is included in Appendix B.

The NBC Language

In this section we'll look at the lexical rules used by the compiler, the structure of programs, statements, and expressions, and the operation of the preprocessor.

Unlike a lot of assembly languages, NBC is a case-sensitive language. This means that the identifier "xYz" is not the same identifier as "Xyz". Similarly, the subtract statement begins with the keyword "sub" but "suB", "Sub", or "SUB" are not keywords, meaning you can use them as valid identifiers.

Comments

Three forms of comments are supported in NBC. You can use traditional C-style block comments that begin with /* and end with */, and as before you cannot nest block comments. The second form of comments is the C++ single line comment, which begins with // and ends with a newline character. The third form of comments supported by NBC is the traditional assembly language single line comment. It begins with a semicolon and ends with a newline.

```
/* a block comment */
// a single line comment
; another single line comment
```

Use comments freely throughout your programs to document the source code. Without comments you will quickly forget why you wrote the code the way you did. Make sure your comments do not simply restate the code. They should describe the purpose of a section of code and explain any portions of the code that might be confusing to another person.

Whitespace

Whitespace is used to separate tokens and to make programs more readable. As long as the tokens are distinguishable, adding or subtracting whitespace has no effect on the meaning of a program. For example, the following lines of code both have the same meaning:

```
set x,2
set   x,   2
```

Generally, whitespace is ignored outside of string constants and constant numeric expressions. However, unlike in C, NBC statements may not span multiple lines. Aside from preprocessor macros, each statement in an NBC program must begin and end on the same line.

```
add x, x, 2 ; okay
add x,       // error
    x, 2     // error

set x, (2*2)+43-12 ; okay
set x, 2 * 2 ; error (expression contains spaces)
```

Numerical Constants

Numerical constants work the same in NBC as they do in NXC. They may be written in either decimal or hexadecimal form. Decimal constants consist of one or more decimal digits. Hexadecimal constants start with 0x or 0X followed by one or more hexadecimal digits.

```
set x, 10 // set x to 10
set x, 0x10 ; set x to 16 (10 hex)
```

String Constants

String constants in NBC, unlike NXC, are delimited with single quote characters. NBC automatically converts a string constant into an array of bytes with the last byte in the array being a zero. The final zero byte is generally referred to as the null terminator for a C-style string.

```
TextOut(0, LCD_LINE1, 'testing')
```

Identifiers and Keywords

Identifiers are used for variable, thread, function, and subroutine names. The first character of an identifier must be an upper or lower case letter or the underscore ('_'). Remaining characters may be letters, numbers, and an underscore.

A number of potential identifiers are reserved for use in the NBC language itself. These reserved words are call keywords and may not be used as identifiers. A complete list of keywords appears in Table 8-1 below.

add	sub	neg	mul
div	mod	and	or
xor	not	cmp	tst
index	replace	arrsize	arrbuild
arrsubset	arrinit	mov	set
flatten	unflatten	numtostr	strtonum
strcat	strsubset	strtoarr	arrtostr
jmp	brcmp	brtst	syscall
stop	exit	exitto	acquire
release	subcall	subret	setin
setout	getin	getout	wait
gettick	thread	endt	subroutine
follows	precedes	segment	ends
typedef	struct	db	byte
sbyte	ubyte	dw	word
sword	uword	dd	dword
sdword	udword	long	slong
ulong	void	mutex	waitv
call	return	abs	sign
strindex	strreplace	strlen	shl
shr	sizeof	compchk	compif
compelse	compend	valueof	isconst

Table 8-1. NBC Keywords

Program Structure

An NBC program is composed of code blocks and global variables in data segments. There are two primary types of code blocks: thread and subroutines. Each of these types of code blocks has its own unique features and restrictions, but they share a common structure.

A third type of code block is the preprocessor macro function. This code block type is used throughout the NBC API. Macro functions are the only type of code block, which use a parameter passing syntax similar to what you might see in a language like C or Pascal.

Data segment blocks are used to define types and to declare variables. An NBC program can have zero or more data segments, which are placed either outside of a code block or within a code block. Regardless of the location of the data segment, all variables in an NBC program are global.

Threads

An NBC thread is like an NXC task in that it directly corresponds to an NXT thread. Threads are defined using the thread keyword with the syntax shown in the code sample below.

```
thread name
   // the thread's code is placed here
endt
```

The name of the thread is any legal identifier. A program must always have at least one thread. If there is a thread named "main" then that thread is the thread that is started whenever the program runs. If none of the threads are named "main" then the very first thread that the compiler encounters in the source code is the main thread. The maximum number of threads and subroutines supported by the NXT is 256.

The body of a thread consists of a list of statements and optional data segments. Threads are started by scheduling dependant threads using the precedes or follows statements. Another thread can stop a thread by using the stop statement, if you have the enhanced NBC/NXC firmware. You can also stop all threads using the stop statement, which is described in the section titled Statements, or by a thread stopping on its own via the exit and exitto statements.

```
thread main
   precedes waiter, worker
   /* thread body goes here */
   exit // all dependants are scheduled
endt

thread waiter
   /* thread body goes here */
   //   exit
   ; exit is optional
endt

thread worker
   precedes waiter
   /* thread body goes here */
   exit // schedule dependant to execute
endt
```

Subroutines

Subroutines allow several different callers to share a single copy of some body of code. This makes subroutines much more space efficient than macro functions. Subroutines are defined using the subroutine keyword with the syntax shown below.

```
subroutine name
   // body of subroutine
   // subroutines must end with a return statement
```

```
   return
ends
```

A subroutine is just a special type of thread that you execute explicitly from within other threads or subroutines. Its name can be any legal identifier. Subroutines are not scheduled to run via the same mechanism that is used with threads. Instead, subroutines and threads execute other subroutines by using the call statement (described in the Statements section below).

```
thread main
  /* body of main thread goes here */
  call mySub // compiler handles the subroutine
          // return address
  exit // finalize execution
      // (details handled by the compiler)
endt

subroutine mySub
  /* body of subroutine goes here */
  return // compiler handles the subroutine
        // return address
ends
```

You can pass arguments into and out of subroutines using global variables. If you design a subroutine to work properly when called by concurrently executing threads, then calls to the subroutine must be protected by acquiring a mutex prior to the subroutine call and releasing the mutex after the call.

You can also call a thread as a subroutine using a slightly different syntax. This technique is required if you want to call a subroutine which executes two threads simultaneously. The subcall and subret statements must be used instead of call and return. You also must provide a global variable to store the return address as shown in the sample code below.

```
dseg segment
  anothersub_returnaddress byte
dseg ends

thread main
  /* thread body goes here */
  acquire ssMutex
  call SharedSub ; automatic return address
  release ssMutex
  // calling a thread as a subroutine
  subcall AnotherSub, anothersub_returnaddress
  exit
endt

subroutine SharedSub
```

```
   /* subroutine body goes here */
   ; return is required as the last operation
   return
ends

thread AnotherSub
   /* threads can be subroutines too */
   ; manual return address
   subret anothersub_returnaddress
endt
```

After the subroutine completes executing, it returns to the calling routine and program execution continues with the next statement following the subroutine call.

Macro Functions

It is often helpful to group a set of statements together into a single function, which you can call as needed. NBC supports macro functions with arguments. Values may be returned from a macro function by changing the value of one or more of the arguments within the body of the macro function. Macro functions are defined using the syntax shown below.

```
#define name(argument_list) \
   // body of the macro function \
// last line in macro function \
// body has no '\' at the end
```

Note that the newline escape character ('\') must be the very last character on the line. If it is followed by any whitespace or comments then the macro body is terminated at that point and the next line is not considered to be part of the macro definition.

The argument list may be empty, or it may contain one or more argument definitions. An argument to a macro function has no type; rather each argument is simply defined by its name. Use commas to separate multiple arguments. Arguments to a macro function are either inputs for the code in the body of the function to process or outputs for the code to modify and return. Inputs are either constants or variables while outputs must be variables. The following sample shows how to define a macro function to simplify the process of drawing text on the NXT LCD screen.

```
#define MyMacro(x, y, msg, berase) \
   mov dtArgs.Location.X, x \
   mov dtArgs.Location.Y, y \
   mov dtArgs.Options, berase \
   mov dtArgs.Text, msg \
   syscall DrawText, dtArgs

MyMacro(0, 0, 'testing', TRUE)
MyMacro(10, 20, 'Please Work', FALSE)
```

NBC macro functions are always expanded inline by the preprocessor. This means that each call to a macro function produces another copy of the function's code in the program. Unless used judiciously, inline macro functions can lead to excessive code size.

Data segments

Data segments contain the type definitions and variable declarations. Data segments are defined using the syntax shown below.

```
dseg segment
// type definitions and variable
// declarations go here
dseg ends

thread main
  dseg segment
    // or here — still global, though
  dseg ends
endt
```

You can have multiple data segments in an NBC program. All variables are global, regardless of where they are declared. Once declared, you can use them within all threads, subroutines, and macro functions. Their scope begins at the declaration and ends at the end of the program.

Type Definitions

Type definitions are always contained within a data segment. They are used to define new type aliases or new aggregate types (known as structures). A type alias is defined using the typedef keyword with the following syntax:

```
type_alias typedef existing_type
```

The new alias name may be any valid identifier. The existing type must be some type already known by the compiler. It can be a native type or a user-defined type. Once a type alias has been defined. You can use it in subsequent variable declarations and aggregate type definitions. The example below demonstrates a simple type alias definition.

```
big typedef dword
; big is now an alias for the dword type
```

Structure definitions must also reside within a data segment. They are used to define a type that aggregates or contains other native or user-defined types. A structure definition is defined using the struct and ends keywords with the syntax shown in the code sample below.

```
TypeName struct
  x byte
  y byte
TypeName ends
```

Structure definitions allow you to manage related data in a single combined type. They can be as simple or complex as the needs of your program dictate. The code sample below is an example of a relatively complex structure.

```
MyPoint struct
  x byte
  y byte
MyPoint ends
ComplexStrut struct
  value1 big              // using a type alias
  value2 sdword
  buffer byte[]           /* array of byte */
  blen word
  extrastuff MyPoint[]    // array of structs
  pt_1 MyPoint            // struct contains struct instances
  pt_2 MyPoint
ComplexStruct ends
```

Variable Declarations

All variable declarations must be contained within a data segment. They are used to declare variables for use in a code block such as a thread, subroutine, or macro function. A variable is declared using the syntax shown below.

```
var_name type_name optional_initialization
```

The variable name may be any valid identifier. The type name must be a type or type alias already known by the compiler. The optional initialization format depends on the variable type, but for non-aggregate or scalar types the format is simply a constant integer or constant expression, which may not contain whitespace. Examples of this appear later in this section.

The NXT firmware supports several different types of variables, which are grouped into two categories: scalar types and aggregate types. Scalar types are a single integer value which may be signed or unsigned and occupy one, two, or four bytes of memory. The keywords for declaring variables of a scalar type are listed in Table 8-2.

Type Name	Information
byte, ubyte, db	8 bit unsigned
Sbyte	8 bit signed
word, uword, dw	16 bit unsigned
Sword	16 bit signed
dword, udword, dd	32 bit unsigned
Sdword	32 bit signed
long, ulong	32 bit unsigned (alias for dword, udword)
Slong	32 bit signed (alias for sdword)
mutex	Special type used for exclusive subroutine access

Table 8-2. Scalar Types

Some examples of scalar variable declarations are shown in the code sample below.

```
dseg segment
  x byte    // initialized to zero by default
  y byte 12          ; initialize to 12
  z sword -2048      /* a signed value */
  myVar big 0x12345 ; use a type alias
  var1 dword 0xFF    ; value is 255
  myMutex mutex
  ; mutexes ignore initialization, if present
  bTrue byte 1
  ; byte variables can be used as booleans
dseg ends
```

Aggregate variables are either structures or arrays of some other type (either scalar or aggregate). Once a user-defined struct type has been defined, you can use it to declare a variable of that type. Similarly, you can use them in array declarations. You can nest arrays and structs as deeply as your program needs dictate, but nesting deeper than 2 or 3 levels may lead to slower program execution due to NXT firmware memory constraints.

A few examples of aggregate variable declarations are shown below.

```
dseg segment
  buffer byte[] // starts off empty
  msg byte[] 'Testing'
  // msg is an array of byte =
// (0x54, 0x65, 0x73, 0x74,
//  0x69, 0x6e, 0x67, 0x00)
```

```
    ; two values in the array
    data long[] {0xabcde, 0xfade0}

    ; declare an instance of a struct
myStruct ComplexStruct

    ; declare an array of a structs
Points MyPoint[]

    ; an array of an array of byte
    msgs byte[][]
dseg ends
```

You can initialize byte arrays either by using braces containing a list of numeric values ({val1, val2, ..., valN}) or by using a string constant delimited with single-quote characters ('Testing'). Embedded single quote characters are not supported. Arrays of any scalar type, other than byte, should be initialized using braces. Arrays of struct and nested arrays cannot be initialized.

Compiler Tokens

NBC supports special tokens, which it replaces on compilation. The tokens are similar to preprocessor #define macros but they are actually handled directly by the compiler rather than the preprocessor. The supported tokens are as shown in Table 8-3.

Token	Usage
__FILE__	This token is replaced with the currently active filename (no path).
__LINE__	This token is replaced with the current line number.
__VER__	This token is replaced with the compiler version number.
__THREADNAME__	This token is replaced with the current thread name.
__I__ , __J__	These tokens are replaced with the current value of I or J. They are both initialized to zero at the start of each thread or subroutine.
__ResetI__ , __ResetJ__	These tokens are replaced with nothing. As a side effect the value of I or J is reset to zero.
__IncI__ , __IncJ__	These tokens are replaced with nothing. As a side effect the value of I or J is incremented by one.
__DecI__ , __DecJ__	These tokens are replaced with nothing. As a side effect the value of I or J is decremented by one.

Table 8-3. Compiler Tokens

The ## preprocessor directive can help make the use of compiler tokens more readable. `__THREADNAME__##_##__I__:` would become something like `main _ 1:`. Without the ## directive it would much harder to read the mixture of compiler tokens and underscores.

Expression Evaluator

Constant expressions are supported by NBC for many statement arguments as well as variable initialization. The compiler evaluates expressions when the program is compiled, not at run time. The compiler will return an error if it encounters an expression that contains whitespace. "4+4" is a valid constant expression but "4 + 4" is not. The expression evaluator supports the operators in Table 8-4 below.

Operator	Meaning
+	Addition
-	subtraction
*	multiplication
/	Division
^	Exponent
%	modulo (remainder)
&	bitwise and
|	bitwise or
~	bitwise xor
<--<--	shift left
-->-->	shift right
()	Grouping subexpressions
PI	constant value

Table 8-4. Constant Expression Operators

The constant expression evaluator also supports the compile-time functions listed in Table 8-5 below.

Compile-time Functions
tan(x), sin(x), cos(x), sinh(x), cosh(x), arctan(x), cotan(x), arg(x), exp(x), ln(x), log10(x), log2(x), logn(x, n), sqr(x), sqrt(x), abs(x), trunc(x), int(x), ceil(x), floor(x), heav(x), sign(x), zero(x), ph(x), rnd(x), random(x), max(x, y), min(x, y), power(x, exp), intpower(x, exp)

Table 8-4. Constant Expression Operators

The code sample below demonstrates how to use a constant expression.

```
// expression value will be truncated
// to an integer
set val, 3+(PI*2)-sqrt(30)
```

IO-Map Address (IOMA) Constants

IOMA constants provide a simplified means of accessing input and output field values without having to use a variable or an input or output statement. The constants are defined in the NBCCommon.h header file, which the compiler includes in your NBC program automatically.

There are IOMA constants for inputs and outputs, defined as preprocessor macros. To specify the port for each IOMA, you must use a constant such as IN_1 or OUT_A. You can often substitute an IOMA constant in statements, which can accept a scalar variable argument. The IOMA macros are shown in Table 8-6 below.

```
InputIOType(port), InputIOInputMode(port),
InputIORawValue(port), InputIONormalizedValue(port),
InputIOScaledValue(port), InputIOInvalidData(port),
OutputIOUpdateFlags(port), OutputIOOutputMode(port),
OutputIOPower(port), OutputIOActualSpeed(port),
OutputIOTachoCount(port), OutputIOTachoLimit(port),
OutputIORunState(port), OutputIOTurnRatio(port),
OutputIORegMode(port), OutputIOOverload(port),
OutputIORegPValue(port), OutputIORegIValue(port),
OutputIORegDValue(port), OutputIOBlockTachoCount(port),
OutputIORotationCount(port)
```

Table 8-6. IOMA Constant Macros

The body of a code block (thread, subroutine, or macro function) is composed of statements. All statements are terminated with the newline character.

Assignment Statements

Assignment statements enable you to copy values from one variable to another or to simply set the value of a variable. In NBC there are two ways to assign a new value to a variable.

The move statement (mov) assigns the value of its second argument to its first argument. The first argument must be the name of a variable. It can be of any valid variable type except mutex. The second argument can be a variable or a constant (numeric or string). If a constant is used, the compiler creates a variable behind the scenes and initializes it to the specified constant value.

Both arguments in the move statement must be of compatible types. You can assign a scalar value to another scalar variable (regardless of type), structs to struct variables (if the structure types are the same), and arrays to an array variable (if the types contained in the arrays are the same). The syntax of the move statement is shown below.

```
mov x, y      // set x equal to y
```

The set statement assigns its first argument to the value of its second argument. The first argument must be the name of a variable, and it must be a scalar type. The second argument must be a numeric constant or constant expression. The syntax of the set statement is shown below.

```
set x, 10     // set x equal to 10
```

Because all arguments must fit into a 2-byte value in the NXT executable, the second argument of the set statement is limited to a 16 bit signed or unsigned value (-32768..65535).

Math Statements

Math statements enable you to perform basic math operations on data in your NBC programs. In high level programming languages like C, mathematical expressions use standard math operators, such as *, -, +, /. However, in NBC (as with other assembly languages), math operations are expressed as statements with the math operation name coming first, followed by the arguments to the operation. All statements in this family have one output argument and two input arguments except the negate statement, the absolute value statement, and the sign statement.

Math statements in NBC differ from traditional assembly math statements because many of the operations can handle arguments of scalar, array, and struct types rather than only scalar types. If, for example, you multiply an array by a scalar then each of the elements in the resulting array equal the corresponding element in the original array multiplied by the scalar value.

Only the absolute value and sign statements require that their arguments are scalar types. If you do not install the enhanced NBC/NXC firmware then the NBC compiler implements these two statements, since the NXT firmware does not have built-in support for them.

The add statement lets you add two input values together and store the result in the first argument. The first argument must be a variable but the second and third arguments can be variables, numeric constants, or constant expressions. The syntax of the add statement is shown below.

```
add x, x, y ; add x and y and store the
            ; result in x
```

The subtract statement (sub) lets you subtract two input values and store the result in the first argument. The first argument must be a variable but the second and third arguments can be variables, numeric constants, or constant expressions. The syntax of the subtract statement is shown below.

```
sub x, x, y ; subtract y from x and store
            ; the result in x
```

The multiply statement (mul) lets you multiply two input values and store the result in the first argument. The first argument must be a variable but the second and third arguments can be variables, numeric constants, or constant expressions. The syntax of the multiply statement is shown below.

```
mul x, x, x ; set x equal to x^2
```

The divide statement (div) lets you divide two input values and store the result in the first argument. The first argument must be a variable but the second and third arguments can be variables, numeric constants, or constant expressions. The syntax of the divide statement is shown below.

```
div x, x, 2 ; set x equal to x / 2
            ; (using integer division)
```

The modulo statement (mod) lets you calculate the remainder of two input values and store the result in the first argument. The first argument must be a variable but the second and third arguments can be variables, numeric constants, or constant expressions. The syntax of the modulo statement is shown below.

```
mod x, x, 4 ; set x equal to x % 4 (0..3)
```

The negate statement (neg) lets you negate an input value and store the result in the first argument. The first argument must be a variable but the second argument can be a variable, a numeric constant, or a constant expression. The syntax of the negate statement is shown below.

```
neg x, y ; set x equal to -y
```

The absolute value statement (abs) lets you calculate the absolute value of an input value and store the result in the first argument. The first argument must be a variable but the second argument can be a variable, a numeric constant, or a constant expression. The syntax of the absolute value statement is shown below.

```
abs x, y ; set x equal to the absolute value of y
```

The sign value statement (sign) lets you calculate the sign value (-1, 0, or 1) of an input value and store the result in the first argument. The first argument must be a variable but the second argument can be a variable, a numeric constant, or a constant expression. The syntax of the sign value statement is shown below.

```
sign x, y ; set x equal to -1, 0, or 1
```

Logic Statements

Logic statements let you perform basic logical operations on data in your NBC program. As with the math statements, the logical operation name begins the statement and it is followed by the arguments to the logical operation. All the statements in this family have one output argument and two input arguments, except the logical not statement. Each statement supports arguments of any type: scalar, array, or struct.

The bitwise AND statement (and) lets you AND together two input values and store the result in the first argument. Each bit in the two input values is compared and the corresponding bit in the output argument is set if both of the input bits are set. The first argument must be a variable but the second and third arguments can be a variable, a numeric constant, or a constant expression. The syntax of the bitwise AND statement is shown below.

```
and x, x, y   // x = x & y
```

The bitwise OR statement (or) lets you OR together two input values and store the result in the first argument. Each bit in the two input values is compared and the corresponding bit in the output argument is set if at least one of the two input bits is set. The first argument must be a variable but the second and third arguments can be a variable, a numeric constant, or a constant expression. The syntax of the bitwise OR statement is shown below.

```
or x, x, y   // x = x | y
```

The bitwise exclusive OR statement (xor) lets you exclusive OR together two input values and store the result in the first argument. Each bit in the two input values is compared and the corresponding bit in the output argument is set if one and only one of the two input bits is set. The first argument must be a variable but the second and third arguments can be a variable, a numeric constant, or a constant expression. The syntax of the bitwise exclusive OR statement is shown below.

```
xor x, x, y   // x = x ^ y
```

The not statement lets you logically NOT its input value and store the result in the first argument. This is not a bitwise operation. The first argument must be a variable but the second argument can be a variable, a numeric constant, or a constant expression. The syntax of the not statement is shown below.

```
not x, x   // x = !x (logical not – not bitwise)
```

Bit Manipulation Statements

The bit manipulation statements perform bit shifts, rotates, and complement operations on data in your NBC program. As with the math statements, the bit manipulation operation name begins the statement, followed by the arguments for the operation. All of the opcodes in this family have one output argument and two input arguments except for the cmnt statement.

The shift right statement (shr) shifts the second argument to the right by the number of bits specified by the third argument, and stores the result in the first argument. Shifting to the right is the same as dividing by two but the shift operation is usually faster than integer division. The first argument must be a variable but the second and third arguments can be a variable, a numeric constant, or a constant expression. The syntax of the shift right statement is shown below.

```
shr x, x, 2   // x = x >> 2
```

The shift left statement (shl) shifts the second argument to the left by the number of bits specified by the third argument, and stores the result in the first argument. Shifting to the left is the same as multiplying by two. Shift operations are usually faster than multiplications. The first argument must be a variable but the second and third arguments can be a variable, a numeric constant, or a constant expression. The syntax of the shift left statement is shown below.

```
shl x, x, y   // x = x << y
```

If you do not have the enhanced NBC/NXC firmware installed then the compiler implements these opcodes with additional code since the standard NXT firmware does not support shift operations. The enhanced firmware also implements the following bit-manipulation statements. These result in a compiler error unless you tell the compiler to target the enhanced firmware.

The complement statement (cmnt) stores the bit-wise complement of the second argument in the first argument. All ones become zeros and all zeros become ones. The syntax of the complement statement is shown below.

```
cmnt x, y   // x = ~y
```

The logical shift right statement (lsr) logically shifts the second argument to the right by the number of bits specified by the third argument, and stores the result in the first argument. When performing a logical shift, any bits that are shifted out are simply discarded, and zeros are shifted in on the other end. This means that logical and arithmetic left shifts are exactly the same. On the other hand, logical shift right operations insert bits with a value of zero instead of copies of the sign bit. Logical shifts are best for unsigned numbers, while arithmetic shifts are best for signed numbers. The syntax of the logical shift right statement is shown below.

```
lsr x, x, 2
```

The logical shift left statement (lsl) logically shifts the second argument to the left by the number of bits specified by the third argument, and then stores the result in the first argument. The syntax of the logical shift left statement is shown below.

```
lsl x, x, y
```

The arithmetic shift right statement (asr) arithmetically shifts the second argument to the right by the number of bits specified by the third argument, and then stores the result in the first argument. The syntax of the arithmetic shift right statement is shown below.

```
asr x, x, y
```

The arithmetic shift left statement (asl) arithmetically shifts the second argument to the left by the number of bits specified by the third argument, and then stores the result in the first argument. The syntax of the arithmetic shift left statement is shown below.

```
asl x, x, y
```

The rotate right statement (rotr) rotates the second argument to the right by the number of bits specified by the third argument, and then stores the result in the first argument. The syntax of the rotate right statement is shown below.

```
rotr x, x, y
```

The rotate left statement (rotl) rotates the second argument to the left by the number of bits specified by the third argument, and then stores the result in the first argument. The syntax of the rotate left statement is shown below.

```
rotl x, x, y
```

Comparison Statements

The comparison statements let you perform compare operations on data in your NBC program. The comparison operation name begins the statement, followed by the arguments for the operation. These statements all take a comparison code constant as their first argument. You can use scalar, array, and aggregate types for the compare or test arguments. Valid comparison codes are listed in Table 8-7.

The compare statement (cmp) compares two sources using the specified comparison code. The boolean result of the comparison is stored in the destination or second argument. The syntax of the compare statement is shown below.

```
cmp EQ, bEqual, x, y  // set bEqual to 1 if x == y
```

The test statement (tst) compares a single source to zero using the specified comparison code. The boolean result of the comparison is stored in the destination or second argument. The syntax of the test statement is shown below.

```
tst GT, bGTZero, x  // set bGTZero to 1 if x > 0
```

Comparison	Constant	Value	Notes
Less than	LT	0x00	'←⋯' may also be used
Greater than	GT	0x01	'⋯→' may also be used
Less than or equal to	LTEQ	0x02	'←⋯=' may also be used
Greater than or equal to	GTEQ	0x03	'⋯→=' may also be used
Equal to	EQ	0x04	'==' may also be used
Not equal to	NEQ	0x05	'!=' or '←⋯⋯→' may also be used

Table 8-7. Comparison Codes

Control Flow Statements

The program flow control statements play a major role in implementing loops and conditional statements that are not natively available to an assembly language. Familiarize yourself with their usage so that you can easily implement any sort of higher-level language construct that you many need to use in your NBC programs.

Each of the control flow statements takes a statement label as one of its arguments. A statement label is any valid identifier followed by a colon. You have seen statement labels before in Chapter 7 when we looked at the syntax of the NXC switch statement.

The jump statement (jmp) takes a label as its only argument and unconditionally jumps to that program location. The syntax of the jump statement is shown below.

```
LoopStart:
// loop body
jmp LoopStart  // jump to LoopStart label
```

There are two conditional branch statements that each takes a comparison code constant as their first argument and a label for the second argument. The branch compare statement (brcmp) compares two source arguments to each other using the specified comparison code. If the comparison is true then the program flow will branch to the specified statement label. The syntax of the branch compare statement is shown below.

```
// jump to EqualValues label if x == y
brcmp EQ, EqualValues, x, y
```

The branch test statement (brtst) compares its source argument to zero using the specified comparison code. If the comparison is true then the program flow will branch to the specified statement label. The syntax of the branch test statement is shown below.

```
// jump to XGtZero label if x > 0
brtst GT, XGtZero, x
```

System Call Statement

The system call statement (syscall) enables execution of various system functions via a constant function ID and an aggregate type variable for passing arguments to and from the system function. The syntax of the system call statement is shown below.

```
syscall SoundPlayTone, ptArgs
// ptArgs is a struct with input/output fields
```

The list of supported function identifiers is as follows:

```
FileOpenRead, FileOpenWrite, FileOpenAppend, FileRead,
FileWrite, FileClose, FileResolveHandle, FileRename,
FileDelete, SoundPlayFile, SoundPlayTone, SoundGetState,
SoundSetState, DrawText, DrawPoint, DrawLine, DrawCircle,
DrawRect, DrawGraphic, SetScreenMode, ReadButton,
CommLSWrite, CommLSRead, CommLSCheckStatus, RandomNumber,
GetStartTick, MessageWrite, MessageRead, CommBTCheckStatus,
CommBTWrite, KeepAlive, IOMapRead, IOMapWrite, IOMapReadByID,
IOMapWriteByID, DisplayExecuteFunction, CommExecuteFunction,
LoaderExecuteFunction
```

The last five of these function identifiers are only supported when you are running the enhanced NBC/NXC firmware on your NXT brick.

Timing Statements

The timing statements perform waits or timing operations in your NBC program. The timing operation name begins the statement, followed by the arguments in the operation. All of the opcodes in this family have one argument.

The wait statement (wait) suspends the current thread for the number of milliseconds specified by its constant argument.

```
wait 1000   // wait for 1 second
```

The wait variable statement (waitv) acts just like the wait statement but it takes a variable argument.

```
// wait for the specified number of ms.
waitv iDelay
```

The get system tick statement (gettick) sets its output argument to the current system tick count.

```
// set x to the current system tick count
gettick x
```

If you do not have the enhanced NBC/NXC firmware installed then NBC implements wait and waitv as thread-specific subroutine calls due to the wait opcode not being implemented in the NXT firmware. If you use a constant argument with waitv the compiler will generate a temporary variable for you. If needed, you can implement simple wait loops using gettick. The code sample below demonstrates how you can perform this function.

```
gettick currTick
add endTick, currTick, waitms
Loop:
  gettick currTick
  brcmp LT, Loop, currTick, endTick
```

Array Statements

The array statements perform array operations in your NBC program. The array operation name begins the statement, followed by the arguments in the operation.

The index statement returns the arraysrc[index] value in the first argument. The source array is the second argument and the index is the third argument. The syntax of the array index statement is shown below.

```
// index output, arraysrc, index
index val, arValues, idx
// extract arValues[idx] and store it in val
```

The replace statement replaces the arraysrc[index] item in arraydest with the fourth argument. The arraysrc argument can be the same variable as arraydest to replace without copying the array. The fourth argument can be an array, in which case multiple items are replaced. The syntax is shown below.

```
// replace arraydest, arraysrc, index, newval
replace arNew, arValues, idx, x
// replace arValues[idx] with x in arNew
```

The array size statement (arrsize) returns the size of the input array in the scalar output argument. The syntax of this statement is shown below.

```
// arrsize output, arraysrc
arrsize nSize, arValues
// nSize == length of array
```

The array initialization statement (arrinit) initializes the output array to the value and size provided. Its syntax is shown below.

```
// arrinit arraydest, value, size
arrinit arValues, nZero, nSize
// initialize arValues with nSize zeros
```

The array subset statement (arrsubset) copies a subset of the input array to the output array. The syntax is shown below.

```
// arrsubset arraydest, arraysrc, index, length
arrsubset arSub, arValues, NA, x
// copy the first x elements to arSub
```

The array build statement (arrbuild) constructs an output array from a variable number of input arrays, scalars, or aggregates. The syntax of the array build statement is shown below.

```
// arrbuild arraydest, src1, src2, …, srcN
arrbuild arData, arStart, arBody, arEnd
// build data array from 3 sources
```

String Statements

The string statements perform string operations in your NBC program. The string operation name begins the statement, followed by the arguments in the operation.

The flatten statement converts its input argument into its string output argument. A resulting string is really a byte array with a null terminator byte at the end. Signed and unsigned longs will flatten into 4 bytes, not counting the null terminator. The bytes are in little-endian order. Aggregate types can be flattened as well. The syntax of the flatten statement is shown below.

```
// flatten strdest, source
flatten strData, args
// copy from the args struct to strData
```

The unflatten statement converts its input string argument to the output argument type. If 'default' does not match the flattened data type exactly (including array sizes) then 'err' is set to TRUE and 'dest' will contain a copy of 'default'. This statement is the inverse of the flatten statement. The syntax is below.

```
// unflatten dest, err, strsrc, default
// convert string to value
unflatten args, bErr, strSource, x
```

The number to string statement (numtostr) converts its scalar input argument to a string output argument. The syntax for this statement is shown below.

```
// numtostr deststr, source
// convert value to a string
numtostr strValue, value
```

The format number statement (fmtnum) formats the numeric value into a string using the provided format string. Use standard printf formatting codes to control output. The syntax of this statement is shown below.

```
// fmtnum deststr, fmtstr, source
// format value to a string
fmtnum strValue, "x = %d", value
```

The string to number statement (strtonum) parses its input string argument into a numeric output argument, advancing an offset past the numeric string. The syntax of the string to number statement is shown below.

```
// strtonum dest, oNext, str, oCur, default
// parse string into num
strtonum value, idx, strValue, idx, nZero
```

The string subset statement (strsubset) copies a subset of the input string to the output string. The syntax is shown below.

```
// strsubset strdest, strsrc, index, length
// copy the first x bytes in strSource to strSub
strsubset strSub, strSource, NA, x
```

The string concatenation statement (strcat) constructs an output string from a variable number of input strings. The syntax for the string concatenation statement is shown below.

```
// strcat strdest, strsrc1, strsrc2, …, strsrcN
strcat strData, strStart, strBody, strEnd
// build data string from 3 sources
```

The array to string statement (arrtostr) copies the input byte array into its output string array and adds a null byte at the end. Its syntax is shown below.

```
// arrtostr strdest, arraysrc
// convert byte array to string
arrtostr strData, arrData
```

The string to array statement (strtoarr) copies the input string array into its output byte array excluding the last byte, which should be a null. The syntax is shown below.

```
// strtoarr arraydest, strsrc
// convert string to byte array
strtoarr arrData, strData
```

The string index statement (strindex) returns the strsrc[index] value in the dest output argument. The syntax for this statement is shown below.

```
// strindex dest, strsrc, index
// extract strVal[idx] and store it in val
strindex val, strVal, idx
```

The string replace statement (strreplace) replaces the strsrc[index] item in strdest with the newval input argument. The strsrc argument can be the same variable as strdest to replace without copying the string. The newval argument can be a string, in which case multiple items are replaced. The syntax is below.

```
// strreplace strdest, strsrc, index, newval
// replace strValues[idx] with newStr in strNew
strreplace strNew, strValues, idx, newStr
```

The string length statement (strlen) returns the size of the input string in the scalar output argument. Its syntax is shown below.

```
// strlen dest, strsrc
strlen nSize, strMsg
// nSize == length of strMsg
```

Scheduling Statements

The scheduling statements control the operation of threads and the calling of subroutines in your NBC program. The scheduling operation name begins the statement, followed by the arguments in the operation.

The start statement starts executing another thread while the current thread keeps running. If the enhanced firmware is not installed on your NXT, the compiler will generate additional code to simulate this operation since the standard firmware does not support this operation natively.

```
// start threadname
start moveThread // start another thread
```

The stopthread statement stops the execution of another thread while the current thread keeps running. This statement requires the enhanced NBC/NXC firmware.

```
// stopthread threadname
stopthread moveThread // stop another thread
```

The priority statement adjusts the priority of the current thread or any other thread. This statement also requires the enhanced NBC/NXC firmware.

```
// priority threadname, priority_level
priority moveThread, 10 // set thread priority
```

The exit statement finalizes the current thread and schedules zero or more dependant threads by specifying start and end dependency list indices. The indices are zero-based and inclusive.

```
// exit [start, end]
exit 0, 2  // schedule this thread's 3 dependants
exit // schedule all this thread's dependants
```

The exitto statement finalizes the current thread and schedules the thread specified by name.

```
// exit now and schedule worker thread
exitto worker
```

The acquire statement acquires the named mutex. If the mutex is already acquired the current thread waits until it becomes available.

```
// acquire mutex for our use
acquire muFoo
```

The release statement releases the named mutex allowing other threads to acquire it.

```
// release mutex for another thread to use
release muFoo
```

The subcall statement calls into the named thread or subroutine and waits for a return, which might not come from the same routine. The second argument is a variable used to store the return address.

```
// call drawText subroutine
subcall drawText, retDrawText
```

The subret statement returns from a thread to the return address value contained in its input argument.

```
subret retDrawText   // return to calling routine
```

The call statement executes the named subroutine and waits for a return. The argument should specify a thread that was declared using the "subroutine" keyword.

```
call MyFavoriteSubroutine  // call routine
```

The return statement returns from a subroutine.

```
return   // return to calling routine
```

The compiler automatically handles the return address for call and return when they are used with subroutines rather than threads

The stop statement stops program execution depending on the value of its boolean input argument. Its syntax is shown in the code sample below.

```
// stop the program if bStop is non-zero
stop bStop
```

Input Statements

The input statements control the NXT inputs in your NBC program, allowing you to read sensors. The input operation name begins the statement, followed by the arguments in the operation.

The set input statement (setin) sets an input field of a sensor on a port to the value specified in its first input argument. The sytax of this statement is shown below.

```
// setin src, port, fieldID
// Configure variables
// set type to switch
set theType, IN_TYPE_SWITCH
// set mode to boolean
set theMode, IN_MODE_BOOLEAN
// set port to #1
set thePort, IN_1
// set sensor on port 1 to switch type
setin theType, thePort, Type
```

```
// set sensor on port 1 to boolean mode
setin theMode, thePort, InputMode
```

Valid input field identifiers for the setin statement are listed below. These fields, along with their values, are documented in Chapter 3.

```
Type, InputMode, RawValue, NormalizedValue, ScaledValue,
InvalidData
```

The get input statement (getin) reads a value from an input field of a sensor on a port and writes the value to its dest output argument. One of the field names from the list above must be used for the third argument of this statement. See the syntax for this statement in the code sample below.

```
// getin dest, port, fieldID
// read raw sensor value
getin rVal, thePort, RawValue
// read scaled sensor value
getin sVal, thePort, ScaledValue
// read normalized sensor value
getin nVal, thePort, NormalizedValue
```

Output Statements

The output statements control the NXT outputs in your NBC program, allowing motor control. The output operation name begins the statement, followed by the arguments in the operation.

The set output statement (setout) sets one or more output fields of a motor on one or more ports to the value specified by the coupled input arguments. While the code may wrap in the sample below, it is important that each statement is on a single line. If you want to use multiple ports with this statement, you must use an array containing some combination of OUT_A, OUT_B, and OUT_C rather than any of the combined motor constants. This statement's syntax is shown in the code sample below.

```
// setout port/portlist, fieldid1, value1, ..., fieldidN, valueN
// set mode to motor on
set theMode, OUT_MODE_MOTORON
// motor running
set rsVal, OUT_RUNSTATE_RUNNING
// set port to #1
// (use an array for multiple ports)
set thePort, OUT_A
// negative power means reverse motor direction
set pwr, -75
// set output values
setout thePort, OutputMode, theMode, RunState, rsVal, Power, pwr
```

Valid output field identifiers for the setout statement are listed below. These fields, along with their values, are documented in Chapter 3.

UpdateFlags, OutputMode, Power, ActualSpeed, TachoCount, TachoLimit, RunState, TurnRatio, RegMode, Overload, Reg-PValue, RegIValue, RegDValue, BlockTachoCount, RotationCount

The get output statement (getout) reads a value from an output field of a sensor on a port and writes the value to its dest output argument. Use one of the field names listed above for the third argument. The syntax for the get output statement is shown in the code sample below.

```
// getout dest, port, fieldID
// read motor regulation mode
getout rmVal, thePort, RegMode
// read tachometer limit value
getout tlVal, thePort, TachoLimit
// read the rotation count
getout rcVal, thePort, RotationCount
```

NBC API

There is a great deal in common between the NBC and NXC programming languages. Most of the motor control API functions are exactly the same in NBC as they are in NXC, while sensor and other API functions are extremely similar.

Controlling Outputs

In NBC the basic motor control functions are OnFwd(ports, power), OnRev(ports, power), Float(ports), and Off(ports). The semantics of these NBC functions are exactly the same as those found in the NXC equivalents.

```
OnFwd(OUT_AB, 75) ; run A and B fwd at 75% power
OnRev(OUT_A, 50) ; run A in reverse at 50% power
```

The OnFwdSync(ports, power, turnratio) and OnRevSync(ports, power, turnratio) functions are also available in NBC as are the OnFwdReg(ports, power, regmode) and OnRevReg(ports, power, regmode) methods.

```
OnFwdSync(OUT_AB, 75, 0); // no turning
OnFwdReg(OUT_A, 75, OUT_REGMODE_SPEED); //reg speed
```

NBC also provides the same NXC API functions for tachometer-limited motion. Use the RotateMotor(ports, power, tacholimit) command or one of its several variants to specify a tachometer limit for one or more motors. This function call will not return until the motors have reached the specified limit or until some other task sets the power level of the first motor to zero.

```
RotateMotor(OUT_A, 75, 720); // 2 full turns
```

For additional information about the motor API functions available in NBC see Appendix B.

Using Sensors

Sensor control in NBC is extremely similar to that found in NXC. Sensor port constants are `IN_1`, `IN_2`, `IN_3`, and `IN_4`. To read a sensor's value in NBC, use the `ReadSensorValue(port, value)` API function.

```
ReadSensorValue(IN_1, x) ; read a sensor value
```

The sensor type and sensor mode constants are all listed in Table 7-9 and Table 7-10 in Chapter 7. NBC also provides some higher-level functions for configuring and controlling the NXT sensor ports. You can configure any port to support any type of NXT or RCX sensor. Each of the API functions shown in the code sample below performs three separate tasks: set the sensor type, set the sensor mode, and wait for the sensor to be ready for reading.

```
SetSensorTouch(S1) ; NXT or RCX touch sensor
SetSensorLight(S2) ; NXT light sensor (active)
SetSensorSound(S3) ; NXT sound sensor (DB)
SetSensorLowspeed(S4) ; NXT I2C sensor (9 volt)
```

If you call `SetSensorType` and `SetSensorMode` or `SetSensor` directly, you should follow the sensor port configuration with a call to the `ResetSensor(port)` API function.

```
ResetSensor(S1) ; wait for sensor to be ready
```

You can also set a sensor value to zero, if you wish, using the `ClearSensor(port)` function. And the NXT motors all provide three separate tachometer values that you can read and reset independently. The RotationCount value is the sensor value that you should use if you want to treat the motor's built-in rotation sensor like the RCX rotation sensor, of course with far greater precision and accuracy. These functions are demonstrated in the code sample below.

```
ClearSensor(S1) ; set port 1 value to zero
ResetTachoCount(OUT_A)
ResetBlockTachoCount(OUT_A)
ResetRotationCount(OUT_A)
ResetAllTachoCounts(OUT_A)
```

The I²C or Lowspeed sensors are a special case since they are not analog sensors. You cannot read an I²C device using the standard sensor routines. Each type of I²C device has its own specific set of commands that you need to send it in order to read data from it. The Ultrasonic sensor that comes with your NXT set has a special NBC API function called `ReadSensorUS(port, value)`, which you use to read its value.

```
ReadSensorUS(S4, x) ; read distance
```

For generic high-level communication with an I²C device, use the
ReadI2CBytes(*port, writebuf, nbytes, readbuf, result*) function.
You have to pass in the port, a buffer containing the data to write to
the device, a variable set to the number of bytes you want to read from
the device, an empty byte array for storing the output data, and a byte
variable for storing the result. The sample program below demonstrates
how to use the ReadI2CBytes function.

```
thread main
dseg segment
  cmdUSReadVersion byte[] {0x02, 0x00}
  returnLength byte 8
  bufin byte[]
dseg ends
  SetSensorLowspeed(IN_1) ; set lowspeed sensor
  ReadI2CBytes(IN_1,
    cmdUSReadVersion,
    returnLength,
    bufin)
  TextOut(0, LCD_LINE1, bufin)
  wait 10000
endt
```

There are three primary low-level API routines for communicating
with I²C devices. The ReadI2CBytes function simply wraps these three
separate calls into a single routine for the sake of simplicity. Each time
you want to read from an I²C device, you must start a transaction using
LowspeedWrite(*port, bytes, buffer, result*), passing in the port,
the number of bytes you want to read, a buffer containing the bytes of
data you are sending to the device, and a byte variable for storing the
result. Then you monitor the state of the transaction, waiting for it to
complete successfully, using LowspeedStatus(*port, bytes, result*),
passing in the port, a variable that will be set to the number of bytes
available to be read, and a variable that will store the status result.
Finally, to read the response, use the LowspeedRead(*port, buflen,
buffer, result*) function, passing in the port, the read buffer length,
the read buffer itself, and a variable for storing the API function results.

Miscellaneous Commands
The NBC API shares with NXC an extremely rich set of functions that provide
you with the power to manipulate and control the NXT in nearly every
way imaginable. See Appendix B for additional information about NBC API
functions. Some of the most frequently used functions are mentioned here.

Timing Functions
The GetFirstTick(*value*) function sets its argument to an unsigned
32-bit value which is the system timing value (called a "tick") in
milliseconds at the time that the program began running.

```
GetFirstTick(x) ; read the start tick value
```

The `GetSleepTimeout(value)` function sets its argument to the number of minutes that the NXT will remain on before it automatically shuts down.

```
GetSleepTimeout(x) ; get the sleep timeout value
```

The `SetSleepTimeout(minutes)` function sets the NXT sleep timeout value to the specified number of minutes. To disable the automatic shutdown, set the sleep time to 0.

```
SetSleepTimeout(8)
```

Math Functions

The `Random(result, value)` function sets its result argument to an unsigned 16-bit random number between 0 and the specified maximum value (exclusive). The value can be a constant or a variable.

```
Random(x, 10) ; return a value of 0..9
```

The `SignedRandom(result)` function, when called without an argument, sets its argument to a signed 16-bit random number.

```
SignedRandom(x) ; return a signed random number
```

The `Sqrt(x, result)` function sets its result argument to the square root of the specified value.

```
Sqrt(100, x) ; x = 10
```

The `Sin(degrees, result)` function sets its result argument to the sine of the specified degrees value. The result is 100 times the sine value (-100..100).

```
Sin(theta, x)
```

The `Cos(degrees, result)` function sets its result argument to the cosine of the specified degrees value. The result is 100 times the cosine value (-100..100).

```
Cos(alpha, x)
```

The `Asin(value, deg)` function sets its result argument to the inverse sine of the specified value (-100..100). The result is degrees (-90..90).

```
Asin(80, deg)
```

The `Acos(value, deg)` function sets its result argument to the inverse cosine of the specified value (-100..100). The result is degrees (0..180).

```
Acos(0, deg)
```

 TRY ME: *In the previous chapter there was a Try Me to output ten random numbers between 50 and 100 to the LCD screen using NXC. Try this problem again, only this time using NBC code instead.*

Sound Functions

The `PlayTone(`*`frequency, duration`*`)` function plays a single tone of the specified frequency and duration. The frequency is in hertz. The duration is in 1000ths of a second. All parameters may be any valid expression.

```
PlayTone(440, 500) ; Play 'A' for 1/2 second
```

The `PlayFile(`*`filename`*`)` function plays the specified sound file (.rso) or a melody file (.rmd). The filename may be any valid string expression.

```
PlayFile('startup.rso')
```

Drawing Functions

Use the `LCD_LINE1` through `LCD_LINE8` constants for drawing numbers and text on the screen. Values other than these will be automatically adjusted to one of the eight fixed LCD lines. Other drawing routines do not have this restriction.

The `NumOut(`*`x, line, value`*`)` function draws a numeric value on the screen at the specified x and line location.

```
NumOut(0, LCD_LINE1, x)
```

The `TextOut(`*`x, line, msg`*`)` function draws a text value on the screen at the specified x and line location.

```
TextOut(0, LCD_LINE3, 'Hello World!')
```

The `GraphicOut(`*`x, y, filename`*`)` function draws the specified NXT picture file on the screen at the specified x and y location. If the file is not found then nothing is drawn and no errors are reported.

```
GraphicOut(40, 40, 'image.ric')
```

The `GraphicOutEx(`*`x, y, filename, vars, clear`*`)` function draws the specified NXT picture file on the screen at the specified x and y location. It uses the values contained in the vars array to transform the parameterized drawing commands contained within the file.

```
GraphicOutEx(4, 4, 'image.ric', variables, 0)
```

The `CircleOut(`*`x, y, radius`*`)` function draws a circle on the screen with its center at the specified x and y location, using the specified radius.

```
CircleOut(40, 40, 10)
```

The `LineOut(`*`x1, y1, x2, y2`*`)` function draws a line on the screen from x1, y1 to x2, y2.

```
LineOut(40, 40, 10, 10)
```

The `PointOut(`*`x, y`*`)` function draws a point on the screen at x, y.

```
PointOut(40, 40)
```

The `RectOut(x, y, width, height)` function draws a rectangle on the screen at x, y with the specified width and height.

```
RectOut(40, 40, 30, 10)
```

The `ClearScreen()` function clears the NXT LCD to a blank screen.

```
ClearScreen()
```

Coming attractions

In the next few chapters we will begin our in-depth look at how to powerfully program the NXT inputs and outputs from the very basic to the most advanced. To get started, we'll build a basic platform that we can reuse and extend as we explore the NXT further.

Building a Basic Robot

Topics in this Chapter

- Building a Basic Mobile Robot

Chapter 9

In this chapter we will build a simple robot named Versa. Although Versa appears simple at first, this robot was designed with expandability in mind. You can extend this robot with different types of modular attachments for all the standard NXT sensors, or even create your own modular attachments. Our basic robot will provide a base for different programs in NXC and NBC that will explore the inputs and outputs provided by the NXT. Let's walk through the steps to build our basic robot.

Building a Basic Mobile Robot

The Versa base platform has two motors with wheels attached directly to the motor. The first module that we will add to the base platform is a simple third wheel in the form of a castor.

The Base Platform

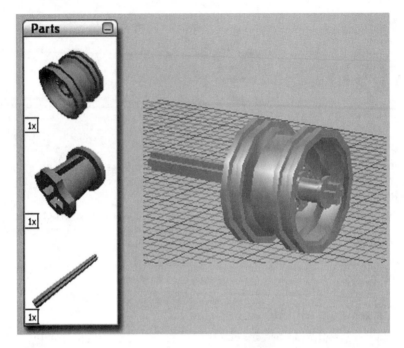

STEP 1. Add parts as shown. The axle is 7 units long.

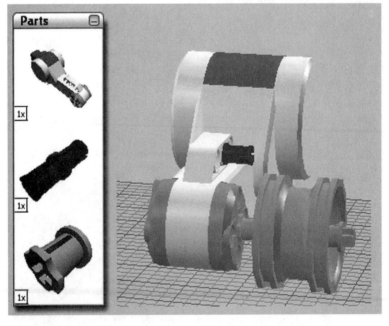

STEP 2. Add parts as shown.

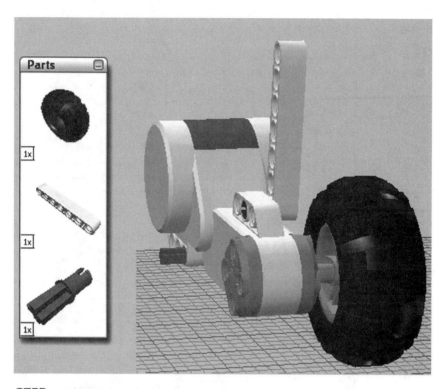

STEP 3. Add parts as shown.

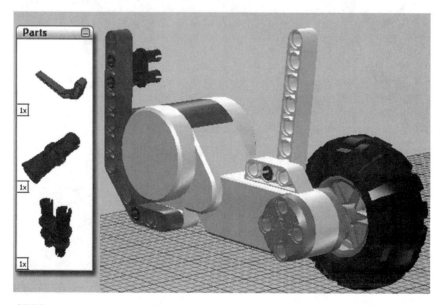

STEP 4. Add parts as shown.

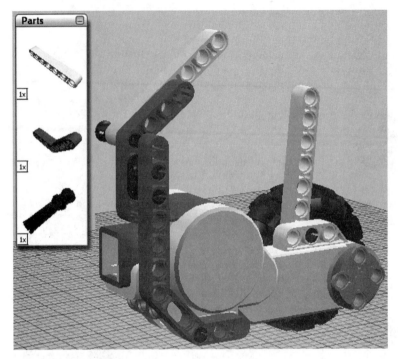

STEP 5. Add parts as shown.

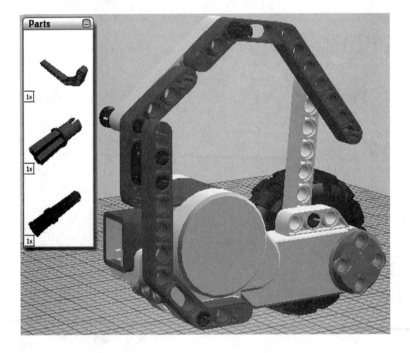

STEP 6. Add parts as shown.

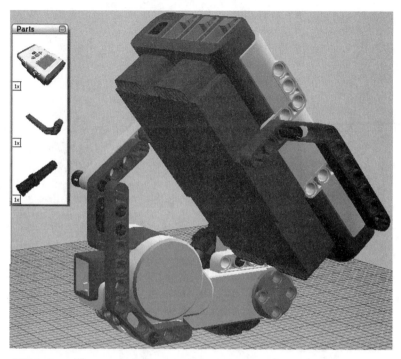

STEP 7. Add parts as shown.

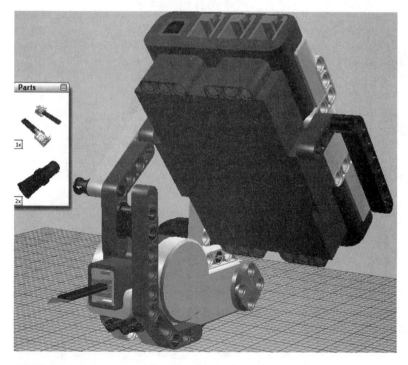

STEP 8. Add pins to the back of the motor. Plug in a medium cable.

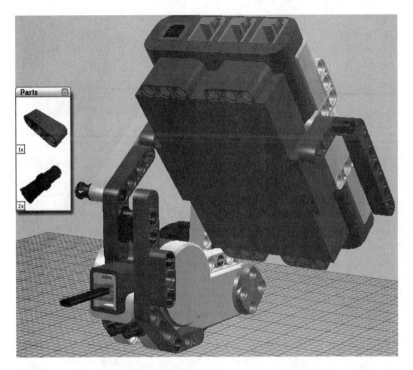

STEP 9. Add parts as shown.

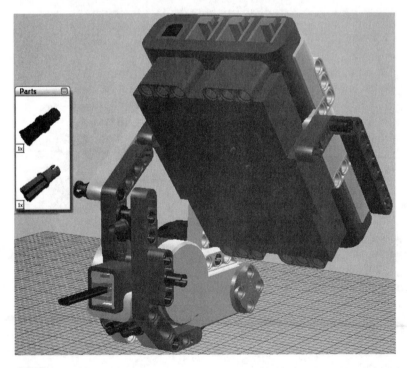

STEP 10. Add parts as shown.

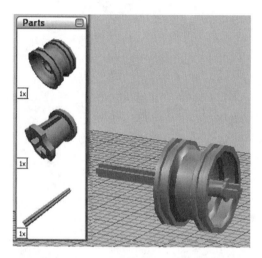

STEP 11. Add parts as shown. The axle is 7 units long.

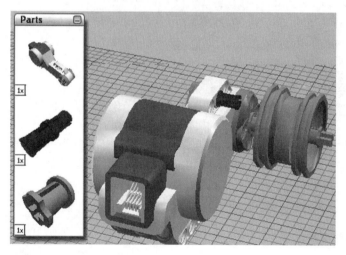

STEP 12. Add parts as shown.

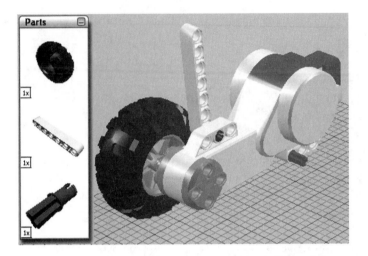

STEP 13. Add parts as shown.

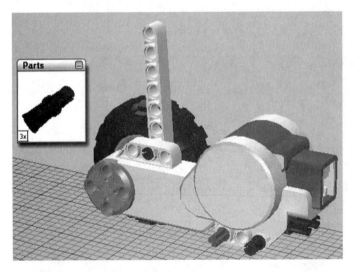

STEP 14. Add parts as shown.

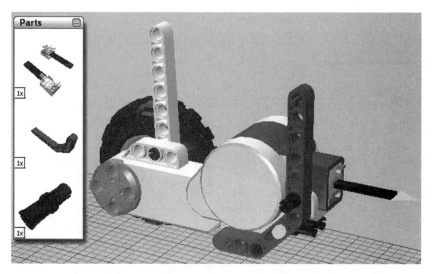

STEP 15. Add parts as shown.

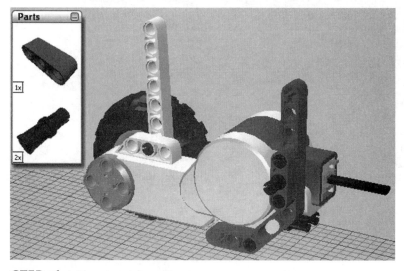

STEP 16. Add parts as shown.

STEP 17. Add parts as shown.

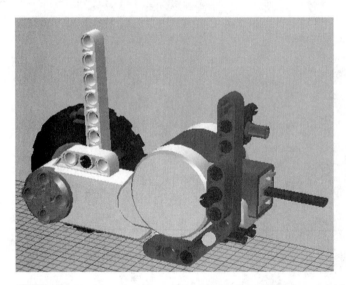

STEP 18. Combine the assembly from step 17 with the assembly from step 16 as shown.

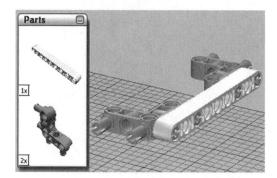

STEP 19. Add parts as shown.

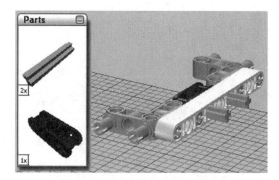

STEP 20. Add parts as shown. The axles are 3 units long.

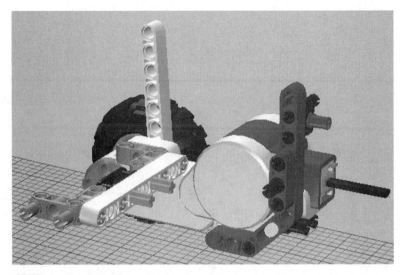

STEP 21. Combine the assembly from step 20 with the assembly from step 18 as shown.

STEP 22. Combine the assembly from step 21 with the assembly from step 10 as shown.

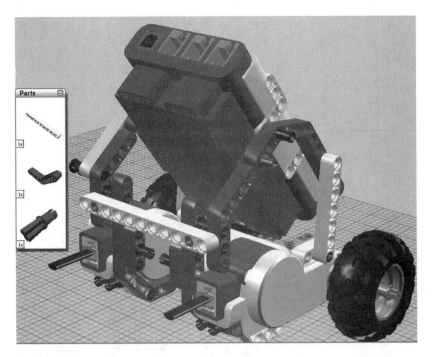

STEP 23. Add parts as shown.

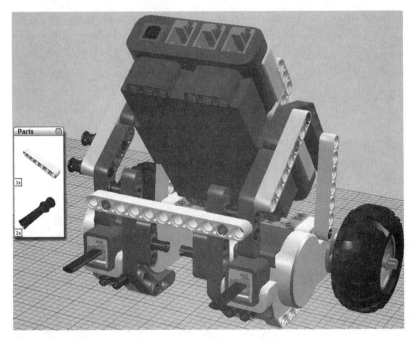

STEP 24. Add parts as shown. The location of the other black pin is more visible in step 25.

STEP 25. Add parts as shown.

STEP 26. Add parts as shown. Attach the right motor cable to port A.

STEP 27. Attach the left motor wire to port C as shown.

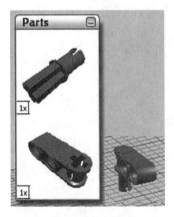

STEP 28. Add parts as shown.

STEP 29. Add the assembly from step 28 to the assembly from step 27 as shown.

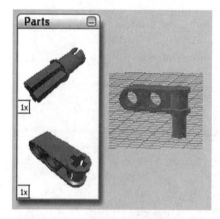

STEP 30. Add parts as shown.

STEP 31. Add the assembly from step 30 to the assembly from step 29 and you are finished with the base platform.

The Castor Wheel Module

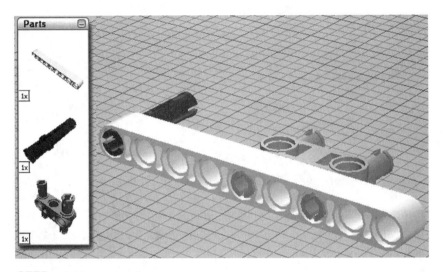

STEP 1. Add parts as shown.

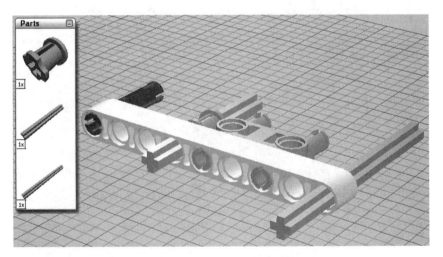

STEP 2. Add parts as shown. The axles are 5 units long and 7 units long.

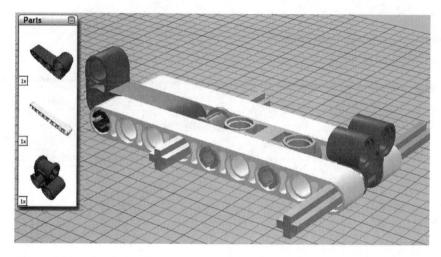

Step 3. Add parts as shown.

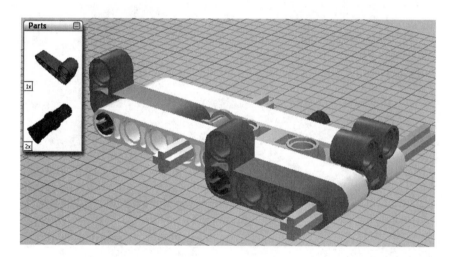

STEP 4. Add parts as shown.

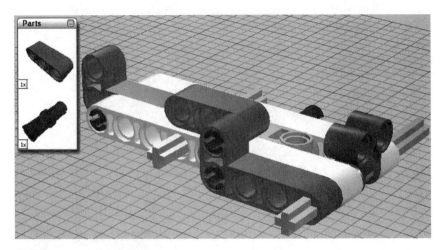

STEP 5. Add parts as shown.

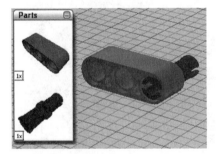

STEP 6. Add parts as shown.

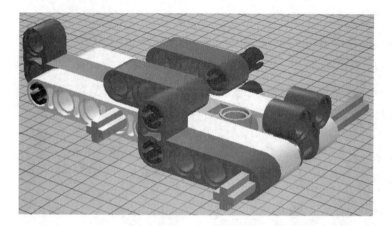

STEP 7. Add the part from step 6 as shown. It rests freely for now.

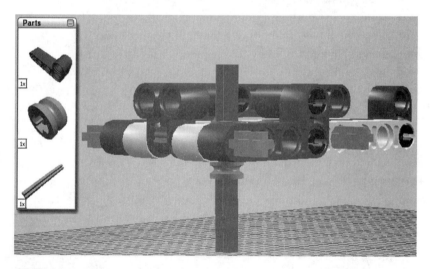

STEP 8. Now the part is connected. The axle is 5 units long.

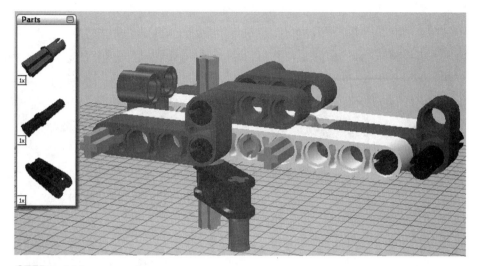

STEP 9. Add parts as shown.

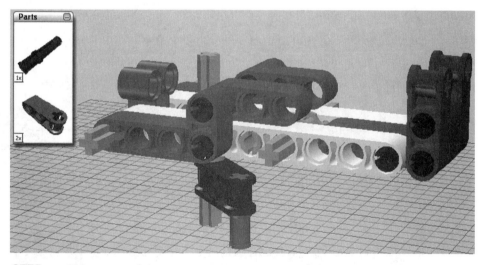

STEP 10. Add parts as shown.

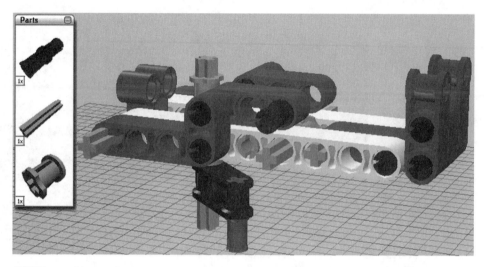

STEP 11. Add parts as shown.

STEP 12. Add parts as shown.

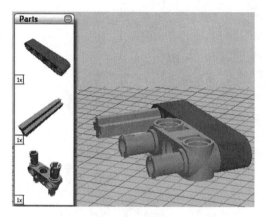

STEP 13. Add parts as shown. The axle is 3 units long.

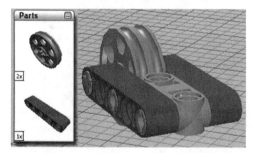

STEP 14. Add parts as shown.

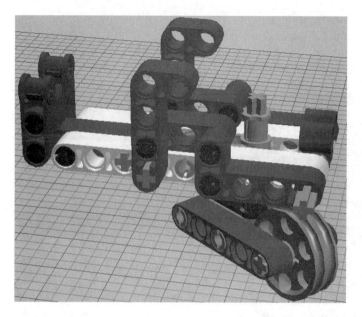

STEP 15. Add the assembly from step 14 to the assembly from step 12 as shown above. Now we are done making the castor wheel module. The next two steps demonstrate how to attach this module to the base platform.

STEP 16. Line up the base platform with the castor wheel.

STEP 17. Attach the two modules together as shown. Use the two 3 unit long axles attached to the black double cross block to lock the castor module into the base platform.

You have now completed the basic robot platform. We'll see how to attach different NXT sensors to the platform in the next few chapters as we begin to write some basic NXC code.

Basic NXT Outputs

Topics in this Chapter

- Motors
- LCD Screen
- Sounds

Chapter 10

One of the keys to power programming is understanding exactly what you can do with a device's inputs and outputs. A program is ultimately all about IO. You input data into a program to process and it returns an output. If we want to get the most out of our NXT, it is crucial to know about the different ways we can give the NXT some type of input to process and the different ways that the NXT can give us back some type of output. In this chapter we will look at the basic NXT outputs.

Motors

Perhaps the most basic outputs on the NXT are its motors, especially for mobile robots. Even a stationary robot needs motors to give the NXT the muscle it needs to physically manipulate objects. Think of a plotter or a robotic arm. Motors are important.

There are four primary types of motor control in the NXT that you can use in your programs. You can turn a motor on and off without any form of feedback or limits on the motor's operation. This is the unregulated mode. You can set a motor's speed and tell the firmware to make sure that it runs at that speed. This is called speed regulation mode. You can pair motors together and operate them in a way that keeps them rotating at the same speed. This is called synchronized motor control. And you can give a motor or set of motors a goal or a destination to rotate to. This is the tachometer-limited mode.

Let's examine how to use all four of these modes using NXC and NBC. We'll use the basic mobile robot we built in the previous chapter along with a new motor attachment. To build the motor attachment just follow the simple steps outlined below.

The Motor Attachment

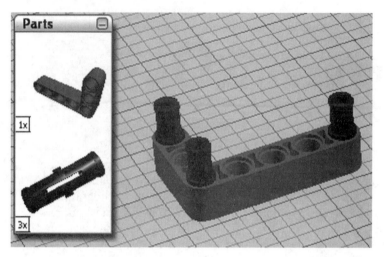

STEP 1. Add parts as shown.

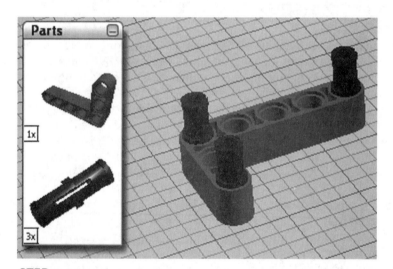

STEP 2. Add parts as shown.

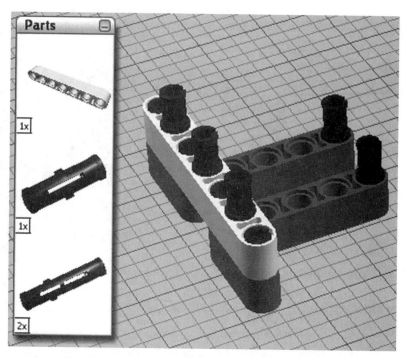

STEP 3. Combine the assemblies from steps 1 and 2, then connect together with the parts as shown.

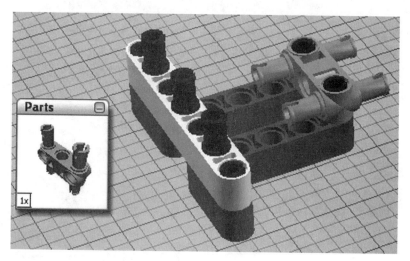

STEP 4. Add part as shown.

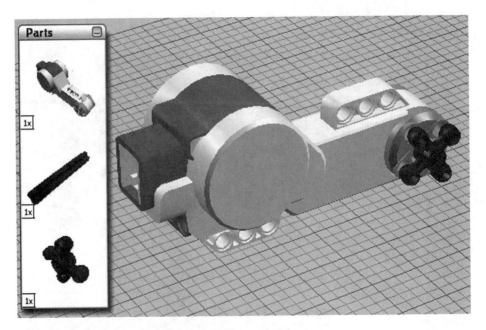

STEP 5. Add parts as shown. The axle is 4 units long.

STEP 6. Add parts as shown.

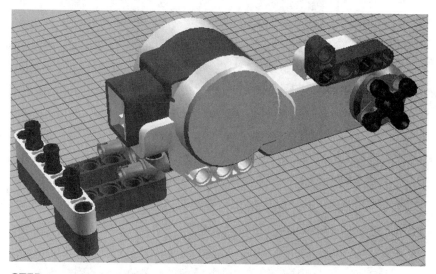

STEP 7. Combine the assembly from step 6 with the assembly from step 4 as shown.

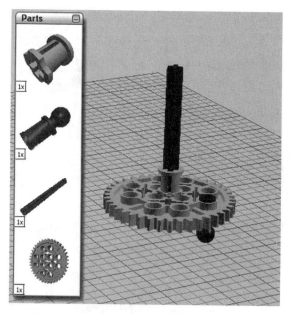

STEP 8. Add parts as shown. The axle is 6 units long.

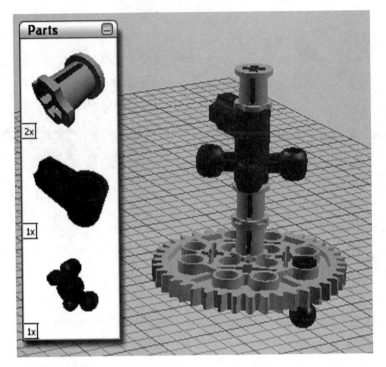

STEP 9. Add parts as shown.

STEP 10. Add parts as shown. The axle is 2 units long.

STEP 11. Add parts as shown.

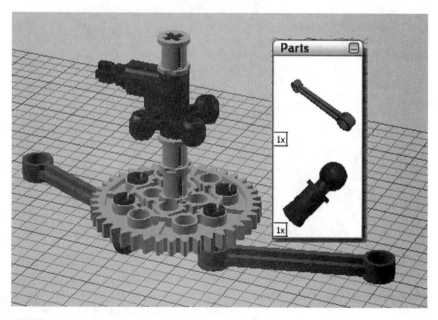

STEP 12. Add parts as shown.

STEP 13. Combine the assembly from step 12 with the assembly from step 7 as shown.

STEP 14. Insert the unit as shown, and then lock it to the available holes in the underside of the robot. Connect the motor to port B using a medium cable.

Unregulated Motors

We previously tried unregulated motor control in NBC and NXC using simple OnFwd and OnRev commands. Within the code, these API functions set the motor mode on with brake enabled so that the motors get a bit more power. They also set the run state to running with a regulation mode of idle, meaning no regulation.

By default, these commands also reset the block and tachometer counters as part of their operation. You can use the OnFwdEx and OnRevEx API functions if you do not want to reset these counters. Just pass in the additional reset constant to indicate which counters, if any, you want to reset. To make sure that none of the counters are reset, use RESET_NONE (0x00).

The NXC program below simply turns on all three motors at 75% power. After five seconds motors A and C are reversed for five more seconds and then all motors are turned off with braking. Five seconds later the program terminates. When you download and run this on our basic mobile robot with the motor attachment on port B the robot should drive forward in a nearly straight line with the "lawn mower" blades spinning. After five seconds the robot should drive backwards toward you, still cutting the grass, and end up close to its original position.

```
1. task main() {
2.    OnFwd(OUT_ABC, 75);
3.    Wait(5000);
4.    OnRev(OUT_AC, 75);
5.    Wait(5000);
6.    Off(OUT_ABC);
7.    Wait(5000);
8. }
```

By way of comparison, here is the same program written in NBC. As you can see, the API functions in NBC make this simple program almost identical with the NXC version.

```
1. thread main
2.    OnFwd(OUT_ABC, 75)
3.    Wait(5000)
4.    OnRev(OUT_AC, 75)
5.    Wait(5000)
6.    Off(OUT_ABC)
7.    Wait(5000)
8. endt
```

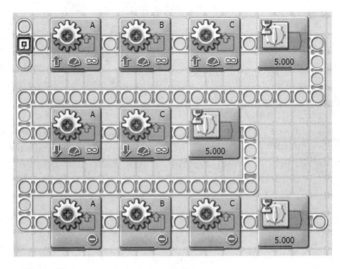

Figure 10-1. The NXT-G unregulated sample program

You can also easily duplicate this simple program in NXT-G. To contrast the simplicity of text-based programming with the complexity of a graphical programming language, we'll recreate the same program in NXT-G (Figure 10-1). There are many other ways I could have written this in NXT-G but I chose these blocks to avoid using regulation.

Table 10-1 summarizes the resulting source code size and compiled executable size of these three programs. As you can see, NXT-G loses in this comparison. Not only is the code size on your PC huge, but you can't tell the motor power levels just by looking at a screenshot of the program. It's also difficult to share NXT-G code in printed form, since you can't copy and paste a screenshot into the LEGO MINDSTORMS NXT software.

Even worse than the source code size is the resulting executable size. Even with a simple NXT-G program like this, the size of the executable has already ballooned to an inexplicably large 2.5 kilobytes. That is almost six times larger than the equivalent NXC or NBC program. The more code in the executable, the slower it executes.

Programming Language	Source Size	Executable Size
NXT-G	1052 kilobytes	2646 bytes
NBC	120 bytes	398 bytes
NXC	125 bytes	454 bytes

Table 10-1. Program sizes

An important thing to remember about the NXT and its motors is that the type of motor control we have just described is the only type of motor control you had available with the RCX brick. Every program ever written for the RCX used nothing more than unregulated motor control. Many amazing programs were written and are still being written in NQC for the RCX, and you can do the same with the NXT using NBC or NXC.

Speed Regulation Motor Control

Speed regulation tells the firmware to actively adjust the power level applied to a motor in order to maintain the desired speed. In order to see this in action, you need to use lower power levels, since that gives the firmware more room to make adjustments. The NXC program below displays the MotorActualSpeed value for outputs A and C on the LCD. As the robot drives forward it uses unregulated motor control with a power level of 40. When it reverses after five seconds it uses the OnRevReg API function and the OUT_REGMODE_SPEED regulation mode constant to drive backward at the same power level.

While going forward, the LCD screen shows that the ActualSpeed of the motors remains at 40, whether the motors achieve that speed or not. Try holding the robot in your hand and force one of the wheels to slow down. In unregulated mode the values on the LCD do not change. While driving backward, however, when you try to slow a motor you will see the ActualSpeed change on the LCD screen. We'll see more examples of using the NXT LCD as an output in the next section.

```
1.  task MotorStatus() {
2.     while (true) {
3.        NumOut(0, LCD_LINE1,
4.           MotorActualSpeed(OUT_A));
5.        NumOut(0, LCD_LINE2,
6.           MotorActualSpeed(OUT_C));
7.     }
8.  }
9.
10. task main() {
11.    start MotorStatus;
12.    OnFwdReg(OUT_ABC, 40, OUT_REGMODE_IDLE);
13.    Wait(5000);
14.    OnRevReg(OUT_AC, 40, OUT_REGMODE_SPEED);
15.    Wait(5000);
16.    Off(OUT_ABC);
17.    Wait(5000);
18.    StopAllTasks();
19. }
```

For unregulated motor control, the OnFwd API function is equivalent to the OnFwdReg function when you use OUT_REGMODE_ IDLE as the regulation mode. So we could replace line 12 with OnFwd(OUT_ABC, 40) and the behavior would be exactly the same. These commands also reset the block and tachometer counters as part of their operation. You can use the OnFwdRegEx and OnRevRegEx API functions if you do not want these counters to be reset.

Another example of using speed regulation involves stopping a motor that is already running or forcing a motor to stay at its current position. This uses a power level of zero with speed regulation enabled. The following example uses this form of stopping instead of using the Off or Coast API functions. The Wait at the end of the program has been increased to ten seconds to give you more time to play with the wheels once it brakes.

```
1. task MotorStatus() {
2.    while (true) {
3.       NumOut(0, LCD_LINE1,
4.          MotorActualSpeed(OUT_A));
5.       NumOut(0, LCD_LINE2,
6.          MotorActualSpeed(OUT_C));
7.    }
8. }
9.
10. task main() {
11.    start MotorStatus;
12.    OnFwdReg(OUT_ABC, 40, OUT_REGMODE_IDLE);
13.    Wait(5000);
14.    OnRevReg(OUT_AC, 40, OUT_REGMODE_SPEED);
15.    Wait(5000);
16.    OnFwdReg(OUT_ABC, 0, OUT_REGMODE_SPEED);
17.    Wait(10000);
18.    StopAllTasks();
19. }
```

Try turning either motor A or motor C during the last ten seconds before the program ends. Notice that any deviation that you cause from the motor's stopped position is corrected. When regulation is enabled the firmware will try to correct any errors or deviations by using a feedback control loop.

The firmware uses a PID control algorithm to try to hold the motors at the current rotation with a speed of zero in our example. The letters P, I, and D stand for proportional, integral, and derivative. The proportional component determines the reaction to the current deviation. The integral component determines the reaction based on the sum of recent deviations. The derivative component determines the reaction using the rate of change of the deviation. These three elements are combined via a weighted sum to determine the corrective response to any error.

WEBSITE: *See the wikipedia article on PID controllers:*
http://en.wikipedia.org/wiki/PID_controller

The NXT firmware sets the default PID weightings at
$P=96$, $I=32$, and $D=32$. You can read the current PID values
using MotorRegPValue(port), MotorRegIValue(port), and
MotorRegDValue(port). Set new values if you want to using
SetOutput(ports, field, value, …).

```
1. SetOutput(OUT_A, RegPValue, 30,
2.    RegIValue, 0, RegDValue, 0,
3.    UpdateFlags, UF_UPDATE_PID_VALUES);
```

Synchronized Motor Control

Synchronized motor control involves a pair of motors. You can't use it
with a single motor or with all three motors. Notice that neither of our
previous programs resulted in the robot driving perfectly straight. With
the RCX you had to carefully select your two driving motors so that
they would turn at as close to the same rate as possible. With the NXT
you can just use the synchronous control mode to keep a pair of motors
driving at the same rate.

```
1. task main() {
2.    OnFwdSync(OUT_AC, 75, 0);
3.    OnFwd(OUT_B, 75);
4.    Wait(5000);
5.    OnRevSync(OUT_AC, 75, 0);
6.    Wait(5000);
7.    Off(OUT_ABC);
8.    Wait(5000);
9. }
```

The NXC program above uses the OnFwdSync and OnRevSync API
functions. These same functions are available in NBC so you could
easily rewrite this with trivial modifications in NBC assembly if you
prefer. This code causes our basic mobile robot to drive forward for five
seconds and then reverse just like in our other simple programs. It will
run the lawn mower attachment without any synchronization.

The last argument to the OnFwdSync and OnRevSync functions
is a turn percentage from -100 to 100. This value controls the relative
speeds of the two synchronized motors. When you run this program you
should see a much straighter course going forward and returning. As
the NXT moves you can observe tiny course corrections along the way.
These commands also reset the block and tachometer counters as part of
their operation. You can use the OnFwdSyncEx and OnRevSyncEx API
functions if you do not want these counters to be reset.

Tachometer-Limited Motor Control

The built-in rotation sensors provide the NXT with feedback regarding how fast and how far the motors have turned. The NXT firmware gives you the ability to use this information when you control the motors. You can tell the firmware to stop powering the motors after a specified number of degrees of rotation have been reached. This goal is called the tachometer limit.

```
1. #define WHEEL_DIAM 56
2. #define METERS     1
3. #define METER2MM   1000
4. #define DEG2ROT    360
5.
6. const int degrees =
7.   (METERS*METER2MM*DEG2ROT)/(PI*WHEEL_DIAM);
8.
9. task main() {
10.   OnFwd(OUT_B, 75);
11.   RotateMotor(OUT_AC, 75, degrees);
12.   RotateMotor(OUT_AC, 75, -degrees);
13.   Off(OUT_B);
14. }
```

This NXC program uses the diameter of the wheels (56 millimeters) and the desired distance (1 meter) to calculate the number of degrees the motors should rotate. Since all these values are constants, the compile-time expression evaluator can calculate the correct value using floating-point precision before converting the final answer into an integer. If you examine the code generated by the compiler, you will see that the degrees constant is simply set to 2046.

There are several other versions of the RotateMotor function, which expose various options. The most flexible version is the RotateMotorExPID API function. You can specify the ports, the power, the desired tachometer limit, the turn percentage, a boolean flag indicating whether or not to synchronize the motors, a boolean flag saying whether or not to brake the motors when the limit is reached, and the proportional, integral, and derivative weights. You can also use RotateMotorEx and RotateMotorPID.

For even lower-level control of the NXT motors in NBC, use the setout and getout statements. NXC provides equivalent support via the GetOutput and SetOutput API functions. See the NXC quick reference guide in Appendix A for more details.

LCD Screen

The LCD screen is another very important basic NXT output. You can draw lines, pixels, rectangles, circles, and text to the LCD. Let's try a few simple projects that exercise each of these options.

Drawing Images

As we have seen, an RICScript picture is a very powerful tool for drawing on the NXT screen. It is not simply a bitmap that you copy to the LCD at a certain location. Using parameters in your NXT pictures makes them even more powerful.

Here's a simple example of drawing an entire tic-tac-toe board using a parameterized NXT picture. First we see the contents of the picture. When the image is output to the screen, it draws the board lines and a 13 by 13 portion of the sprite into each of the nine positions on the board.

```
1.  // TTT.ric
2.  // the sprite defines an O, a space, and an X
3.  // each sub-image is 13x13
4.  sprite(1, 0x0F8000180C, 0x1040002412,
5.             0x2720003226, 0x489000194C,
6.             0x9048000C98, 0xA028000630,
7.             0xA028000220, 0xA028000410,
8.             0x90480009C8, 0x4890001364,
9.             0x2720002632, 0x1040002C1A,
10.            0x0F8000180C);
11. // Always draw the tic-tac-toe board
12. line(0, 15, 46, 15, 0);
13. line(0, 31, 0, 31, 46);
14. line(0, 0, 15, 46, 15);
15. line(0, 0, 31, 46, 31);
16. // draw the 9 positions with either an O,
17. // a space, or an X.
18. copybits(0, 1, arg(0), 0, 13, 13, 1, 1);
19. copybits(0, 1, arg(1), 0, 13, 13, 17, 1);
20. copybits(0, 1, arg(2), 0, 13, 13, 33, 1);
21. copybits(0, 1, arg(3), 0, 13, 13, 1, 17);
22. copybits(0, 1, arg(4), 0, 13, 13, 17, 17);
23. copybits(0, 1, arg(5), 0, 13, 13, 33, 17);
24. copybits(0, 1, arg(6), 0, 13, 13, 1, 33);
25. copybits(0, 1, arg(7), 0, 13, 13, 17, 33);
26. copybits(0, 1, arg(8), 0, 13, 13, 33, 33);
```

Save the file as ttt.rs, compile the NXT picture using the RICScript compiler and download it to your brick. Here's a simple NXC program that uses this picture. Notice the use of the asm statement in the BoardToField function. This shows how easy it is to drop down into NBC assembly language when needed. In this case, the NXC parser does not allow some of the polymorphic array operations that are supported by the NBC statements. If you add a scalar to an array it automatically iterates through all the elements in the array and adds the value to each element. The same rule applies to the NBC multiply statement and all of the other native math statements.

```
1.  #define TTT_O       0
2.  #define TTT_EMPTY 13
3.  #define TTT_X       26
4.
5.  int field[];
6.  int Board[9];
7.  int players[] = {1, -1};
8.
9.  void BoardToField() {
10.    asm {
11.       add field, Board, 1
12.       mul field, field, 13
13.    }
14. }
15.
16. void DrawBoard() {
17.    BoardToField();
18.    GraphicOutEx(0, 0, "ttt.ric", field, true);
19.    Wait(1000);
20. }
21.
22. int MoveCount() {
23.    int empty = 0;
24.    for (int i=0; i < 9; i++)
25.       if (Board[i] == 0)
26.          empty++;
27.    return 9-empty;
28. }
29.
30. void Move(const int p) {
31.    Board[p] = players[MoveCount() % 2];
32.    DrawBoard();
33. }
34.
35. task main()
36. {
37.    DrawBoard();
38.    Move(4); // first move
39.    Move(0); // second move
40.    Move(2); // third move
41.    Move(6); // fourth move
42.    Move(3); // fifth move
43.    Move(5); // sixth move
44.    while(true);
45. }
```

Another thing you can easily do using NXT pictures is simple animation or image scrolling on the LCD. Here's a simple RICScript for a smiley face NXT picture.

```
1. // smile.ric
2. sprite(1, 0x07C0, 0x1830, 0x2008, 0x4004, 0x4C64,
3.           0x8C62, 0x8002, 0x8002, 0x8002, 0x8002,
4.           0x4824, 0x47C4, 0x2008, 0x1830, 0x07C0);
5. copybits(0, 1, 0, 0, 15, 15, 0, 0);
```

Use this in a very trivial NBC program to bounce a smiley face around the LCD screen. The slightly tricky part about this code is the way it handles bouncing off the screen edges. The image is 15x15 so we subtract 15 from 64 when checking the maximum y value and we subtract 15 from 100 when checking the maximum x value. The wait statement is required since the loop executes so quickly. Without it, the smiley face is never visible long enough to see it clearly.

```
1.  thread main
2.    dseg segment
3.      x sword
4.      y sword
5.      dy sbyte
6.      dx sbyte
7.    dseg ends
8.    set x, 0
9.    set y, 0
10.   set dx, 2
11.   set dy, 2
12. Loop:
13.   ClearScreen()
14.   GraphicOut(x, y, 'smile.ric')
15.   wait 100
16.   add x, x, dx
17.   add y, y, dy
18.   brtst LT, FixXMin, x
19.   brcmp GT, FixXMax, x, 85
20.   brtst LT, FixYMin, y
21.   brcmp GT, FixYMax, y, 49
22.   jmp Loop
23. FixXMin:
24.   set x, 0
25.   Random(dx, 3)
26.   add dx, dx, 1
27.   jmp Loop
28. FixXMax:
29.   set x, 85
30.   Random(dx, 3)
31.   add dx, dx, 1
32.   mul dx, dx, -1
33.   jmp Loop
```

```
34. FixYMin:
35.    set y, 0
36.    Random(dy, 3)
37.    add dy, dy, 1
38.    jmp Loop
39. FixYMax:
40.    set y, 49
41.    Random(dy, 3)
42.    add dy, dy, 1
43.    mul dy, dy, -1
44.    jmp Loop
45. endt
```

There are lower-level functions available in NBC and NXC for drawing NXT pictures. Ultimately these functions end up using the syscall NBC statement with the DrawGraphic syscall function ID. In NBC we can use the syscall statement directly along with the TDrawGraphic syscall argument structure type. NXC exposes this same low-level access via the SysCall or SysDrawGraphic API functions. Use the DrawGraphicType structure along with these routines.

```
struct DrawGraphicType {
    char Result;
    LocationType Location;
    string Filename;
    int Variables[];
    unsigned long Options;
};
```

Simply define a variable of the DrawGraphicType type, set the fields appropriately, and call the SysDrawGraphic function. The bouncing smiley is recreated here in NXC using the low-level SysDrawGraphic function. One benefit of using SysDrawGraphic is that within the loop you can set only those fields that have to change rather than all of the fields, which is what happens in the higher-level GraphicOut API function.

```
1. task main() {
2.    int dx = 2;
3.    int dy = 2;
4.    DrawGraphicType dgt;
5.    dgt.Filename = "smile.ric";
6.    dgt.Options = 0x0001;
7.    dgt.Location.X = 0;
8.    dgt.Location.Y = 0;
9.    while (true) {
10.       SysDrawGraphic(dgt);
11.       Wait(100);
12.       dgt.Location.X += dx;
13.       dgt.Location.Y += dy;
14.       if (dgt.Location.X < 0) {
```

```
15.        dgt.Location.X = 0;
16.        dx = Random(3)+1;
17.     }
18.     else if (dgt.Location.X > 85) {
19.        dgt.Location.X = 85;
20.        dx = (Random(3)+1)*(-1);
21.     }
22.     if (dgt.Location.Y < 0) {
23.        dgt.Location.Y = 0;
24.        dy = Random(3)+1;
25.     }
26.     else if (dgt.Location.Y > 49) {
27.        dgt.Location.Y = 49;
28.        dy = (Random(3)+1)*(-1);
29.     }
30.   }
31. }
```

Drawing Shapes

It is easy to draw shapes on the NXT. Just call the API function and it works. You can use these drawing operations to spruce up your NXT programs with a nice user interface or to graph sensor data. The sample NXC program below draws a few shapes with random sizes.

```
1. task main() {
2.    for (int i=0;i<10;i++) {
3.       CircleOut(Random(100), Random(64),
4.                 Random(20));
5.    }
6.    for (int i=0;i<10;i++) {
7.       int x = Random(100);
8.       int y = Random(64);
9.       RectOut(x, y, Random(100-x), Random(64-y));
10.   }
11.   for (int i=0;i<10;i++) {
12.      LineOut(Random(100), Random(64),
13.              Random(100), Random(64));
14.   }
15.   Wait(10000);
16. }
```

Carl Sturmer's dot tail NXC program is an effective demonstration of the PointOut API function.

```
1. #define SCREEN_WIDTH   100
2. #define SCREEN_HEIGHT  64
3. #define TAIL_LENGTH    20
4. int xDir=1;
5. int yDir=1;
6. int X[]={1, 1, 1, 1, 1, 1, 1, 1, 1, 1,
7.          1, 1, 1, 1, 1, 1, 1, 1, 1, 1};
```

```
8.  int Y[]={1, 1, 1, 1, 1, 1, 1, 1, 1, 1,
9.              1, 1, 1, 1, 1, 1, 1, 1, 1, 1};
10.
11. // move the elements in the tail
12. void moveTail() {
13.   for( int i = TAIL_LENGTH-1; i > 0; i-- ) {
14.     X[i] = X[i-1];
15.     Y[i] = Y[i-1];
16.   }
17. }
18.
19. // move the X[0] and Y[0] to a new position,
20. // keeping them on the screen
21. void move() {
22.   X[0] += xDir;
23.   Y[0] += yDir;
24.   if( ( X[0] < 2 ) ||
25.       ( X[0] >= SCREEN_WIDTH-1 ) )
26.     xDir *= -1;
27.   if( ( Y[0] < 2 ) ||
28.       ( Y[0] >= SCREEN_HEIGHT-1 ) )
29.     yDir *= -1;
30. }
31.
32. task main() {
33.   while( true ) {
34.     // move each element in the array
35.     // out one slot so that the value
36.     // in slot 0 goes to slot 1, etc.
37.     moveTail();
38.     // move the 0 slot in the array to
39.     // its new position
40.     move();
41.     ClearScreen();
42.     // draw the array in reverse order
43.     for( int i = TAIL_LENGTH-1; i > 0; i-- )
44.       PointOut( X[i], Y[i] );
45.   }
46. }
```

The enhanced NBC/NXC firmware provides an additional ability when it comes to drawing shapes on the LCD. Not only can you set pixels by calling the shape drawing API functions but you can also clear pixels using these same functions. Each drawing function takes the optional "clear screen" argument as a final parameter. When the value is 0x01 or true, it tells the firmware to clear the screen before drawing the shape. With the enhanced firmware you can use a value of 0x04 (DRAW_OPT_CLEAR_PIXELS) to cause the function to clear the pixels rather than set them.

Use the SysDrawPoint, SysDrawLine, SysDrawRect, and SysDrawCircle API functions for lower-level access to the shape drawing features of the NXT firmware. The arguments to these drawing functions are passed via the DrawPointType, DrawLineType, DrawRectType, and DrawCircleType structures.

```
struct DrawPointType {
  char Result;
  LocationType Location;
  unsigned long Options;
};

struct DrawLineType {
  char Result;
  LocationType StartLoc;
  LocationType EndLoc;
  unsigned long Options;
};

struct DrawCircleType {
  char Result;
  LocationType Center;
  byte Size;
  unsigned long Options;
};

struct DrawRectType {
  char Result;
  LocationType Location;
  SizeType Size;
  unsigned long Options;
};
```

Drawing Text

Using the LCD for text and numeric output couldn't be simpler. As with NXT picture files and basic shapes, you can use a high-level API function or drop down to a lower level if you desire. The NBC sample code below uses the TDrawText structure and the syscall statement to show off the low level access. Then it uses TextOut and NumOut to demonstrate the higher level API.

```
1. thread main
2.   dseg segment
3.     dt TDrawText
4.     msg byte[] 'Hello World!'
5.     val dword
6.     sval dword
7.     hh word
8.     mm word
9.     ss word
```

```
10.     mmm word
11.   deg ends
12.   gettick sval
13.   ClearScreen()
14.   set dt.Location.X, 0
15.   set dt.Location.Y, LCD_LINE4
16.   mov dt.Text, msg
17.   syscall DrawText, dt
18.   TextOut(0, LCD_LINE6, 'Elapsed time:')
19. Loop:
20.   gettick val
21.   sub val, val, sval
22.   div hh, val, 3600000
23.   div mm, val, 60000
24.   mod mm, mm, 60
25.   div ss, val, 1000
26.   mod ss, ss, 60
27.   mod mmm, val, 1000
28.   NumOut(10, LCD_LINE8, hh)
29.   NumOut(30, LCD_LINE8, mm)
30.   NumOut(50, LCD_LINE8, ss)
31.   NumOut(70, LCD_LINE8, mmm)
32.   jmp Loop
33. endt
```

Direct Screen Drawing

In NBC and NXC you have low-level access to the LCD screen buffer
memory. There are actually two buffers for the LCD. The popup buffer
is used for drawing the low battery message but you can use it in your
own program if you are careful. The other buffer is the normal LCD
buffer. All of the drawing functions manipulate the contents of this
buffer. You can read the contents of this buffer or write to this buffer.
Any changes you make to the screen buffer contents are reflected on
the LCD screen.

```
1. #define NOTPOP ~DISPLAY_POPUP
2. task main() {
3.   byte data[];
4.   byte d;
5.   long fred;
6.
7.   // set to normal display
8.   d = DisplayFlags();
9.   d &= NOTPOP;
10.  SetDisplayFlags(d);
11.
12.  ClearScreen();
13.  // draw on normal screen memory
14.  d = 0xaa; // 10101010
15.  ArrayInit(data, d, 4);
```

```
16.   SetDisplayNormal(30, TEXTLINE_5, 4, data)
17.   Wait(1000);
18.   // draw on popup screen memory
19.   d = 0xaa; // 10101010
20.   ArrayInit(data, d, 4);
21.   SetDisplayPopup(30, TEXTLINE_6, 4, data);
22.
23.   long flipthen = CurrentTick();
24.   d = DisplayFlags();
25.   while (CurrentTick() < (flipthen+10000)) {
26.      //  SwapDisplayMemory
27.      d ^= DISPLAY_POPUP;
28.      SetDisplayFlags(d);
29.      Wait(5);
30.   }
31.   // set to normal display
32.   d = DisplayFlags();
33.   d &= NOTPOP;
34.   SetDisplayFlags(d);
35.   Wait(8000);
36. }
```

The two buffers are switched using the DisplayFlags and SetDisplayFlags API functions. You write to the two buffers using the SetDisplayNormal and SetDisplayPopup API functions. To read screen data you would use either GetDisplayNormal or GetDisplayPopup.

Writing directly to screen memory is not trivial but it does give you the ability to do interesting things that are difficult to do any other way. If you ever need to save the current screen and then bring it back again after performing some other operation then direct screen memory access is probably the best way to do it.

Sounds

The NXT sound system gives you another basic output that you can put to good use. Audible feedback is a powerful tool to keep a user informed about what is going on in a program. You might use it to let a user know that you are processing a button press or when a sensor value goes below or above thresholds.

There are three basic types of sound output that you can use with the NXT. You can play simple tones, melody files containing multiple tone definitions, and sound files, which contain PCM samples.

Playing Tones
Use the PlayTone or PlayToneEx API functions in both NBC and NXC to play simple tones. You specify the frequency and the duration of the tone and the NXT will play it for you. If you want to play multiple tones one after the other, you need to add Wait statements between each call to PlayTone so that the previous tone has a chance to play. The duration

of the Wait should be as long as the duration you specified in the PlayTone call.

```
1.  #define WHEEL_DIAM  56
2.  #define METERS      2.25
3.  #define METER2MM    1000
4.  #define DEG2ROT     360
5.
6.  bool programComplete = FALSE;
7.
8.  const int degrees =
9.    (METERS*METER2MM*DEG2ROT)/(PI*WHEEL_DIAM);
10.
11. task backingBeep() {
12.   while (!programComplete) {
13.     while (MotorPower(OUT_A) < 0) {
14.       PlayTone(440, 500);
15.       Wait(750);
16.     }
17.   }
18. }
19.
20. task main() {
21.   start backingBeep;
22.   OnFwd(OUT_B, 75);
23.   RotateMotor(OUT_AC, 75, degrees);
24.   RotateMotor(OUT_AC, 75, -degrees);
25.   Off(OUT_B);
26.   programComplete = true;
27. }
```

The code above causes our basic lawn-mowing robot to drive forward 2.25 meters. Then it backs up while emitting a repeating tone to let all the other vehicles in the area know that it is reversing. There are many interesting uses for the PlayTone API function. You could play tones where the frequency is dependant on the speed of the motors or the amount of light in a room. Use your imagination!

Melody Files

A melody file gives you a simple way to collect a series of tones and durations and play them all back in sequence, without having to execute calls to PlayTone and Wait. You can create melody files using the tools and utilities discussed in chapter 6. Playing a melody file is exactly like playing a sound file. Just call the PlayFile API function in either NBC or NXC.

```
1. task main() {
2.   PlayFileEx("fatherjohn.rmd", 3, true);
3.   while(true) {
4.     OnFwd(OUT_AC, 75); Wait(3000);
5.     OnRev(OUT_AC, 75); Wait(3000);
```

```
6.   }
7. }
```

If you need to loop the melody file or play it at a volume that is different from the user interface volume setting then use the PlayFileEx API function instead. The sample program above shows that the song plays and repeats in the background while the motor loop executes. The fatherjohn.rmd file was created using the Brick Piano utility.

Sound Files

If you know how to play a melody file then you know how to play a sound file. Internally, the NXT firmware sees what type of sound file it is based on the first two bytes of data in the file. If it is a sound file the firmware plays the PCM samples and if it is a melody file the firmware plays simple tones.

```
1. task main() {
2.    PlayFile("T56.rso");
3.    while (SoundFlags() != SOUND_FLAGS_IDLE);
4.    PlayFile("MINDSTORMS.rso");
5.    until (SoundFlags() == SOUND_FLAGS_IDLE);
6. }
```

When you play a sound or melody file you probably want to avoid calling other sound functions while it is playing. If you try to play a tone or another sound or melody file too soon it will start your new command before it has finished the previous one. With the PlayTone API function you know exactly how long the tone is going to last but with a sound or melody file you have no idea. This is where checking the SoundFlags comes in handy. As you can see above, you can either wait while the value is not SOUND_FLAGS_IDLE or you can use the until-loop construct in NXC to wait until it becomes idle again.

Sound Control

You can examine the NXT sound module using a variety of functions. These let you check the current state of the module to find out not only whether it is busy with a previous task but also what type of task it is working on. You can ask for the SoundFlags, the SoundMode, and the SoundModuleState. These values help you determine whether the NXT is ready to play another sound or not.

In addition to being able to check the current state of the sound module, you have the ability to modify its state using a number of low-level sound module API functions. The sample program below demonstrates how you can manipulate the sample rate while it plays back a sound file. Under normal circumstances it uses the value stored within the sound file itself but right at the start of the playback process you can tell the sound module to use a different sample rate.

```
1. task main() {
2.    for (int i=2000; i < = 16000; i += 2000) {
3.       PlayFile("T56.rso");
4.       SetSoundSampleRate(i);
5.       while (SoundFlags() != SOUND_FLAGS_IDLE);
6.    }
7. }
```

You can also do the same thing with tones by changing the tone frequency or duration right after a PlayTone call. All of the API functions are described in the NXC quick reference guide in Appendix A. Complete details regarding all the sound module constants and functions can be found in the *NXC Programmer's Guide,* which can be found in the NXC documentation section of the NBC website.

Basic NXT Inputs

Topics in this Chapter

- Buttons
- Basic Sensors

Chapter **11**

The NXT has four input ports to which you can connect sensors. In this chapter we will limit our focus to the three analog sensors that come with the NXT: the sound sensor, the light sensor, and the touch sensor. Another basic sensor that we will examine here is the rotation sensor built into the NXT motor. Counting the three built-in rotation sensors, the NXT retail set gives you a total of six basic sensors. All six of these sensors can be at use in a single robotics project with an extra sensor port to spare. Before we get to the analog sensors, though, we'll take a look at the four NXT buttons and see how they can be put to use.

Buttons

The first thing to know about the four buttons on the NXT is that with the standard firmware you do not have access to the dark gray escape button. The NXT virtual machine that executes your code is also checking to see if the escape button (BTN1 or BTNEXIT) is pressed. As soon as it is pressed, the virtual machine aborts your program. So your code never has a chance to do anything when BTN1 is pressed. The good news is that the other three buttons are completely at your disposal.

Reading Buttons

There are two different mechanisms you can use to check for button presses in your programs. The first involves the ReadButton system call statement or its high level wrapper functions in NBC and NXC. In NBC you can use the ReadButtonEx API function. In NXC you can use the ReadButtonEx, ButtonCount, and ButtonPressed API functions. The reset boolean input argument to each of these functions tells the firmware whether or not it should reset to zero all the button release counters for the specified button.

A constant or variable is allowed for the button index and the reset arguments. When using ReadButtonEx the output arguments must be variables since the function modifies those arguments. With NBC the output arguments are the boolean Pressed value, the byte Count value, and char Result value. In NXC the result is returned by the API function itself rather than via a function argument.

```
1.  void playSound(const byte btn, string file) {
2.    byte cnt;
3.    bool bPressed;
4.    byte x;
5.    switch(btn) {
6.      case BTNLEFT   : x =  0; break;
7.      case BTNCENTER : x = 40; break;
8.      case BTNRIGHT  : x = 80; break;
9.    }
10.   until(!ButtonPressed(btn, false));
11.   ReadButtonEx(btn, false, bPressed, cnt);
12.   NumOut(x, LCD_LINE8, cnt);
13.   PlayFile(file);
14. }
15.
16. task main() {
17.   bool bDone = false;
18.   TextOut(5, LCD_LINE4, "Press Buttons!");
19.   while (!bDone) {
20.     // loop until all three sounds have
21.     // been played at least twice
22.     bDone =
23.       (ButtonCount(BTNLEFT, false) >= 2) &&
24.       (ButtonCount(BTNRIGHT, false) >= 2) &&
25.       (ButtonCount(BTNCENTER, false) >= 2);
26.     if (ButtonPressed(BTNCENTER, false)) {
27.       playSound(BTNCENTER, "mindstorms.rso");
28.     }
29.     else if (ButtonPressed(BTNLEFT, false)) {
30.       playSound(BTNLEFT,
31.                 "have a nice day.rso");
32.     }
33.     else if (ButtonPressed(BTNRIGHT, false)) {
34.       playSound(BTNRIGHT, "snore.rso");
35.     }
36.   }
37.   until (SoundFlags() == SOUND_FLAGS_IDLE);
38. }
```

The sample program above shows how to use the ButtonPressed, ButtonCount, and ReadButtonEx NXC API functions. The counts are not reset in any of these calls since the program wants to know how many times a button was pressed and released. The statement on line 37 is necessary so that the last sound will play until it is finished even after the loop has exited. Without it the last sound doesn't play at all because the program ends before it has a chance to begin playing. Be sure to upload the three required sound files located in your LEGO NXT engine\ sounds directory.

Advanced Buttons

You can also directly access the button state value and the counts that are stored in the Button module IOMap structure for each button. These values are read using the ButtonState, ButtonPressCount, ButtonLongPressCount, ButtonReleaseCount, ButtonLongReleaseCount, and ButtonShortReleaseCount API functions. Each of these functions requires you to use a constant button index as the parameter to the function call. In NBC these functions all start with Get (as in GetButtonState) and they require a variable as the second parameter so that the requested value is returned. In NXC you just use the API function return value.

In both NBC and NXC you can set these values using the SetButtonState, SetButtonPressCount, SetButtonLongPressCount, SetButtonShortReleaseCount, SetButtonLongReleaseCount, and SetButtonReleaseCount API functions. All of these require a constant button index as the first argument. The second argument is either a constant or a variable.

Note that the enhanced NBC/NXC firmware provides one additional capability when it comes to the NXT buttons. If you install the enhanced firmware, you can tell the virtual machine to use a long button press on the escape button to abort a running program. Just call SetLongAbort(true) at the start of your program. After turning on this feature you will be able to use the escape button in your programs just like the other buttons with the sole exception of using the long press state of the escape button.

 TRY IT: *Write a program that uses the buttons to select different menu options in your program like the NXT user interface does. Use parameterized NXT pictures to simplify the drawing code.*

Basic Sensors

Let's turn our attention to the analog sensors and the built-in rotation sensors. Using sensors is a major part of nearly every robotic program. As we progress through the rest of this chapter, we'll build a few sensor attachments for our mobile robot platform and use those sensors in some simple projects.

Touch Sensor

The NXT touch sensor is like your fingers or an insect's antennae. It can provide your program with feedback from a physical event that presses or releases the button on the touch sensor. Mobile robots often have one or more bumpers that give them an awareness of obstacles they encounter along their path. The very popular Roomba robotic vacuum cleaner uses its bumper to help it vacuum a room.

We'll start our exploration of the touch sensor by building a front bumper attachment for our robot. Follow the steps outlined below to build the bumper using the NXT touch sensor.

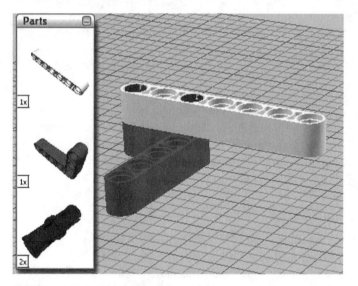

STEP 1. Add parts as shown.

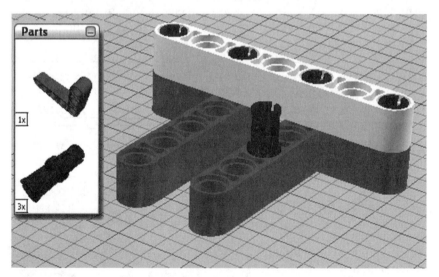

STEP 2. Add parts as shown.

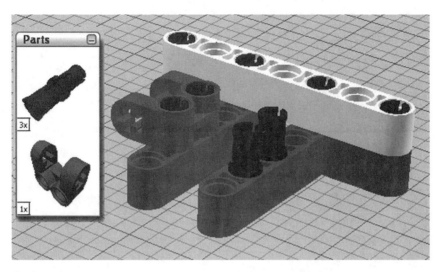

STEP 3. Add parts as shown.

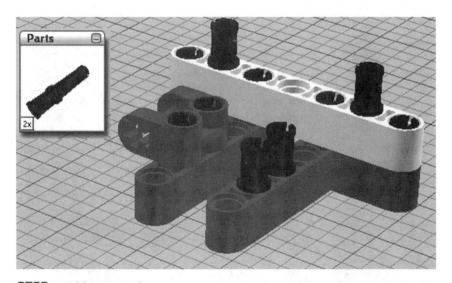

STEP 4. Add parts as shown.

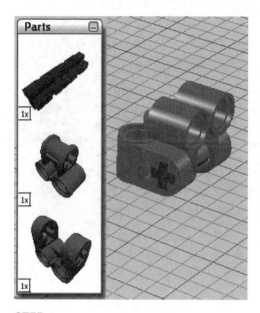

STEP 5. Add parts as shown.

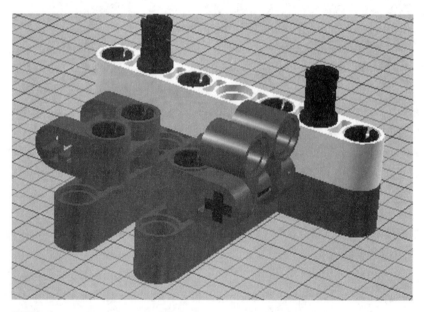

STEP 6. Combine the assembly from step 5 with the assembly from step 4 as shown.

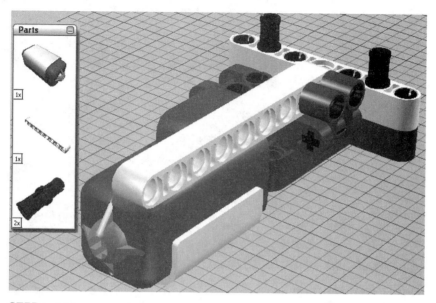

STEP 7. Add parts as shown. The touch sensor is not yet attached to the beam so hold it in place until the next step.

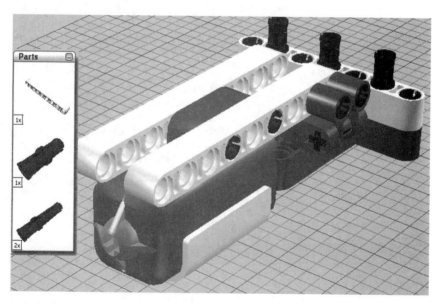

STEP 8. Add parts as shown. The 3-unit long pins go through the touch sensor and both beams.

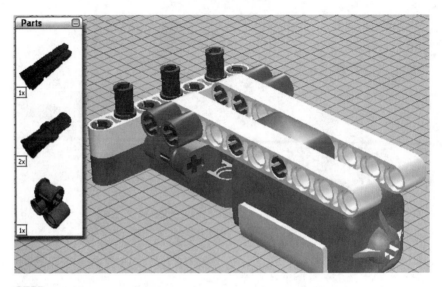

STEP 9. Add parts as shown.

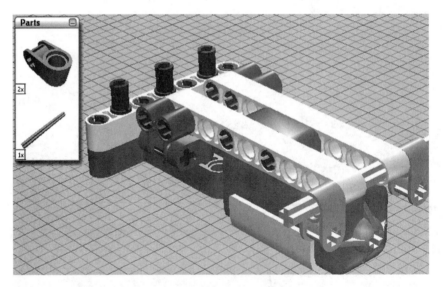

STEP 10. Add parts as shown. The axle is 7 units long.

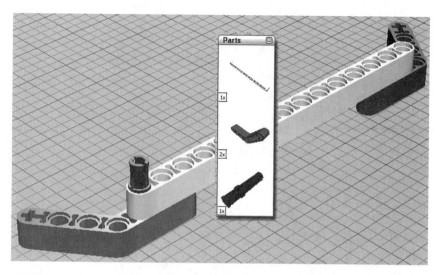

STEP 11. Add parts as shown. The axle is 7 units long.

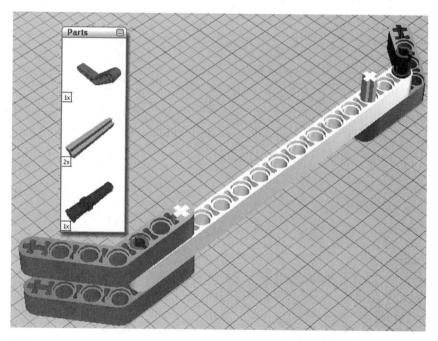

STEP 12. Now start with the bumper. The beam is 15 units long.

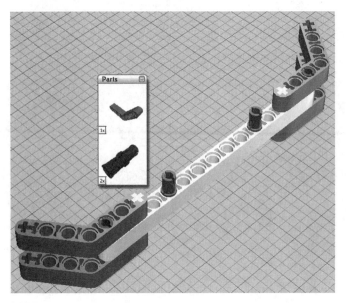

STEP 13. Add parts as shown.

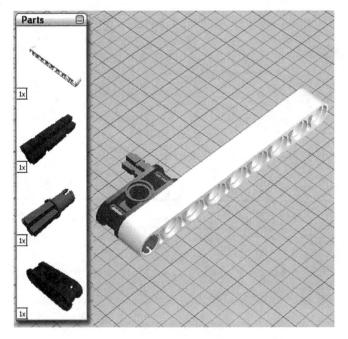

STEP 14. Add parts as shown. The axle is 2 units long.

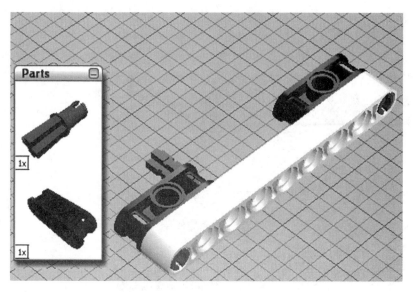

STEP 15. Add parts as shown.

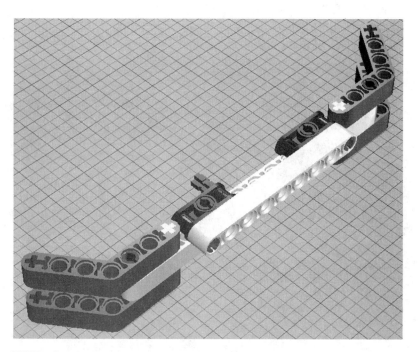

STEP 16. Clip the assembly from step 15 to the bumper.

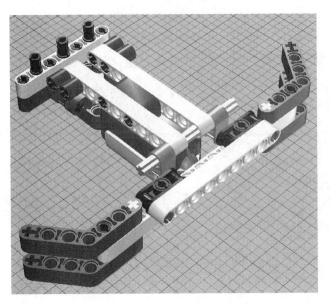

STEP 17. Combine the assembly from step 16 with the assembly from step 10 as shown.

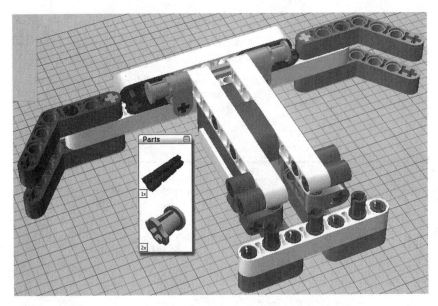

STEP 18. Add the bushes and 2 unit axle to secure the part.

STEP 19. Now attach the bumper to the base using the three black friction pins at the rear of the attachment. Connect a cable from the touch sensor to port 1 on the NXT.

In order to attach the bumper you will need to detach the motor attachment we added back in Chapter 10. We'll remove the bumper unit later in this chapter to make room for other projects. It easily snaps in and out using the three black friction pins.

The sample NXC program produces behavior similar to a Roomba vacuum cleaner. It wanders around a room and if it runs into something, it backs up and turns a random amount and then starts moving forward again. The sensor port and the macro for reading the touch sensor value are used in two preprocessor declarations at the start of the program so that if you decide to change the port you can reconfigure the entire program in one place.

This program also demonstrates how you can safely share a motor resource between two tasks that are running at the same time. The motor resources are protected using the mutex defined on line 4. If a task wants to use the motor, it has to acquire the motor mutex and then release it when it has finished using the motors.

```
1. #define BUMPER_PORT S1
2. #define BUMPER        SENSOR_1
3.
4. mutex motorMutex;
5.
6. task Move() {
7.   while(true) {
8.     Acquire(motorMutex);
9.     OnFwdSync(OUT_AC, 75, 0);
```

```
10.        Release(motorMutex);
11.        Wait(500);
12.      }
13.  }
14.
15.  task WatchBumper() {
16.    while (true) {
17.      if (BUMPER) {
18.        // sensor is pressed so back up and turn
19.        Acquire(motorMutex);
20.        OnRev(OUT_A, 40+Random(60));
21.        OnRev(OUT_C, 60+Random(40));
22.        Wait(500+Random(1000));
23.        Release(motorMutex);
24.      }
25.    }
26.  }
27.
28.  task main() {
29.    SetSensorTouch(BUMPER_PORT);
30.    Precedes(WatchBumper, Move);
31.  }
```

Both the WatchBumper task and the Move task are scheduled to run concurrently by using the Precedes statement in task main. The only other thing that main does is configure the bumper port as a touch sensor. As soon as the main task exits, the two tasks that it precedes will begin to execute.

TRY IT: *See if you can build a robot that can safely drive around on a table without falling off the edge. In this case the touch sensor would need to detect the table edge – preferably from both the front and the back of the robot. With one sensor this might be difficult, but it is possible using a clever bumper design.*

Light Sensor

The NXT light sensor is similar to your eyes in some ways. It provides your program with a limited form of vision and you can measure the amount of ambient light in a room. This might come in handy if you want your robot to act differently during the day than it does at night. You can also use this sensor to measure how much light is reflected off a surface near the light sensor. This mode of operation is commonly used when you want your robot to follow a line or gather data about the reflectivity of the surface it is passing over. Phillipe Hurbain used the NXT light sensor in its active reflection mode to build a Segway-like robot using the NXT.

WEBSITE: *Visit Philo's Homepage for information about using the NXT light sensor to build your own NXTway. www.philohome.com/nxtway/ nxtway.htm*

We'll start exploring the light sensor just as we did with the touch sensor. Let's build a simple light sensor attachment for our mobile robot platform. Follow the steps outlined below to add the NXT light sensor to the base.

Building the front sensor attachment

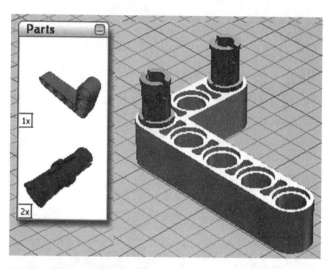

STEP 1. Add parts as shown.

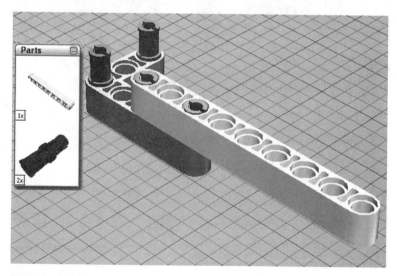

STEP 2. Add parts as shown.

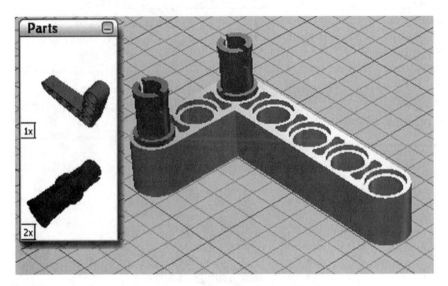

STEP 3. Add parts as shown.

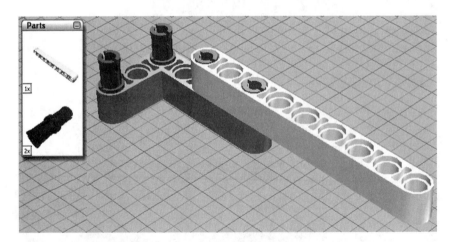

STEP 4. Add parts as shown.

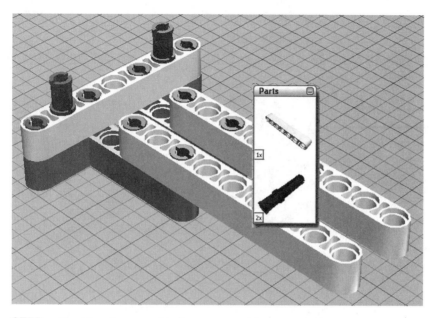

STEP 5. Combine the assembly from step 4 with the assembly from step 2 as shown and add the parts as shown.

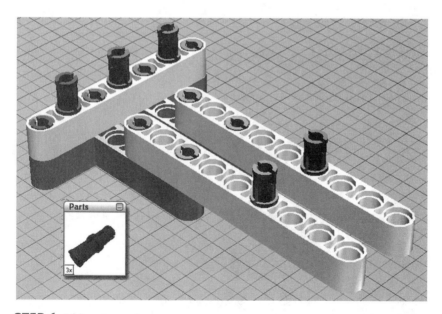

STEP 6. Add parts as shown.

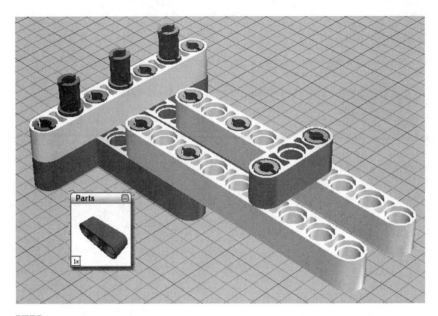

STEP 7. Add parts as shown.

Building the light sensor attachment

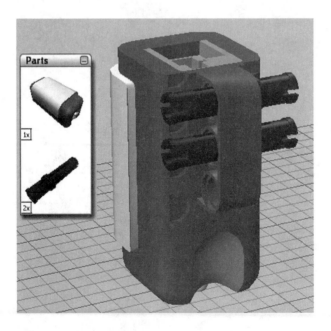

STEP 1. Add parts as shown.

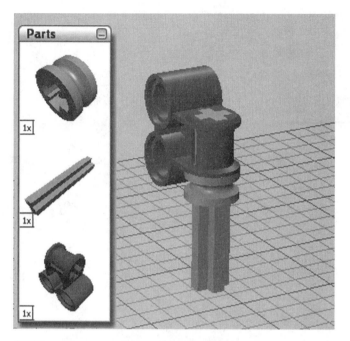

STEP 2. Add parts as shown. The axle is 3 units.
Make two copies of this assembly.

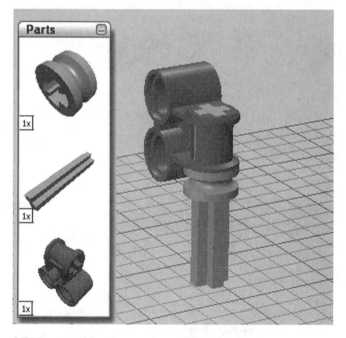

STEP 3. Combine the two assemblies from step 2
with the assembly from step 1 as shown.

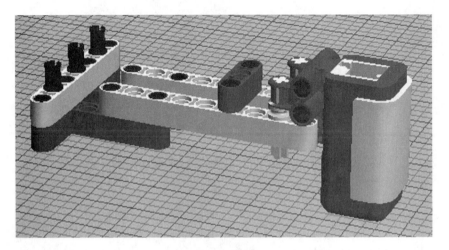

STEP 4. Drop the axles from the assembly in step 3 into the second hole in both beams as shown.

STEP 5. Now connect the light sensor to the base using the three black friction pins at the rear of the attachment. Connect a cable from the light sensor to port 1 on the NXT.

Now let's see if we can get our mobile robot to follow the line on the test pad that came with the NXT. You can also make your own line for the robot to follow using black electricians tape. Just make sure you have permission to put tape on the floor or you may find yourself stuck in your room without your NXT for a while.

The NXC program below allows you to calibrate the light sensor using buttons. First it configures port 1 to work with an active type of light sensor and to return a raw value. Then it starts waiting for button presses. At first the program runs in a while loop waiting for a left button press. During this phase you should move the light sensor over the white background of the test pad and over the black line on the test pad. The lowest and highest light sensor readings are being calculated. Press the left arrow button to exit this phase.

```
1.  #define SLOW_SPEED  25
2.  #define FAST_SPEED  75
3.  #define EYE_PORT    S1
4.  #define EYE_VALUE   SENSOR_1
5.  #define DELTA       200
6.  inline int min(int v1, int v2) {
7.    return (v1 < v2) ? v1 : v2;
8.  }
9.
10. inline int max(int v1, int v2) {
11.   return (v1 > v2) ? v1 : v2;
12. }
13.
14. task main() {
15.   SetSensorType(EYE_PORT,
16.                 SENSOR_TYPE_LIGHT_ACTIVE);
17.   SetSensorMode(EYE_PORT, SENSOR_MODE_RAW);
18.   ResetSensor(EYE_PORT);
19.   int BlackThreshold, WhiteThreshold, SV;
20.   int lowest = 2000, highest = -2000;
21.   while (!ButtonPressed(BTNLEFT, false)) {
22.     SV = EYE_VALUE;
23.     lowest  = min(lowest, SV);
24.     highest = max(highest, SV);
25.     NumOut(0, LCD_LINE1, SV);
26.   }
27.   BlackThreshold = lowest+DELTA;
28.   WhiteThreshold = highest-DELTA;
29.   NumOut(0, LCD_LINE1, BlackThreshold, true);
30.   NumOut(0, LCD_LINE2, WhiteThreshold);
31.   until(ButtonPressed(BTNCENTER, false));
32.   while (true) {
33.     SV = EYE_VALUE;
34.     if (SV < WhiteThreshold)
35.       OnFwd(OUT_A, FAST_SPEED);
```

```
36.      else
37.         OnFwd(OUT_A, SLOW_SPEED);
38.      if (SV > BlackThreshold)
39.         OnFwd(OUT_C, FAST_SPEED);
40.      else
41.         OnFwd(OUT_C, SLOW_SPEED);
42.   }
43. }
```

The black and white thresholds are set based on the low and high light sensor readings along with a DELTA value that you will probably have to tweak to get the robot to work right. Light sensor readings are very sensitive to a number of different factors such as the ambient light level in your room, and the NXT battery level. Tuning a line follower is more of an art than a science.

The program displays the selected threshold values and waits for one more center button press and then it starts trying to follow the line. The basic line following algorithm goes straight fast so long as the reading is between the two thresholds. If the sensor reading exceeds the white threshold then motor A slows down causing the robot to turn to the right. If the sensor reading drops below the black threshold then motor C slows down causing the robot to turn to the left. That means this program is designed to follow the outside edge of the line traveling clockwise around the line on the test pad. It will not handle counter-clockwise turns since it thinks that it will find a blacker color by turning toward the right.

TRY IT: *Using the light sensor in inactive mode (ambient) write a program for your robot so that it can find the brightest spot in your room. You will have to modify the light sensor attachment so that it points straight up.*

Try this same project with the light sensor attached to a modified version of our motor attachment so that the sensor itself is turned around to look for the brightest light.

Sound Sensor

The sound sensor is a great addition to the suite of analog sensors, though it probably won't get used as often as the light sensor or the touch sensor. It definitely is not as critically important to a robotic application as the rotation sensor. But it can be useful for adding more feedback to your program.

It gives your NXT robot a very crude form of hearing much like the light sensor provided a form of seeing. You won't be able to do things like distinguish frequencies of sound but you can at least measure sound pressure level or loudness within the sound sensor's range of sensitivity.

We'll start exploring the sound sensor just as we did with the touch sensor and the light sensor. Let's build a simple sound sensor attachment for our basic mobile robot. Follow the steps outlined below to add the NXT sound sensor to the base.

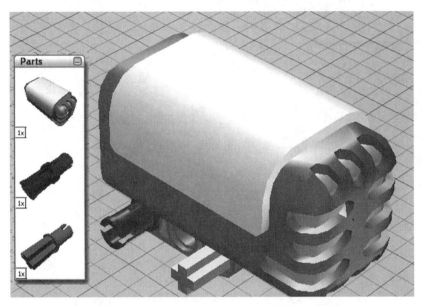

STEP 1. Add parts as shown.

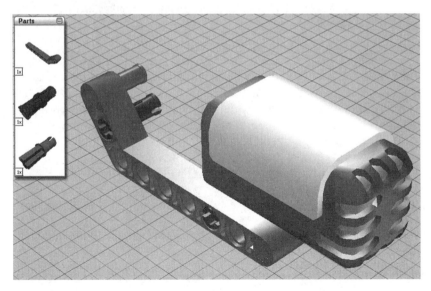

STEP 2. Add parts as shown.

STEP 3. Now attach the sound sensor to the robot platform, as shown here. The blue axle pin fits into the slot in the double bent beam on the side of the robot.

Our sample program is a clapper robot. It waits until it hears a loud sound like a clap and starts moving randomly. It stops moving the next time it hears a loud clap. This cycle keeps repeating until you abort the program. The Wait in the checkSound task's main loop is important since it gives the sensor time to read a new, presumably lower, sound level after the clap. Without it the bMoving boolean value might get switched on and then switched back off or vice versa.

```
1.  #define EARS        SENSOR_1
2.  #define EAR_PORT    S1
3.  #define THRESHOLD   45
4.
5.  bool bMoving = FALSE;
6.
7.  task checkSound() {
8.    while (true) {
9.      int value = EARS;
10.     NumOut(0, LCD_LINE1, value);
```

```
11.       if (value > THRESHOLD)
12.          bMoving = !bMoving;
13.       Wait(50);
14.    }
15. }
16.
17. task main() {
18.    SetSensorSound(EAR_PORT);
19.    start checkSound;
20.    while (true) {
21.       if (bMoving) {
22.          if (Random() < 0)
23.            OnFwdSync(OUT_AC,
24.                        Random(100), -Random(100));
25.          else
26.            OnFwdSync(OUT_AC,
27.                        Random(100), Random(100));
28.          Wait(100);
29.       }
30.       else
31.          Off(OUT_AC);
32.    }
33. }
```

Sometimes the random values for the power value in the OnFwdSync calls may be too low for the motors to actually turn but if you leave it alone it should start moving in a rather jerky sort of way. You may also need to fiddle with the threshold value so that it doesn't take a really loud clap to start it moving or, conversely, so that it doesn't keep turning on and off due to the ambient noise level.

Rotation Sensor

As previously mentioned, each NXT motor has a built-in rotation sensor. It is such a useful sensor that it has been hard to avoid using it until now. Of course, the firmware uses it implicitly when you use any form of regulated motor control. But we want to use it directly in our programs too. The sample NXC program below demonstrates this.

This program uses the active braking technique that we saw in the previous chapter. We'll use the low-level SetOutput API call instead of OnFwdReg, as we did before. This will make it easy for us to tweak the PID parameters during our experimentation. You can see where these values are set on lines 8, 9, and 10 in the program below. In addition, our program uses the ClearScreen, PointOut, MotorRotationCount, and ResetRotationCount API functions.

```
1. task main() {
2.    ResetRotationCount(OUT_A);
3. //    OnFwdReg(OUT_A, 0, OUT_REGMODE_SPEED);
4.    SetOutput(OUT_A, Power, 0,
```

```
 5.      OutputMode, OUT_MODE_MOTORON+
    OUT_MODE_BRAKE+OUT_MODE_REGULATED,
 6.      RegMode, OUT_REGMODE_SPEED,
 7.      RunState, OUT_RUNSTATE_RUNNING,
 8.      RegPValue, 96,
 9.      RegIValue, 32,
10.      RegDValue, 32,
11.      UpdateFlags, UF_UPDATE_SPEED+
    UF_UPDATE_MODE+UF_UPDATE_PID_VALUES);
12.    int x = 0;
13.    while (true) {
14.      int y = MotorRotationCount(OUT_A) + 32;
15.      PointOut(x, y);
16.      Wait(10);
17.      x++;
18.      if ((x % 100) == 0)
19.        ClearScreen();
20.      x %= 100;
21.    }
22. }
```

This little program resets the rotation sensor on OUT_A back to zero and then it starts active braking on that output. It then starts an infinite loop where it uses PointOut to draw a graph showing the motor's rotation count as you turn the wheel attached to the motor. The graph gives you a visual representation of the effect that different PID values have on the motor control loop.

You can perform a small experiment by turning the motor and letting it go repeatedly. See if you can plot a sine wave on the LCD screen. Configure a Watch in BricxCC on Motor A's rotation count and turn on regular polling at a fast polling rate. Then bring up the Graph window and collect some data points while you turn and release the wheel. All the data is stored in memory and can be copied to the clipboard so that you can paste it into Excel for further analysis. Repeat this process with different values of RegPValue, RegIValue, and RegDValue to see what impact these weights have on the NXT motor control algorithm.

 TRY IT: *Using the rotation sensor, the diameter of the NXT wheels, and the center-to-center distance between the wheels on our mobile robot platform, write a program that can accurately turn the robot in place by a specified number of degrees. Take advantage of the compile-time expression evaluator for maximum accuracy.*

Coming Up Next

In Chapter 12 we'll look at some of the advanced outputs available for use in NXC and NBC programs. If custom I²C sensors, writing to files, or sending Bluetooth messages sounds interesting then read on. We'll even take a peek at writing to the USB and Hi-Speed output buffers directly.

Advanced NXT Outputs

Topics in this Chapter

- Creating and Writing to Files
- Sending Messages
- Custom I^{2}C Output Devices

Chapter 12

In addition to all the basic outputs we looked at in Chapter 10, the NXT has additional outputs that are more advanced. The goal of this chapter is to describe how to use these, and spark your curiosity about the different ways to use these outputs in your own programs. I want to challenge you to try using LEGO MINDSTORMS in ways no one has tried before.

Creating and Writing to Files

Something you may not immediately think of as an output is file writing. But just like the other outputs, file writing is yet another way our programs can provide a response to the user. In this section we'll learn how to use the NBC and NXC API functions to create files, open existing files, delete files, and rename files.

Files on the NXT have a fixed size, which is specified at the time the file is created. When you write to a file you gradually fill up all the space available within the file. Once a file is completely full you cannot write any more data to it.

The NXT-G File Access block contains code that handles the situation when the file has reached its capacity. When you use this block, it writes data to the specified file until it runs out of room. Then it actually renames the file, creates a new larger file using the old filename, copies all the data from the old file into the new file, deletes the old file, and then writes the remaining bytes into the new file.

The NXC API functions all operate at a much lower level, so they don't perform this resize operation for you. Using the knowledge you gain in this chapter and the next, you will be empowered to go out and create your own NXC version of the File Access block to use in your programs.

Let's examine the API functions to find out how to create files in our code. The sample program below uses CreateFile to create a new file. The return value of the file API functions is a Loader module status code. They are listed in Table 12-1. Make sure that you check the return value since it will tell you if the file operation was a success.

Loader Result Codes	Value
LDR_SUCCESS	0X0000
LDR_INPROGRESS	0X0001
LDR_REQPIN	0X0002
LDR_NOMOREHANDLES	0x8100
LDR_NOSPACE	0x8200
LDR_NOMOREFILES	0x8300
LDR_EOFEXPECTED	0x8400
LDR_ENDOFFILE	0x8500
LDR_NOTLINEARFILE	0x8600
LDR_FILENOTFOUND	0x8700
LDR_HANDLEALREADYCLOSED	0x8800
LDR_NOLINEARSPACE	0x8900
LDR_UNDEFINEDERROR	0x8A00
LDR_FILEISBUSY	0x8B00
LDR_NOWRITEBUFFERS	0x8C00
LDR_APPENDNOTPOSSIBLE	0x8D00
LDR_FILEISFULL	0x8E00
LDR_FILEEXISTS	0x8F00
LDR_MODULENOTFOUND	0x9000
LDR_OUTOFBOUNDARY	0x9100
LDR_ILLEGALFILENAME	0x9200
LDR_ILLEGALHANDLE	0x9300
LDR_BTBUSY	0x9400
LDR_BTCONNECTFAIL	0x9500
LDR_BTTIMEOUT	0x9600
LDR_FILETX_TIMEOUT	0x9700
LDR_FILETX_DSTEXISTS	0x9800
LDR_FILETX_SRCMISSING	0x9900
LDR_FILETX_STREAMERROR	0x9A00
LDR_FILETX_CLOSEERROR	0x9B00

Table 12-1. Loader result codes

CreateFile requires you to pass in the name of the file along with the desired file size. You also pass in a byte variable to store the handle for the file you are creating. All further access to the file within a session is performed using this file handle, so make sure your code retains it. On line 6 the program checks to see if the Loader module reported that the file already exists. If the file could not be created because it already exists then it tries to open the existing file for writing using the OpenFileAppend API function.

```
int CreateFile(string filename, unsigned long fsize, byte &
handle);
int OpenFileAppend(string filename, unsigned long & fsize,
byte & handle);
```

OpenFileAppend takes a filename as its sole input. The last two arguments are outputs from the function call. It returns the number of bytes of empty space left in the file as well as the file handle. If the file exists but it is completely full then OpenFileAppend will return `LDR_FILEISFULL`. The program checks for this value on line 23 and outputs an appropriate error message.

```
1.  task main() {
2.     byte handle;
3.     unsigned int result;
4.     int fileSize = 256;
5.     result = CreateFile("test.dat", fileSize, handle);
6.     if (result == LDR_FILEEXISTS)
7.        result = OpenFileAppend("test.dat", fileSize, handle);
8.     if (result == LDR_SUCCESS) {
9.        TextOut(0, LCD_LINE1, "writing to file");
10.       for (byte i=0; i<fileSize;i++) {
11.          result = Write(handle, i);
12.          if (result == LDR_FILEISFULL) {
13.             TextOut(0, LCD_LINE1, "no more room");
14.             break;
15.          }
16.          else if (result != LDR_SUCCESS) {
17.             NumOut(0, LCD_LINE1, result);
18.             break;
19.          }
20.       }
21.       CloseFile(handle);
22.    }
23.    else if (result == LDR_FILEISFULL)
24.       TextOut(0, LCD_LINE1, "File is full");
25.    else {
26.       TextOut(0, LCD_LINE1, "Open write failed");
27.       NumOut(0, LCD_LINE2, result);
28.    }
29.    Wait(3000);
30. }
```

The two other API functions used in the above sample are CloseFile and Write. The program uses CloseFile on line 21 if it was able to either create or open the file for writing. Closing the file as soon as you are done writing is important, since the NXT can only have sixteen files open at the same time.

```
int CloseFile(byte handle);
int Write(byte handle, <Type> n);
```

The Write API function writes binary data to a file. You have to pass in the handle and a variable containing the data you want to write. The number of bytes that are written to the file depends on the type of the variable you pass into the function. Scalar types write 1, 2, or 4 bytes. With structures and arrays the number of bytes depends on the structure definition or the number and type of elements in the array. This makes it trivially easy to save persistent data to a file that can be loaded the next time your program is executed. A binary file format is often a better choice than a text format since it is almost always more compact – it takes six bytes to write -32767 to a file as a string but it only takes two bytes to write it to a file using a binary format.

To write data to a file as text, use WriteString instead of Write. You need to use NumToStr to convert numeric data to a string before using WriteString to write it to a file. With the enhanced NBC/NCX firmware you can also use FormatNum to perform this conversion. When writing using a text format, you may want to use WriteLnString, which appends a pair of DOS-style line endings (0x0D 0x0A) after the data so that each write is on a separate line in the file. Both of these functions set the length parameter to the number of bytes actually written.

```
int WriteString(byte handle, string str, unsigned long &
length);
int WriteLnString(byte handle, string str, unsigned long &
length);
```

The Write API function suffers from the problem of not telling you that a partial write has occurred due to reaching the end of the file in the middle of a write operation. If you are not sure there is enough space left in a file that you are writing to, you should use WriteBytes or WriteBytesEx instead. These two routines take a byte array buffer as the data you want to write. You can still write either binary or text files but you are responsible for the format of the data in your buffer.

```
int WriteBytes(byte handle, byte buffer[], unsigned long &
length);
int WriteBytesEx(byte handle, unsigned long & length, byte
buffer[]);
```

When you call either of these functions, you pass in an unsigned long variable in addition to the handle and the data buffer. In WriteBytes the variable is an output only, so it is the last parameter in the function. It is set to the actual number of bytes written. If the function's return value is `LDR_FILEISFULL` then you can use the value of length to tell whether some of the data still needs to be written to a file. The WriteBytesEx function takes length as an input parameter, which lets you specify that you only want to write a portion of the buffer contents to the file. The value of length is updated just like it is with the WriteBytes, WriteString, and WriteLnString API functions.

In NBC and NXC you can also rename or delete existing files. Simply use the RenameFile and DeleteFile API functions. RenameFile takes two string values as arguments. The first is the old filename and the second is the new filename. If a file is found with the old name you provided, then it is renamed using the new name. The DeleteFile function simply takes a filename as its sole parameter. If a file by that name exists on the NXT it is deleted.

```
int RenameFile(string oldname, string newname);
int DeleteFile(string fname);
```

If you need to write to a file that is already open but you don't have the handle available for some reason, you can look it up. The ResolveHandle API function gives you this ability. Pass in a filename and it tries to find an open file that matches your filename. If found, it returns the handle and a boolean value indicating whether the handle is for reading or for writing.

```
int ResolveHandle(string filename, byte & handle, bool & writeable);
```

Data Logging

The sample program below demonstrates a number of concepts we just discussed. It uses a binary file format to save rotation sensor values. The motor on OUT_A is stopped with active braking. For ten seconds you can rotate the motor by hand and record the motor rotation sensor value every five milliseconds. The data is stored in a structure and inserted into an in-memory array.

After ten seconds have elapsed, the contents of the array (and a couple of header structures) are all written to the file, which we have named "motor.rdt". If the file already exists, it is deleted and recreated.

```
1. #define DATALOG_FILENAME "motor.rdt"
2.
3. struct DatalogFileHeader {
4.    unsigned int Format;
5.    unsigned int DataBytes;
6.    unsigned long TotalTime;
7. };
8.
9. struct DatalogSensorHeader {
10.    byte Combined;
11.    byte Sensor_Type;
12.    byte Sensor_Mode;
```

```
13.    byte CustPctFullScale;
14.    byte CustActiveStatus;
15.    unsigned long SampleTime;
16. };
17.
18. struct DatalogSyncSensorData {
19.    byte Combined;
20.    unsigned long Sensor_Value;
21. };
22.
23. struct DatalogAsyncSensorData {
24.    byte Combined;
25.    unsigned long Sensor_Value;
26.    unsigned long Sensor_Time;
27. };
28.
29. const int HeaderType    = 0x80;
30. const int SyncDataType  = 0x00;
31. const int AsyncDataType = 0x40;
32.
33. byte CombinedValue(byte dType, byte port) {
34.    return dType + (port & 0x3F);
35. }
36.
37. byte HeaderCombined(byte port) {
38.    return CombinedValue(HeaderType, port);
39. }
40.
41. byte SyncDataCombined(byte port) {
42.    return CombinedValue(SyncDataType, port);
43. }
44.
45. byte AsyncDataCombined(byte port) {
46.    return CombinedValue(AsyncDataType, port);
47. }
48.
49. task main() {
50.    DatalogFileHeader hdr;
51.    hdr.Format    = 0x0090;
52.    hdr.TotalTime = 10000;
53.    hdr.DataBytes = 9+(2000*5);
54.
55.    DatalogSensorHeader sHead;
56.    sHead.Combined = HeaderCombined(OUT_A);
57.    sHead.Sensor_Type = 0;
58.    sHead.Sensor_Mode = 0;
59.    sHead.CustPctFullScale = 0;
60.    sHead.CustActiveStatus = 0;
61.    sHead.SampleTime = 5;
62.
```

```
63.    DatalogSyncSensorData sData;
64.    sData.Combined = SyncDataCombined(OUT_A);
65.
66.    DatalogSyncSensorData storage[2000];
67.
68.    TextOut(0, LCD_LINE1, "Press Orange");
69.    TextOut(0, LCD_LINE2, "to start");
70.
71.    ResetRotationCount(OUT_A);
72.    OnFwdReg(OUT_A, 0, OUT_REGMODE_SPEED);
73.    until(ButtonPressed(BTNCENTER, false));
74.    // collect data
75.    int idx = 0;
76.    while (idx < 2000) {
77.      sData.Sensor_Value =
78.        MotorRotationCount(OUT_A);
79.      storage[idx] = sData;
80.      idx++;
81.      Wait(5);
82.    }
83.    // save it to a file
84.    byte handle;
85.    unsigned int result;
86.    unsigned long fsize = hdr.DataBytes+8;
87.    result = CreateFile(DATALOG_FILENAME,
88.      fsize, handle);
89.    if (result == LDR_FILEEXISTS) {
90.      DeleteFile(DATALOG_FILENAME);
91.      result = CreateFile(DATALOG_FILENAME,
92.        fsize, handle);
93.    }
94.    if (result = LDR_SUCCESS) {
95.      Write(handle, hdr);
96.      Write(handle, sHead);
97.      for (int i=0; i<ArrayLen(storage);i++) {
98.        sData = storage[i];
99.        Write(handle, sData);
100.     }
101.     CloseFile(handle);
102.   }
103. }
```

The motor.rdt file format is well documented by the structure definitions within the program itself. Writing a program on a Mac or a PC to process this data is trivial. Keep in mind that multi-byte values in each structure are stored in little-endian order.

The primary limitation of this approach to data logging is that a running program has a maximum of 32 kilobytes of memory available. Our ten-second experiment, with one sensor recording data every five milliseconds, uses 10k of memory. If we used the asynchronous data structure it would have taken 18k. Of course, the set of structures we used above are not the only way you could design an in-memory data-logging mechanism. A more data-efficient approach might suit your needs better.

Low-level File Output

You can perform every file operation described above using NBC syscall statements or the NXC SysCall function. In NXC there are also low-level wrapper functions for each of the file system calls. Instead of calling DeleteFile, for example, you could use a FileDeleteType structure and call SysFileDelete.

The low-level NXC SysCall structures are all documented in the *NXC Programmer's Guide*. All of the file output system call wrapper functions are listed below. See Appendix A for additional information about these API functions.

```
void SysFileOpenWrite(FileOpenType & args);
void SysFileOpenAppend(FileOpenType & args);
void SysFileWrite(FileReadWriteType & args);
void SysFileClose(FileCloseType & args);
void SysFileResolveHandle(FileResolveHandleType & args);
void SysFileRename(FileRenameType & args);
void SysFileDelete(FileDeleteType & args);
```

Sending Messages

Another advanced form of output at your command is the ability to send messages from the NXT to other devices. Let's take an in-depth look at the different options available for inter-device messaging.

Bluetooth

The most obvious form of message sending via the NXT is using its built-in support for Bluetooth. You can pair up to four NXT bricks with one primary and three secondary. The primary brick controls all the communication back and forth in the standard form of this configuration. The secondary bricks do not actually initiate any messages via Bluetooth.

With the standard firmware you must establish your connections manually using the NXT menu system before running your programs on the primary and secondary bricks. The brick that initiates the connection is the primary brick.

The standard NXT communication protocol (described in the LEGO MINDSTORMS NXT Bluetooth Developer Kit) is the easiest way to communicate. Built on top of this protocol are the NXT Direct and System commands. At the highest level, the NXT provides a mailbox system that is used in conjunction with Bluetooth for exchanging data between two bricks.

WEBSITE: *The Bluetooth Developer Kit is available on the LEGO website.* *http://mindstorms.lego.com/Overview/NXTreme.aspx*

Each brick has ten incoming message queues or mailboxes that can hold up to five messages of at most 59 characters each. There are also ten outbound mailboxes that are used on a secondary brick to reply to a message received from a primary brick. Secondary bricks simply write their reply message to the outbound mailbox corresponding to the inbound mailbox. The primary brick retrieves this message using a MessageRead direct command sent to the secondary brick. Outbound mailboxes numbers are always 10 plus the inbound mailbox number.

Messages are always stored as a string with a null terminator, even though the data may represent a numeric or boolean value. The NXT-G block for reading a mailbox message automatically tries to convert the value into all three types: string, numeric, and boolean.

In NXC you can write a message to a local mailbox using the SendMessage API function. A very similar function is the SendResponseString API function. Both of these simply write a string to a mailbox or queue using the MessageWrite syscall statement. SendResponseString automatically adds 10 to the queue value, since a response message is supposed to be written to the outbound queue associated with an inbound queue. With SendMessage you have to offset the queue yourself in order for the message to go to an outbound mailbox. Both of these functions are best used on a secondary brick using the standard NXT communication protocol.

```
int SendMessage(byte queue, string msg);
int SendResponseString(byte queue, string str);
```

NXC and NBC also provide two functions for sending a boolean or numeric message as a response to a message from the primary NXT. These functions are SendResponseBool and SendResponseNumber. As with SendResponseString, they both offset the queue by ten automatically. These functions format the boolean or numeric value according to the standard protocol requirements and write the null terminated result to the outbound queue. Numeric values are always written as a four byte long in little-endian order followed by the null terminator. Boolean values are written as a single byte, which is either zero or one, again followed by a null byte.

```
int SendResponseBool(byte queue, bool bval);
int SendResponseNumber(byte queue, long val);
```

On the primary NXT, message sending via the NXT communication protocol involves sending a MessageWrite direct command to a secondary NXT over a specific Bluetooth connection. NXC wraps this procedure into three simple API functions: SendRemoteBool, SendRemoteNumber, and SendRemoteString. It also provides the RemoteMessageWrite API function, which is an alias for SendRemoteString.

```
int SendRemoteBool(byte conn, byte queue, bool bval);
int SendRemoteNumber(byte conn, byte queue, long val);
int SendRemoteString(byte conn, byte queue, string str);
int RemoteMessageWrite(byte conn, byte queue, string msg);
```

The return value of each of these functions is the result of the CommBTWrite syscall statement. If the operation succeeds it returns STAT_COMM_PENDING. This means the Bluetooth subsystem has started to process your request. You will need to make sure you do not interrupt a previous Bluetooth write request by following any API function that performs a Bluetooth write operation with a while loop containing a call to the BluetoothStatus API function. When BluetoothStatus returns NO_ERR (0) it is safe to begin another write operation.

```
int BluetoothStatus(byte conn);
```

NXC also provides a number of high-level routines that wrap several of the direct commands into simple to use API functions. Only direct commands that do not require a response can be used, with one exception. Since the MessageRead direct command is part of the NXT firmware's basic mailbox messaging system the Bluetooth response to this command will be successfully received and processed by the firmware. If you send any other direct command that expects a response, the firmware will ignore it. The direct command API functions are listed below. Do not forget to use the BluetoothStatus wait loop following calls to any of these API functions.

```
int RemoteMessageRead(byte conn, byte queue);
int RemoteStartProgram(byte conn, string filename);
int RemoteStopProgram(byte conn);
int RemotePlaySoundFile(byte conn, string filename, bool
bloop);
int RemotePlayTone(byte conn, unsigned int frequency,
unsigned int duration);
int RemoteStopSound(byte conn);
int RemoteKeepAlive(byte conn);
int RemoteResetScaledValue(byte conn, byte port);
int RemoteResetMotorPosition(byte conn, byte port, bool
brelative);
int RemoteSetInputMode(byte conn, byte port, byte type, byte
mode);
int RemoteSetOutputState(byte conn, byte port, char speed,
byte mode, byte regmode, char turnpct, byte runstate, long
tacholimit);
```

At a lower level NXC and NBC both provide direct access to the CommBTWrite syscall statement. Here there is no direct link to the standard NXT communication protocol. You are simply writing a string of bytes to a Bluetooth connection. Devices other than NXT bricks do not know anything about the structure and format of a standard NXT message. For non-NXT devices there is really no compelling reason to adhere to this protocol. If you are writing to a paired mobile device such as a smart phone or a PDA, you may need to either adhere to a specific protocol implemented by the device, such as AT commands for your Bluetooth-enabled telephone, or you may just want to implement a simpler outbound protocol to improve the overall data transfer rate.

```
int BluetoothWrite(byte conn, byte buffer[]);
```

For data coming back to the NXT from your non-NXT device the restrictions of the standard firmware make it difficult to use a protocol other than the standard. Difficult does not mean impossible, however. For quite a while many of the supposed NXT experts believed it was simply impossible for the standard firmware to receive and process GPS data sent to the NXT via Bluetooth. This data definitely does not follow the NXT communication protocol standards. Yet a relatively simple NXC program was written that uses low-level API functions to extract GPS data from Bluetooth input buffers directly.

WEBSITE: *To view an NXC program that reads and validates Bluetooth GPS data using the standard NXT firmware, visit the NXC sample programs section of the NBC website at http://bricxcc.sourceforge.net/nbc/*

Access to the Bluetooth buffers, configuration, and state data is provided in NBC and NXC via IOMapRead and IOMapWrite syscall statements. If you have the enhanced firmware installed on your NXT, make sure you enabled the Enhanced Firmware option in the compiler. Use the -EF command line switch or in BricxCC simply check the Enhanced Firmware option on the NBC/NXC compiler page of the Preferences dialog. When this option is enabled, all of the NXC and NBC API functions that use IOMapRead and IOMapWrite are automatically converted to use the enhanced firmware's IOMapReadByID and IOMapWriteByID syscall statements. Accessing IOMap structure members by the module ID is much faster than accessing them by the module name due to expensive string comparisons required when looking up the module ID using the name provided. The enhanced NBC/ NXC firmware also supports reading and writing up to 800 bytes in a single IOMapRead or IOMapWrite operation. The standard firmware uses a 64-byte limit. The Bluetooth buffers are 128 bytes long and it is much faster to read the entire buffer in a single operation.

A few of the API functions that provide access to low-level Bluetooth data using the Comm module IOMap structure are listed below.

```
void GetBTInputBuffer(byte offset, byte count, byte & data[]);
byte BTInputBufferInPtr();
byte BTInputBufferOutPtr();
void SetBTInputBufferInPtr(byte pointer);
void SetBTInputBufferOutPtr(byte pointer);
```

```
byte BluetoothState();
void SetBluetoothState(byte state);
void SetBTInputBuffer(byte offset, byte count, byte data[]);
```

NXC also provides low-level API function for sending messages, which are thin wrappers around the syscall statement. The SysCall structures for these functions as well as all of the low-level IOMap-based Bluetooth routines are fully documented in the NXC Programmer's Guide. You can find additional information about these functions in Appendix A.

```
void SysMessageWrite(MessageWriteType & args);
void SysCommBTWrite(CommBTWriteType & args);
void SysCommBTCheckStatus(CommBTCheckStatusType & args);
```

Using the enhanced NBC/NXC firmware you can also directly call the Comm module's low-level functions using the SysCommExecuteFunction API routine. Using this feature, you can establish connections to known Bluetooth devices, disconnect devices, and turn on or off the Bluetooth radio.

```
void SysCommExecuteFunction( CommExecuteFunctionType &
args);
```

USB and Hi-speed Output

The NXT has USB and Hi-speed (port 4) output buffers. These are not exposed at a high level within NXC and NBC. However, they are accessible using the IOMapWrite syscall statement and the low-level NXC API functions that wrap this functionality. Writing directly to the USB output buffer or to the Hi-speed port's output buffer is a cutting edge mechanism for sending data to another device.

There hasn't been a lot of experimentation in these areas so far by NXT developers. As of this writing, no one has used the hi-speed output port in a project. In order to write code for these output ports, you might have to pore over the firmware source code. If you have success, you may become semi-famous for being the first to delve the depths of these features and return with something amazing. All of the low-level functions that you can use in your experimentation are listed in Appendix A and are more fully documented in the *NXT Programmer's Guide.*

Custom I²C output devices

Another advanced form of NXT output is a certain type of I²C device that is not a sensor in the purest sense of the word. Rather than requesting information from these devices, they actually send data or control an output such as a motor.

An example of this type of device is the mindsensors.com I²C motor multiplexer. You can send commands to the multiplexer that turn on or off up to four legacy motors with 255 different power levels. The multiplexers allow you to run the motors forward or reverse and stop them in either brake or float mode. All of this is controlled using simple I2CWrite operations. The low-level I²C API routines in NXC are listed below.

```
int I2CWrite(byte port, byte retlen, byte buffer[]);
int I2CCheckStatus(byte port);
int I2CRead(byte port, byte buflen, byte buffer[]);
bool I2CBytes(byte port, byte inbuf[], byte & count, byte &
outbuf[]);
```

If you use the low-level I²C routines in a program, the standard pattern is to first call I2CCheckStatus in a while loop until it returns a status of zero, indicating that the I²C bus is idle. Then call I2CWrite followed by another while loop with I2CCheckStatus waiting for the write operation to complete. Finally, call I2CRead to retrieve the requested data. This pattern is performed for you by the I2CBytes API function in order to make using I²C devices easier. When you do not need to read any response data, you should just call I2CWrite followed by the I2CCheckStatus loop.

The default device address for the mindsensors motor multiplexer is 0xB4. To control motor 1, write to register address 0x42. Likewise, motors 2, 3 and 4 are controlled via register addresses 0x44, 0x46 and 0x48 respectively. At each register you simply write the direction command byte followed by the speed byte. The direction commands are 0x00 for float, 0x01 for forward, 0x02 for reverse, and 0x03 for brake.

```
byte buf[];
SetSensorLowspeed(S1);
// motor 1 forward at half power
ArrayBuild(buf, 0xB4, 0x42, 0x01, 127);
I2CWrite(S1, 0, buf);
while(I2CCheckStatus(S1) > 0);
```

You can also read the current control values to find out the state of any of the four motors. For example, just write to {0xB4, 0x42} and request two bytes to get the direction and speed of motor 1.

```
ArrayBuild(buf, 0xB4, 0x42);
byte outputBuf[];
byte count = 2;
if (I2CBytes(S1, buf, count, outputBuf)) {
  byte m1Dir = outputBuf[0];
  byte m1Pwr = outputBuf[1];
}
```

HiTechnic IrLink

The iRLink I²C device from HiTechnic is an advanced input and output device. It uses infrared (IR) to communicate with an RCX brick running the LEGO RCX 2.0 firmware. It can also control the LEGO Power Function motors and the LEGO IR train motors using their specific IR protocols.

With the Power Function motors and IR trains, the iRLink is strictly an output device. It sends IR messages to an IR receiver and the motors respond appropriately. With the RCX, however, you can send any direct opcode to the RCX brick and it will perform the requested command. Most commands do not have a response but there are some that send an IR message back to the iRLink. If you send the Poll opcode to the RCX, for example, you will receive a 12-byte IR message containing the two-byte signed word value you requested.

The HiTechnic iRLink is so powerful that many of its capabilities have been wrapped up into official NBC and NXC API functions. To control the Power Function motors you can simply use the HTPowerFunctionCommand API function.

```
int HTPowerFunctionCommand(const byte port, byte pfchannel,
byte mtrcmd1, byte mtrcmd2);
```

The API defines constants for the power function channels. These are listed in Table 12-1 below.

Power Function Channel Constants	Value
HTPF_CHANNEL_1	0
HTPF_CHANNEL_2	1
HTPF_CHANNEL_3	2
HTPF_CHANNEL_4	3

Table 12-1. Power Function Channels

There are also constants defined for the power function motor commands. These are listed in Table 12-2 below.

Power Function Motor Commands	Value
HTPF_CMD_STOP	0
HTPF_CMD_FWD	1
HTPF_CMD_REV	2

Table 12-2. Power Function Motor Commands

The sample program below shows how simple it is to use this function. The iRLink device is attached to port 1. The power function IR receiver is set to channel 1. After properly configuring the port for I2C communication, the program begins an infinite loop in which it repeatedly starts and stops the two power function motors that are attached to the receiver.

```
1.  task main() {
2.     SetSensorLowspeed(IN_1);
3.     while (true) {
4.        HTPowerFunctionCommand(IN_1,
5.           HTPF_CHANNEL_1, HTPF_CMD_FWD,
6.           HTPF_CMD_STOP);
7.        Wait(500);
8.        HTPowerFunctionCommand(IN_1,
9.           HTPF_CHANNEL_1, HTPF_CMD_STOP,
10.          HTPF_CMD_FWD);
11.       Wait(500);
12.       HTPowerFunctionCommand(IN_1,
13.          HTPF_CHANNEL_1, HTPF_CMD_STOP,
14.          HTPF_CMD_REV);
15.       Wait(500);
16.       HTPowerFunctionCommand(IN_1,
17.          HTPF_CHANNEL_1, HTPF_CMD_STOP,
18.          HTPF_CMD_STOP);
19.       Wait(500);
20.    }
21. }
```

For communicating with an RCX brick, the NBC and NXC API includes functions that wrap all of the commonly used RCX opcodes. Only a few of them are listed here but you can find complete documentation of all the HiTechnic iRLink RCX functions in the *NXT Programmer's Guide*.

```
int HTRCXPoll(byte src, byte value);
int HTRCXBatteryLevel(void);
void HTRCXPing(void);
void HTRCXDeleteTasks(void);
void HTRCXStopAllTasks(void);
void HTRCXPBTurnOff(void);
void HTRCXDeleteSubs(void);
void HTRCXClearSound(void);
void HTRCXClearMsg(void);
void HTRCXLSCalibrate(void);
void HTRCXMuteSound(void);
void HTRCXUnmuteSound(void);
void HTRCXClearAllEvents(void);
```

The output control functions listed below are modeled after the NQC functions with similar names. For the outputs you can use RCX_OUT_A, RCX_OUT_B, RCX_OUT_C, RCX_OUT_AB, RCX_OUT_AC, RCX_OUT_BC, and RCX_OUT_ABC. The output mode constants are RCX_OUT_FLOAT, RCX_OUT_OFF, and RCX_OUT_ON. The output direction constants are RCX_OUT_REV, RCX_OUT_TOGGLE, and RCX_OUT_FWD.

```
void HTRCXSetOutput(byte outputs, byte mode);
void HTRCXSetDirection(byte outputs, byte dir);
void HTRCXSetPower(byte outputs, byte pwrsrc, byte pwrval)
void HTRCXOn(byte outputs);
void HTRCXOff(byte outputs);
void HTRCXFloat(byte outputs);
void HTRCXToggle(byte outputs);
void HTRCXFwd(byte outputs);
void HTRCXRev(byte outputs);
void HTRCXOnFwd(byte outputs);
void HTRCXOnRev(byte outputs);
void HTRCXOnFor(byte outputs, unsigned int ms);
```

The sample program below demonstrates how to use the HTRCXSetOutput and HTRCXSetDirection API functions. The port is configured for I²C communication first. Then the IRLink port is defined so that we don't have to pass it into each of the subsequent HiTechnic RCX function calls.

```
1. task main() {
2.    SetSensorLowspeed(S1);
3.    // set the IRLink port global
4.    HTRCXSetIRLinkPort(S1);
5.    // turn on OUT_A
6.    HTRCXSetOutput(RCX_OUT_A, RCX_OUT_ON);
7.    repeat (5) {
8.      HTRCXSetDirection(RCX_OUT_A, RCX_OUT_FWD);
9.      Wait(200);
10.     HTRCXSetDirection(RCX_OUT_A, RCX_OUT_REV);
11.     Wait(200);
12.   }
13.   // turn OUT_A off
14.   HTRCXSetOutput(RCX_OUT_A, RCX_OUT_OFF);
15. }
```

In the next chapter we will examine the advanced NXT inputs. We'll learn how to use the legacy RCX sensors and we'll continue exploring the world of I²C devices from the input perspective.

Advanced NXT Inputs

Topics in this Chapter

- Advanced Sensors
- Opening and Reading Files
- Receiving Messages
- Custom Analog Sensors
- Custom I^{2}C

Chapter 13

Now it is time to turn our attention to the advanced NXT inputs. In this chapter we'll look at using the ultrasonic I2C sensor and legacy RCX sensors. We'll examine reading from files and receiving messages. Finally, we'll look at how you can use custom third-party analog and I2C sensors in your NXT programs.

Advanced Sensors

The two types of advanced sensors we'll look at are all made by LEGO. The first is the ultrasonic sensor that comes with the NXT set. It is classed as an advanced sensor because it is not an analog sensor like every other sensor in the set. We'll also examine all of the legacy RCX sensors. Although they are made by LEGO, they do not come with the NXT set. You will have to get a LEGO MINDSTORMS RIS set or purchase these legacy sensors off the Internet if you want to use them with your NXT.

The Ultrasonic Sensor

The ultrasonic sensor acts like both a pair of eyes and a pair of ears. It doesn't actually see anything. Rather, it bounces ultra-high frequency sound waves off distant objects to measure how far away they are. This is how a blind bat can fly around without colliding into things. Likewise, the ultrasonic sensor will help your robots navigate without bumping into obstacles.

To begin our examination of the ultrasonic sensor we will build a simple attachment for the mobile robot platform. Follow the steps outlined below to attach this powerful sensor to your robot.

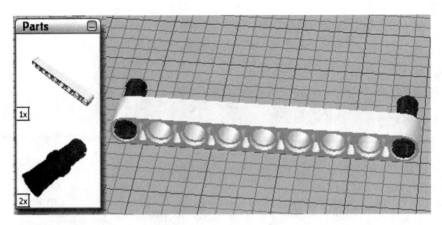

STEP 1. Add parts as shown.

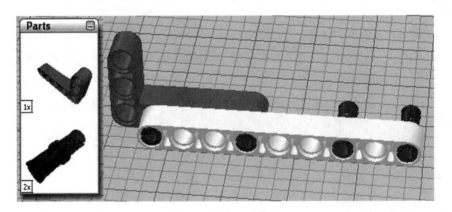

STEP 2. Add parts as shown.

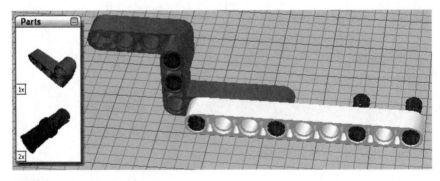

STEP 3. Add parts as shown.

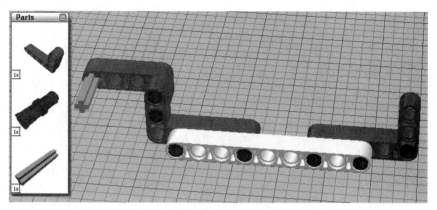

STEP 4. Add parts as shown. The axle is 3 units long.

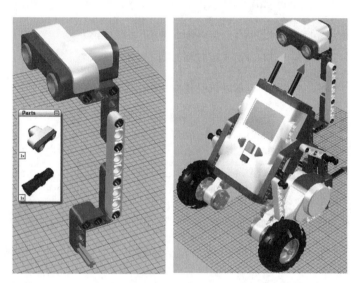

STEP 5. Add parts as shown. Now connect the robot to the back of the mobile platform by pushing the axle into the axle hole at the top of the double bent beam on the left side of the robot.

Reading Distances

You can't just read the value of the ultrasonic sensor using the Input IOMap or the normal NXC sensor API functions and macros. Since it is an I²C digital sensor, reading its value requires that you use the I2CWrite, I2CCheckStatus, and I2CRead API functions or the I2CBytes wrapper routine that we used in the previous chapter.

NXC and NBC also provide an easy to use API function that makes it seem as if the ultrasonic sensor is just like the other sensors. In NXC you can simply use the SensorUS API function. NBC uses ReadSensorUS and it requires a second parameter, which is used to return the ultrasonic sensor's value. NXC uses the API function's return value instead. Both of these functions require that you pass the sensor port as the first parameter.

```
byte SensorUS(byte port); // NXC
ReadSensorUS(byte port, byte & distance) // NBC
```

A power programmer clearly understands the differences between analog sensors and I²C digital sensors like the ultrasonic sensor. All I²C sensors have a device address and a standard set of registers addresses to which you can write data or from which you can read data. The ultrasonic sensor uses 0x02 as its device address. All of this sensor's register addresses are documented in Appendix 7 of the LEGO MINDSTORMS Hardware Developer Kit. The most important register address is 0x42. When you call I2CWrite with an input buffer containing {0x02, 0x42} and a return length equal to 1 then the sensor will respond with the distance value, which you can read using I2CRead.

WEBSITE: *The Hardware Developer Kit is downloadable from the LEGO MINDSTORMS website. http://mindstorms.lego.com/Overview/ NXTreme.aspx*

Ultrasonic Sensor Modes

The ultrasonic sensor allows three different modes. To change the mode, write to register address 0x41. Writing 0x01 to this register address puts the sensor into single shot mode. When the sensor is in this mode it sends out a single ultrasonic ping and records the distances of up to eight different objects. All eight of the distance values can be read in a single transaction by writing to register address 0x42 and requesting 8 bytes. To obtain a new set of measurements just write 0x01 to address 0x41 again.

The normal operating mode is mode 0x02. It is the continuous measurement mode. In this mode the sensor continually sends out ultrasonic pings and updates a single distance value at register address 0x42. You can specify the measurement interval for this mode by writing a value from 0x01 to 0x0f to register address 0x40. This is accomplished using I2CWrite and an input buffer containing {0x02, 0x40, interval} where interval is a byte value from 1 to 15.

The third operating mode is called event capture mode. You can use this mode by writing {0x02, 0x41, 0x03} using I2CWrite. The documentation says that in this mode, the sensor measures whether other ultrasonic sensors are in the area and that this information can be used to determine when to make a measurement so that it doesn't interfere with the other sensors. You can also turn off the ultrasonic sensor by writing {0x02, 0x41, 0x00} to the sensor and reset it by writing {0x02, 0x41, 0x04}.

Most of the I2C sensors available for the NXT follow the standard set by the ultrasonic sensor with respect to providing a set of three register addresses that can be read. These provide you with the sensor version, the product ID and the sensor type. In each case, it returns a string of eight characters. You can read the sensor version by writing {DeviceAddress, 0x00}, the product ID by writing {DeviceAddress, 0x08}, and the sensor type by writing {DeviceAddress, 0x10}. As previously mentioned, the device address of the ultrasonic sensor is 0x02 but there are I2C devices for the NXT that use a different device address. Some even allow you to change the device address programmatically.

Legacy RCX Sensors

All of the LEGO sensors that work with the RCX can be used with the NXT. You will need to purchase a set of adapter wires from LEGO or make your own using the procedure outlined by Phillipe Hurbain.

 WEBSITE: *Instructions for a homemade cable are available on the web. www.philohome.com/nxtcables/nxtcable.htm.*

Since these are analog sensors, they are as easy to use as the basic sensors that we examined two chapters ago. The only difference is the sensor type and sensor mode values that you use when you configure the sensor port. NXC includes a number of NQC compatibility macros, which make configuring and using the legacy sensors a breeze. The simplest approach is to use the SetSensor API function with the sensor configuration constants listed in Table 7-8 in Chapter 7. To manually set the sensor type and sensor mode, use the SetSensorType and SetSensorMode API functions with the constants for the RCX legacy sensors that are listed in Table 7-9 and Table 7-10 in Chapter 7.

The sample program below uses the ultrasonic sensor to help our mobile robot avoid obstacles. It also uses a legacy RCX touch sensor. If you have an RCX touch sensor and an adapter cable you can very easily swap out the NXT touch sensor in our bumper attachment and use the RCX sensor in its place. If not, this program will work exactly the same with the NXT touch sensor.

```
1. #define BUMPER_PORT S1
2. #define BUMPER      SENSOR_1
3.
4. #define BAT_PORT    S4
5. #define BAT_THRESHOLD 30
6.
7. mutex motorMutex;
8.
9. task Move() {
```

```
10.    while(true) {
11.       Acquire(motorMutex);
12.       OnFwdSync(OUT_AC, 75, 0);
13.       Release(motorMutex);
14.       Wait(500);
15.    }
16. }
17.
18. task WatchUltra() {
19.    while (true) {
20.       int dist = SensorUS(BAT_PORT);
21.       if (dist < BAT_THRESHOLD) {
22.          Acquire(motorMutex);
23.          PlayTone(440, 500);
24.          OnRevSync(OUT_AC, 40+Random(60), 0);
25.          Wait(500+Random(500));
26.          OnFwdSync(OUT_AC, 40+Random(60), 100);
27.          Wait(500+Random(1000));
28.          Release(motorMutex);
29.       }
30.    }
31. }
32.
33. task WatchBumper() {
34.    while (true) {
35.       if (BUMPER) {
36.          Acquire(motorMutex);
37.          PlayTone(880, 500);
38.          OnRev(OUT_A, 40+Random(60));
39.          OnRev(OUT_C, 60+Random(40));
40.          Wait(500+Random(1000));
41.          Release(motorMutex);
42.       }
43.    }
44. }
45.
46. task main() {
47.    SetSensorTouch(BUMPER_PORT);
48.    SetSensorLowspeed(BAT_PORT);
49.    Precedes(WatchBumper, WatchUltra, Move);
50. }
\
```

We have added the WatchUltra task to the original wanderer program that we saw two chapters ago. This task performs a similar function to the WatchBumper task. Instead of checking a simple boolean value, though, it compares the distance in millimeters that it reads from the ultrasonic sensor on port four to a configurable threshold. If 30 centimeters is either too sensitive or not sensitive enough, you can change this value on line five. The ultrasonic sensor is at the back of the robot so you will need to allow for the distance from the front of the bumper attachment to the front of the ultrasonic sensor in any adjustments you make.

Since there are two different collision avoidance behaviors, the program also plays a different tone depending on which behavior is activated. Notice that the approach used to turn in the WatchUltra task is different from the one used in WatchBumper. The robot will back straight up for anywhere from half to one full second, then it will spin in place using the turn ratio parameter of 100 on line 26 above.

TRY IT: *If you have some RCX sensors, try incorporating one or more of them into your next NXT robot. They are easy to use in a program and the added flexibility you get from having more sensors at your disposal is a huge bonus.*

Use the ultrasonic sensor in a data-logging experiment of your own design. Keep in mind that each call to SensorUS can take as long as 30 milliseconds to complete due to the characteristics of the ultrasonic sensor.

Experiment with the ultrasonic sensor using the low-level I2C API routines and the different modes of operation.

Opening and reading from files

We learned how to create files and write to them in the previous chapter. Now we will look at how you can use files as an advanced form of input. To open an existing file for reading, use the OpenFileRead API command.

```
int OpenFileRead(string filename, unsigned int & fsize, byte &
handle);
```

When you call OpenFileRead, it returns one of the Loader module status codes we have previously examined. If the file does not exist, you will get back `LDR_FILENOTFOUND`. If the file can be opened, the function returns `LDR_SUCCESS`. Also returned by way of two function arguments are the number of empty bytes in the file and the handle. As before, keep a reference to the handle so that you can use it in the Read routines.

Reading data from a file is done in binary or text mode just like writing. To read binary data from a file, use the Read function. The type of the second parameter determines how many bytes are read from the file. Scalar types are 1, 2, or 4 bytes in length. Structure types each have a unique length depending on the structure definition. The Read API function in NBC and NXC is overloaded so that it can handle almost any variable type. A successful read will have a function return value of `LDR_SUCCESS`. When you reach the end of the file, the Read routines will return `LDR_ENDOFFILE`.

```
int Read(byte handle, <Type> & n);
```

You can also read text files using the ReadLnString API function. This command reads up to the next 0x0D, 0x0A DOS-style line ending pair. The data is returned as a string.

```
int ReadLnString(byte handle, string & output);
```

NBC and NXC also provide a function called ReadBytes that lets you read any kind of data stored in a file into a byte array. You pass the number of bytes that you want to read and a reference to a byte array to store the data. In response, you get the data. It sets the length argument to the number of bytes actually read.

```
int ReadBytes(byte handle, unsigned int & length, byte &
buf[]);
```

NBC and NXC also have a function that uses the basic file IO operations to resize a file. If you have filled up an existing file and want to keep writing to it, you can use the ResizeFile API function to add free space to the file. This actually creates a new larger file, copies the data from the old file, and then deletes the old file.

```
int ResizeFile(string filename, unsigned int newsize);
```

You can also use the low-level syscall functions that are provided by NBC and NXC. To open a file, simply use the SysFileOpenRead API function and pass in a FileOpenType structure. To read from a previously opened file you can use the SysFileRead API function with a FileReadWriteType structure.

```
SysFileOpenRead(FileOpenType & args);
SysFileRead(FileReadWriteType & args);
```

The sample program below demonstrates how you can read a structure containing a game's previous state so that it can be used during this run of the program. When you use the Read and Write API functions with a struct, the size must be a constant. That is why the program uses three char fields for storing the initials of the player with the high score. As a simulation of a real game, each time you run the program it increments the GamesPlayed and LastPlayedLevel fields in the game state structure. It also calculates a random score for the current game. If the score is greater than the current HighScore value, it replaces the high score in the global gState structure and generates a random set of initials.

```
1. #define STATE_FILE      "xyzzy.dat"
2. #define STATE_FILESIZE 8
3.
4. struct GameState {
5.    unsigned int HighScore;
6.    unsigned int GamesPlayed;
7.    char Initial1;
```

```
8.    char Initial2;
9.    char Initial3;
10.   byte LastPlayedLevel;
11. };
12.
13. GameState gState;
14.
15. void InitGameState() {
16.   gState.Initial1 = 'J';
17.   gState.Initial2 = 'C';
18.   gState.Initial3 = 'H';
19.   gState.HighScore = 0;
20.   gState.GamesPlayed = 0;
21.   gState.LastPlayedLevel = 0;
22. }
23.
24. void LoadGameState() {
25.   // if file doesn't exist don't do anything
26.   unsigned int result, fsize;
27.   byte handle;
28.   result = OpenFileRead(STATE_FILE, fsize, handle);
29.   if (result == LDR_SUCCESS) {
30.     Read(handle, gState);
31.     CloseFile(handle);
32.   }
33. }
34.
35. void PlayGame() {
36.   // TODO write game
37.   NumOut(0, LCD_LINE1, gState.GamesPlayed);
38.   NumOut(0, LCD_LINE2, gState.LastPlayedLevel);
39.   NumOut(0, LCD_LINE3, gState.HighScore);
40.   string Initials = "    ";
41.   Initials[0] = gState.Initial1;
42.   Initials[1] = gState.Initial2;
43.   Initials[2] = gState.Initial3;
44.   TextOut(0, LCD_LINE4, Initials);
45.   unsigned int score = Random(1000);
46.   if (score > gState.HighScore) {
47.     gState.HighScore = score;
48.     gState.Initial1 = Random(26)+'A';
49.     gState.Initial2 = Random(26)+'A';
50.     gState.Initial3 = Random(26)+'A';
51.   }
52.   gState.GamesPlayed++;
53.   gState.LastPlayedLevel++;
54.   Wait(1000);
55. }
56.
57. void SaveGameState() {
```

```
58.    // delete and re-create game state file
59.    unsigned int result, fsize;
60.    byte handle;
61.    DeleteFile(STATE_FILE);
62.    result = CreateFile(STATE_FILE, STATE_FILESIZE, handle);
63.    if (result == LDR_SUCCESS) {
64.       Write(handle, gState);
65.       CloseFile(handle);
66.    }
67. }
68.
69. task main() {
70.    InitGameState();
71.    LoadGameState();
72.    PlayGame();
73.    SaveGameState();
74. }
```

Receiving Messages

We saw in the previous chapter that NBC and NXC provide a wealth of capabilities when it comes to sending messages from NXT to NXT, or to other devices. It can use Bluetooth messaging with the NXT communication protocol or with raw unformatted data streams. It can also directly write to outbound USB and Hi-speed port buffers.

On the receiving end, there are fewer API functions but there is no reduction in functionality. The high-level API functions for message reception are all based around the standard NXT communication protocol – specifically the mailbox system and MessageRead direct commands sent via Bluetooth. On a secondary brick, a call to ReceiveMessage simply reads a string from the head of the specified message queue or mailbox. But on a primary brick, if there aren't any messages waiting in the queue, the brick will issue a MessageRead direct command to one of its secondary bricks.

```
int ReceiveMessage(byte queue, bool clear, string & msg);
```

In addition to the ReceiveMessage API function, you have at your disposal four other easy to use routines for retrieving data from the mailbox system. If you want to read a Boolean value from a mailbox, use ReceiveRemoteBool. A 4-byte numeric value can be read using the ReceiveRemoteNumber API function. And for simple text messages, use ReceieveRemoteString. This last function is just another name for the ReceiveMessage routine. The fourth API function is ReceiveRemoteMessageEx. This function acts somewhat like the NXT-G mailbox read block in that it outputs all three types of data.

```
int ReceiveRemoteBool(byte queue, bool clear, bool & bval);
int ReceiveRemoteNumber(byte queue, bool clear, long & val);
int ReceiveRemoteString(byte queue, bool clear, string &
```

```
str);
int ReceiveRemoteMessageEx(byte queue, bool clear, string &
str, long & val, bool & bval);
```

The basic functionality of all these routines is built on top of the MessageRead syscall statement in NBC. You can drop down to this low level if you desire. In NXC this system call is exposed as the SysMessageRead API function. Pass data into and out of the system call via a MessageReadType structure.

```
void SysMessageRead(MessageReadType & args);
```

In the previous chapter we discussed using the Comm module IOMap structure to directly access buffers used for sending and receiving messages to and from other devices. To receive messages, use IOMapRead or IOMapReadByID calls into the appropriate offsets of the Comm module IOMap. The primary low-level routines for receiving raw message data from the Comm module IOMap are GetBTInputBuffer, BTInputBufferInPtr, BTInputBufferOutPtr, SetBTInputBufferInPtr, and SetBTInputBufferOutPtr.

```
void GetBTInputBuffer(byte offset, byte count, byte & data[]);
byte BTInputBufferInPtr(void);
byte BTInputBufferOutPtr(void);
void SetBTInputBufferInPtr(byte n);
void SetBTInputBufferOutPtr(byte n);
```

USB and Hi-speed Input

If you crave more exploration, try reading data directly from the NXT USB and Hi-speed (port 4) input buffers. The API functions that may come in handy are listed below. These and other low-level IOMap-based API functions are documented in the *NXC Programmer's Guide.*

```
void GetUSBInputBuffer(byte offset, byte count, byte &
data[]);
byte USBInputBufferInPtr(void);
byte USBInputBufferOutPtr(void);
void SetUSBInputBufferInPtr(byte n);
void SetUSBInputBufferOutPtr(byte n);
void GetHSInputBuffer(byte offset, byte count, byte & data[]);
byte HSInputBufferInPtr(void);
byte HSInputBufferOutPtr(void);
void SetHSInputBufferInPtr(byte n);
void SetHSInputBufferOutPtr(byte n);
```

Custom Sensors

Advanced NXT inputs are not limited to just I²C devices. There are also third-party analog sensors available for the NXT. We'll have a brief look at two such devices in this section.

HiTechnic Gyro Sensor

The HiTechnic Gyro sensor allows you to measure the rate of rotation about a single axis. The value the sensor returns represents the number of degrees of rotation per second. It is sensitive enough to report as much as +/- 360 degrees rotation per second. It's also fast. You can read values as often as 300 times each second.

NXC and NBC provide an API function for configuring a sensor port for the HiTechnic Gyro and a function for reading the sensor's value. The NBC gyro read function is called ReadSensorHTGyro and it returns the sensor value via an extra result parameter rather than via a function return value.

```
void SetSensorHTGyro(const byte port);
int SensorHTGyro(const byte port, int offset);
```

The offset value that you pass into SensorHTGyro is used to "zero" the sensor. This value should be calculated by sampling the gyro several times while it is completely still. The program below demonstrates how you can use these API functions and the HiTechnic Gyro sensor in your own application. The offset calculation is performed in the for-loop starting on line 6. Ten readings are taken with an offset of zero and then the average reading is calculated from the total.

```
1.  task main() {
2.    SetSensorHTGyro(S1);
3.    // calculate an average offset while
4.    // sensor is perfectly still
5.    int total = 0, offset;
6.    for (int i=0; i<10; i++) {
7.      total += SensorHTGyro(S1, 0);
8.      Wait(10);
9.    }
10.   offset = total / 10;
11.
12.   while (true) {
13.     NumOut(0, LCD_LINE1,
14.       SensorHTGyro(S1, offset));
15.   }
16. }
```

TRY IT: *Starting with Phillipe Hurbain's NXTway robot, which uses a light sensor, see if you can incorporate the HiTechnic Gyro instead. Use the NBC code at www.philohome.com/nxtway/nxtway.htm as inspiration for your own control system to keep the NXTway upright and stable.*

Mindsensors Pressure Sensor

The mindsensors pneumatic pressure sensor for the NXT is another advanced input you can use in your robotic creations. Like other sensors from mindsensors.com, the sensor is not packaged in a fancy shell. The pressure sensor itself has a plastic tube perfectly sized so that you can attach a standard LEGO pneumatic hose to it. The PPS35-Nx sensor can read from zero to 35 PSI.

Based on the official NXT-G block for this sensor, you should configure it in NXC as SENSOR_TYPE_LIGHT. This is the legacy RCX light sensor type. Set the sensor mode to raw. Read the raw sensor value, subtract the raw value from 1024, and then divide the difference by 25. The sample code below demonstrates how you would read this sensor's value in NXC.

```
1. task main() {
2.    SetSensorType(S2, SENSOR_TYPE_LIGHT);
3.    SetSensorMode(S2, SENSOR_MODE_RAW);
4.    ResetSensor(S2);
5.
6.    while (true) {
7.       int pressure = (1024-SensorRaw(S2))/25;
8.       NumOut(0, LCD_LINE4, pressure);
9.       Wait(50);
10.   }
11. }
```

Custom I²C Sensors

Another advanced form of NXT input is the I²C sensor. There are many different I²C sensors available for the NXT. HiTechnic and mindsensors.com are the two main companies that sell I²C sensors but more companies are offering NXT-compatible I²C sensors every day. In this section we will examine how to read various I²C sensors in our NXC programs.

Compass

Both HiTechnic and mindsensors sell a compass sensor. The mindsensors compass is read via a two-byte read at {0x02, 0x42}. The first byte is the least significant byte (LSB) and the second is the most significant byte (MSB). You can take a compass reading using the sample code below.

```
1. #define cmpAddr 0x02;
2. byte bufCmd[]     = {cmpAddr, 0x41, 0x49};
3. byte bufLSWrite[] = {cmpAddr, 0x42};
4.
5. task main() {
6.    SetSensorLowspeed(S2);
7.    I2CWrite(S2, 0, bufCmd);
```

```
8.    string outbuf;
9.    byte count;
10.   while (true) {
11.      byte bLo, bHi;
12.      count = 2;
13.      int value;
14.      if (I2CBytes(S2, bufLSWrite, count, outbuf)) {
15.         bLo = outbuf[0];
16.         bHi = outbuf[1];
17.         value = bHi*256 + bLo;
18.      }
19.      else
20.         value = 0;
21.      NumOut(10, LCD_LINE5, value);
22.   }
23. }
```

With the HiTechnic compass you can just use the SensorHTCompass API function in NXC or the ReadSensorHTCompass function in NBC. These functions encapsulate the I2C read transaction so that you don't need to remember the correct write buffer contents or how to calculate the compass direction using the output buffer contents. The sample program below shows how simple it is to use the SensorHTCompass API function.

```
1. task main() {
2.    int value;
3.    SetSensorLowspeed(S2);
4.    while (true) {
5.       ClearScreen();
6.       value = SensorHTCompass(S2);
7.       NumOut(10, LCD_LINE1, value);
8.       Wait(500);
9.    }
10. }
```

Compass sensors like the ones produced by HiTechnic and mindsensors.com are affected by magnetic fields close to the sensor. The NXT and its motors both generate small magnetic fields that can adversely affect the sensor readings. To lessen effects of the magnetic fields you should try to mount the compass sensor as far away from the NXT and the motors as possible. HiTechnic recommends between four and six inches. Fortunately, both compass sensors provide a calibration mode that lets you configure the sensor so that it can compensate for the local magnetic interference. The program below shows how to calibrate the HiTechnic compass.

```
1. byte buf[] = {0x02, 0x41, 0x43};
2.
3. task main()
4. {
```

```
5.    int result;
6.    SetSensorLowspeed(IN_4);
7.    result = LowspeedWrite(IN_4, 0, buf);
8.
9. // now rotate slowly 1.5 to 2 times,
10. // one turn in about 20 seconds
11.
12.    buf[2] = 0;
13.    result = LowspeedWrite(IN_4, 0, buf);
14.    // done calibrating
15. }
```

TRY IT: *Use a compass sensor to help keep your mobile robot moving in a specific direction.*

HiTechnic & Mindsensors Accelerometer

Both mindsensors and HiTechnic market an I²C acceleration sensor for the NXT. They are both easy to use and can add a great deal of power to your robot. Several projects utilizing these sensors can be found on the Internet.

Phillipe Hurbain used the mindsensors ACCL3x-Nx acceleration sensor in his NXTiimote, which is modeled after the Wiimote. A slightly modified version of the NXC program he wrote for his NXTiimote is listed below. It reads the x and y tilt values from the mindsensors device by writing {0x02, 0x42} and requesting two bytes.

WEBSITE: *For more information about the NXTiimote visit www.philohome.com/nxtiimote/ni.htm.*

At the start of the program at line 13 it checks for a connection to another NXT on connection number one. This is so that when it tries to write the x and y tilt deltas to the other device at lines 33 through 36, the connection has already been verified. If the Bluetooth connection has not been established then the until-loop on line 34 will never exit.

```
1. #define ACCLadr 0x02
2. #define TiltAddr 0x42
3. #define port IN_2
4.
5. byte bufout[]={ACCLadr, TiltAddr};
6. byte bufin[];
7. byte nbytes;
8. int x0, y0;
9. int x,y;
10.
11. task main () {
12.    // check for Bluetooth connection first
```

```
13.    if (BluetoothStatus(1) != NO_ERR) {
14.       TextOut(0, LCD_LINE1, "No connection");
15.       Wait(2000);
16.       Stop(true);
17.    }
18.    SetSensorLowspeed(port);
19.    nbytes = 2;
20.    if (I2CBytes(port, bufout, nbytes, bufin)) {
21.       x0 = bufin[0];
22.       y0 = bufin[1];
23.    }
24.    while (true) {
25.       nbytes = 2;
26.       if (I2CBytes(port, bufout, nbytes, bufin)) {
27.          x=x0-bufin[0];
28.          y=bufin[1]-y0;
29.          TextOut (0, 16, "y-tilt" , true);
30.          NumOut (20, 8, y);
31.          TextOut (0, 40, "x-tilt");
32.          NumOut (20, 32, x);
33.          SendRemoteNumber(1, 0, x);
34.          until(BluetoothStatus(1)==NO_ERR);
35.          SendRemoteNumber(1, 1, y);
36.          until(BluetoothStatus(1)==NO_ERR);
37.       }
38.    }
39. }
```

Using the HiTechnic acceleration sensor is even easier since NBC and NXC have an API function designed just for this device. After configuring the sensor using SetSensorLowspeed, simply call the ReadSensorHTAccel API function. Like other API functions, in NBC the boolean function return value is accessed via an additional result parameter.

```
bool ReadSensorHTAccel(const byte port, int & x, int & y, int
& z);
```

The sample program below demonstrates how easy it is to add the power of a HiTechnic acceleration sensor to your next project. If the function returns true then the values of x, y, and z are okay to use. In this simple example the values are written to the LCD screen. You could log these values to a file as part of a data logging experiment or use the data to help you make crucial decisions about controlling the NXT motors. You can output the values as part of a Bluetooth message to another device. Use your imagination and come up with a creative project that you can build to try out this great sensor.

```
1. task main() {
2.    int x, y, z;
3.    SetSensorLowspeed(S2);
```

```
 4.    while (true) {
 5.       if (ReadSensorHTAccel(S2, x, y, z)) {
 6.          NumOut(0, LCD_LINE1, x);
 7.          NumOut(0, LCD_LINE2, y);
 8.          NumOut(0, LCD_LINE3, z);
 9.       }
10.    }
11. }
```

HiTechnic Color Sensor

The HiTechnic color sensor is another advanced input you may want to consider adding to your next project. Like the other sensors from HiTechnic it is reliable, easy to operate, and looks great. HiTechnic also offers great customer service in case you have any problems. Following the standard first established by the LEGO ultrasonic sensor, the HiTechnic color sensor uses a device address of 0x02 and the start of sensor data is at register address 0x42. With NBC and NXC, however, you don't need to worry about the low-level I2C details. Just use the ReadSensorHTColor API function instead.

```
bool ReadSensorHTColor(const byte port, byte & ColorNum, byte
& Red, byte & Green, byte & Blue);
```

The sample program below demonstrates how easy it is to use the HiTechnic color sensor in an NXC program. If the ReadSensorHTColor function returns true then the color number and the red, green, and blue values are valid. As with our previous sample we are simply writing the color data to the LCD screen. The color sensor extends the possibilities for robotic vision far beyond the capabilities provided by the analog light sensor that comes with the NXT set.

```
 1. task main() {
 2.    int c, r, g, b;
 3.    SetSensorLowspeed(S2);
 4.    while (true) {
 5.       if (ReadSensorHTColor(S2, c, r, g, b)) {
 6.          NumOut(0, LCD_LINE1, c);
 7.          NumOut(0, LCD_LINE2, r);
 8.          NumOut(0, LCD_LINE3, g);
 9.          NumOut(0, LCD_LINE4, b);
10.       }
11.    }
12. }
```

The red, green, and blue values range from 0 to 255, while color numbers range from 0 to 17. The color sensor comes with printed documentation that shows what colors are associated with which color numbers.

> **TRY IT:** *See if you can use the HiTechnic color sensor to build a simple brick sorter. Have the robot drop bricks into four different containers based on the brick's color.*
>
> *Build a simple photo scanner that writes the red, green, and blue color values to a file as it passes the color sensor back and forth across the image.*

HiTechnic IRSeeker

HiTechnic also markets a device that detects infrared transmissions and returns direction data that you can use to adjust the robot's course. The IR Seeker sensor reports a direction value from 0 to 9 where zero means that no IR was detected. When the sensor value is equal to five it means that the IR is directly in front of the sensor. In NXC and NBC you can read the six bytes of sensor data using the ReadSensorHTIRSeeker API function.

```
bool ReadSensorHTIRSeeker(const byte port, byte & dir, byte
& s1, byte & s3, byte & s5, byte & s7, byte & s9)
```

The five sensor element values can be used to calculate approximately how far away the IR source is. If the IR source is coming from a direction that is between two of the five sensors you will need to average the two adjacent values to estimate the distance to the source. The sample below demonstrates calling the ReadSensorHTIRSeeker function and averaging the relative signal strength values.

```
1.  task main() {
2.     int dir, se1, se2, se3, se4;
3.     int se5, se6, se7, se8, se9;
4.     SetSensorLowspeed(S2);
5.     while (true) {
6.        if (ReadSensorHTIRSeeker(S2, dir, se1, se3, se5, se7,
   se9)) {
7.           se2 = (se1+se3)/2;
8.           se4 = (se3+se5)/2;
9.           se6 = (se5+se7)/2;
10.          se8 = (se7+se9)/2;
11.          int deviation = 5-dir;
12.          if (dir > 5) {
13.             // turn to the right
14.          }
15.          else if (dir < 5) {
16.             // turn to the left
17.          }
18.          // go straight
19.       }
20.    }
21. }
```

TRY IT: *Use the IRSeeker with our mobile robot platform and your television IR remote. See if you can write a simple program in NXC that makes the robot follow you around the room as you press buttons on the remote*

Mindsensors DIST-Nx Distance Sensor

The mindsensors DIST-Nx sensor is an optical distance sensor that can measure the distance to an obstacle. There are different models of this sensor available from mindsensors.com. Depending on the model, the sensor can measure accurate distances from a few millimeters to a few meters. The code sample below demonstrates how you can properly configure and use this sensor in your own NXC programs.

```
1.  byte cmdBufGP2D120[] = {0x02, 0x41, 0x32};
2.  byte cmdBufEnergize[] = {0x02, 0x41, 0x45};
3.  byte cmdBufRead[] = {0x02, 0x42};
4.
5.  task main() {
6.    SetSensorLowspeed(S2);
7.    I2CWrite(S2, 0, cmdBufGP2D120);
8.    int status = I2CCheckStatus(S2);
9.    while (status > NO_ERR)
10.     status = I2CCheckStatus(S2);
11.   Stop(status < NO_ERR);
12.
13.   I2CWrite(S2, 0, cmdBufEnergize);
14.   status = I2CCheckStatus(S2);
15.   while (status > NO_ERR)
16.     status = I2CCheckStatus(S2);
17.   Stop(status < NO_ERR);
18.
19.   while (true) {
20.     byte bLo, bHi;
21.     byte count = 2;
22.     string outBuf;
23.     int value;
24.     if (I2CBytes(S2, cmdBufRead, count, outBuf)) {
25.       value = outBuf[1]*256 + outBuf[2];
26.     }
27.     else {
28.       value = 0;
29.     }
30.     NumOut(0, LCD_LINE1, value);
31.   }
32. }
```

Mindsensors RTC-Nx Real-time Clock

The mindsensors RTC-Nx sensor is real-time clock for the NXT that can be used to keep track of dates, times, and the day of week. The clock's date and time information is stored in binary coded decimal (BCD) within the sensor's registers. After retrieving the eight bytes of data from the real-time clock we have to decode each byte into a decimal value. The RTC device address is hard-coded as 0xD0. Have a look at the sample code below in order to see how easy it is to configure and use this sensor.

```
1.  #define REG_SEC  0
2.  #define REG_MIN  1
3.  #define REG_HRS  2
4.  #define REG_DOW  3
5.  #define REG_DATE 4
6.  #define REG_MON  5
7.  #define REG_YEAR 6
8.
9.  byte cmdRTC[] = {0xD0, 0x00};
10.
11. byte bcd2dec(byte bcd) {
12.   byte decTens = bcd / 16;
13.   byte decOnes = bcd % 16;
14.   return (decTens*10)+decOnes;
15. }
16.
17. string BuildStr(byte b1, byte b2, byte b3, string sep) {
18.    string result = NumToStr(b1);
19.    result += sep;
20.    result += NumToStr(b2);
21.    result += sep;
22.    result += NumToStr(b3);
23.    return result;
24. }
25.
26. task main() {
27.    SetSensorLowspeed(S2);
28.
29.    while (true) {
30.      byte count = 8;
31.      byte outBuf[];
32.      string dateStr, timeStr;
33.      byte dow;
34.      if (I2CBytes(S2, cmdRTC, count, outBuf)) {
35.        // eight bytes of data
36.        // ss mm hh dow dd MMM yy xx
37.        byte ss  = bcd2dec(outBuf[REG_SEC]);
38.        byte mm  = bcd2dec(outBuf[REG_MIN]);
39.        byte hh  = bcd2dec(outBuf[REG_HRS]);
40.        dow      = bcd2dec(outBuf[REG_DOW]);
```

```
41.        byte dd  = bcd2dec(outBuf[REG_DATE]);
42.        byte MMM = bcd2dec(outBuf[REG_MON]);
43.        byte yy  = bcd2dec(outBuf[REG_YEAR]);
44.        dateStr  = BuildStr(dd, MMM, yy, "//");
45.        timeStr  = BuildStr(hh, mm, ss, ":");
46.      }
47.      else {
48.        dateStr = "";
49.        timeStr = "";
50.        dow     = 0;
51.      }
52.      TextOut(0, LCD_LINE1, dateStr);
53.      TextOut(0, LCD_LINE2, timeStr);
54.      NumOut(0, LCD_LINE3, dow);
55.    }
56. }
```

Intruder Alert

Topics in this Chapter

- Building a Bedroom Security System

Chapter 14

A teenager's bedroom is a place of refuge. It's a sanctuary to listen to MP3's in peace or play video games without being bothered by mom or little sister. But in spite of the keep-out signs, someone's always barging in for no good reason. And when the room is vacant, there's a good chance a nosey brother will go snooping around where he shouldn't. A teenager needs a way to keep intruders at bay – to protect the borders of a private space. In this chapter, we'll build and program a simple, yet effective, device to help keep your room safe whether you are there or not.

Building a security system

Let's get started by building the platform and turret, and then we'll attach three sensors at the end. The security system uses two motors and the bulk of the parts in the NXT set.

The Platform and Turret

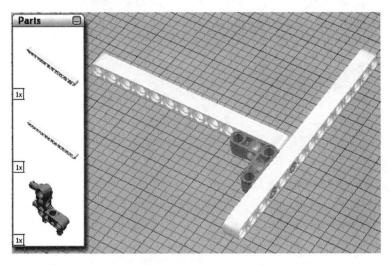

STEP 1. Add parts as shown.

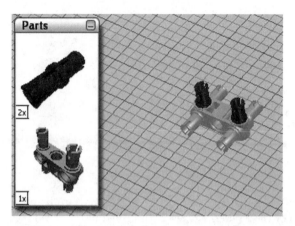

STEP 2. Add parts as shown.

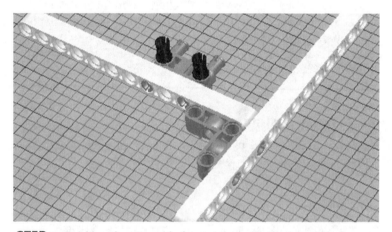

STEP 3. Combine the assembly from step 2 with the assembly from step 1 as shown.

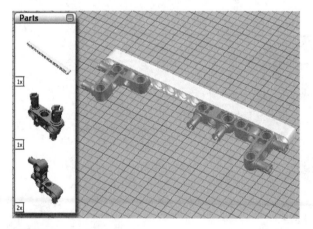

STEP 4. Add parts as shown.

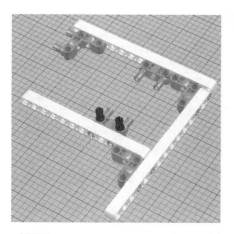

STEP 5. Combine the assembly from step 4 with the assembly from step 3 as shown.

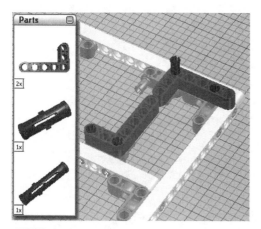

STEP 6. Add parts as shown.

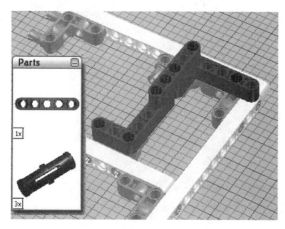

STEP 7. Add parts as shown.

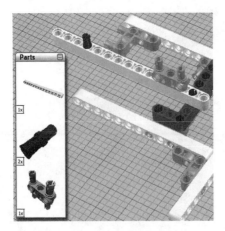

STEP 8. Add parts as shown.

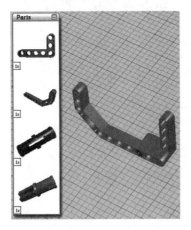

STEP 9. Add parts as shown.

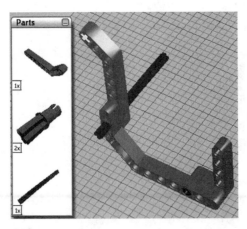

STEP 10. Connect the bent beam using two blue pins. The axle is 8 units long.

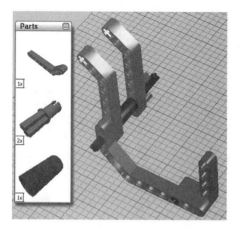

STEP 11. Add parts as shown.

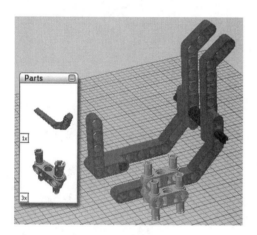

STEP 12. Add parts as shown.

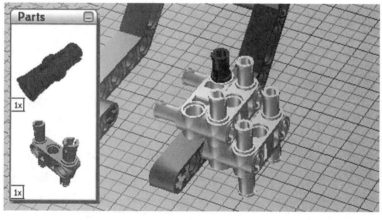

STEP 13. Add parts as shown.

STEP 14. Combine the assembly from step 13 with the assembly from step 8 as shown.

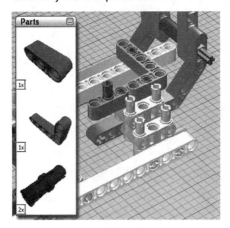

STEP 15. Add parts as shown.

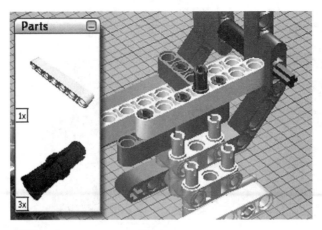

STEP 16. Add parts as shown.

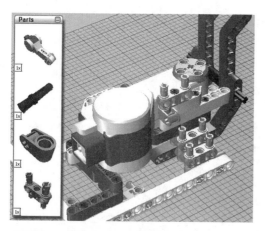

STEP 17. Combine the assembly from step 13 with the assembly from step 8 as shown.

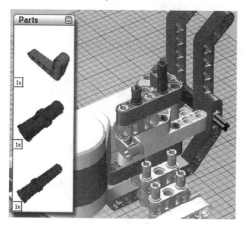

STEP 18. Add parts as shown.

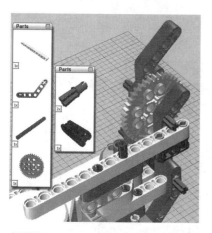

STEP 19. Add parts as shown. The axle is 6 units long. The bent beam attaches to the axle beside the large gear.

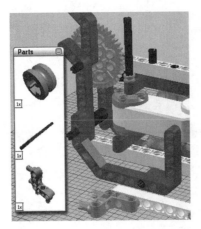

STEP 20. Add parts as shown. The axle is 10 units long.

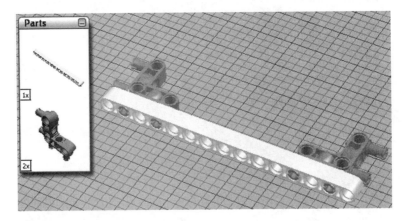

STEP 21. Add parts as shown.

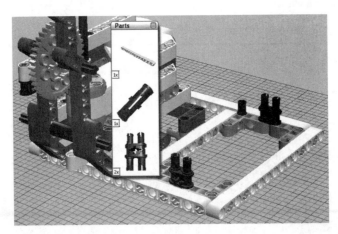

STEP 22. Combine the assembly from step 21 with the assembly from step 20 as shown.

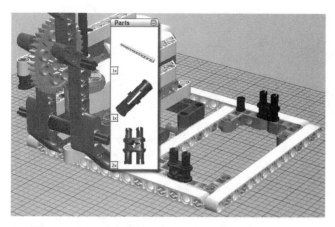

STEP 23. Add parts as shown.

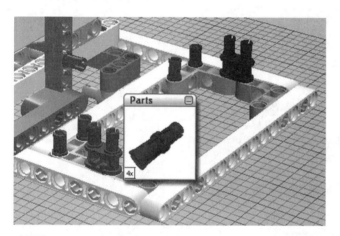

STEP 24. Add parts as shown.

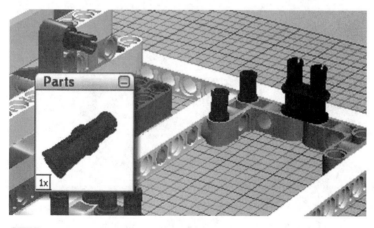

STEP 25. Add parts as shown.

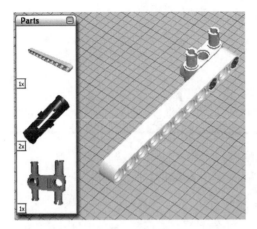

STEP 26. Add parts as shown.

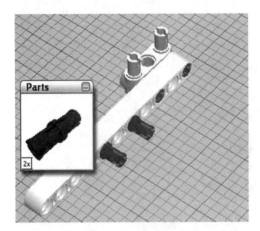

STEP 27. Add parts as shown.

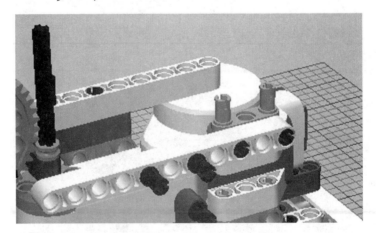

STEP 28. Combine the assembly from step 27 with the assembly from step 25 as shown.

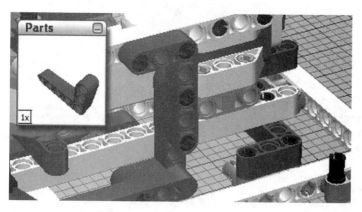

STEP 29. Add parts as shown.

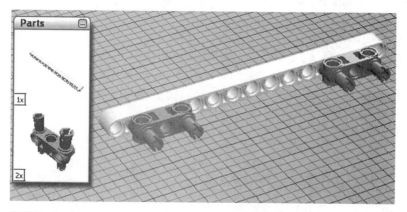

STEP 30. Add parts as shown.

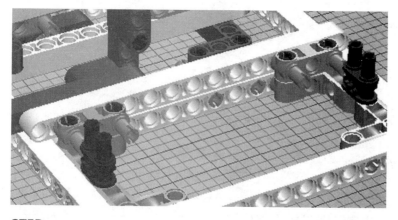

STEP 31. Combine the assembly from step 30 with the assembly from step 29 as shown.

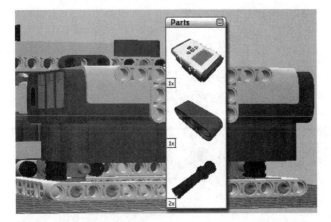

STEP 32. Add parts as shown. Step 39 shows the position of the NXT brick from an above view.

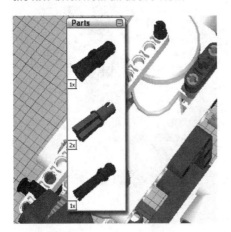

STEP 33. Add parts as shown.

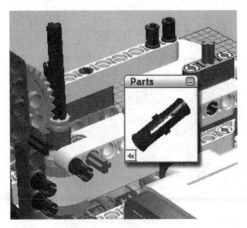

STEP 34. Add parts as shown.

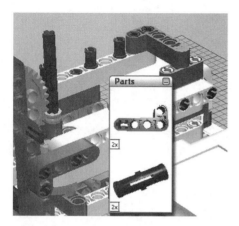

STEP 35. Add parts as shown.

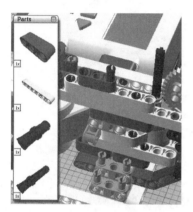

STEP 36. Add parts as shown. The double pin is on the right.

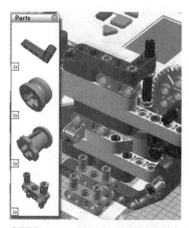

STEP 37. Add parts as shown. Before adding the half-bush to the axle, slide a worm gear (not pictured) onto the axle so that it meshes with the 40-tooth gear.

STEP 39. Now we have finished the platform and turret.

A Ball Launcher

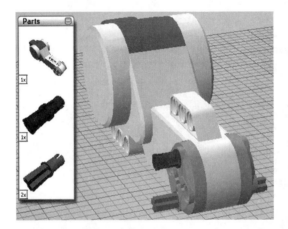

STEP 40. Add parts as shown.

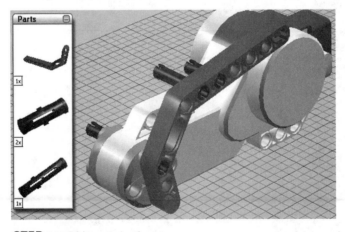

STEP 41. Add parts as shown.

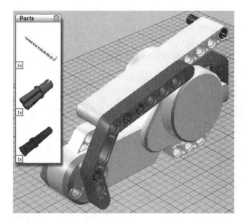

STEP 42. Add parts as shown.

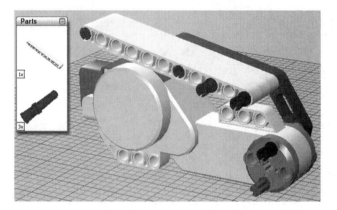

STEP 43. Add parts as shown.

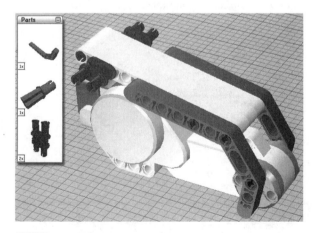

STEP 44. Add parts as shown.

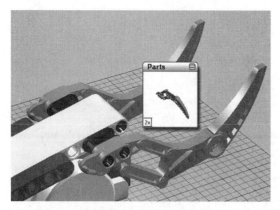

STEP 45. Add parts as shown.

STEP 46. Using a 3-unit long gray axle and a 5-unit long gray axle, mount the launcher to the bent beam attached to the 40-tooth gear. Use two full bush pieces to lock the longer axle in place.

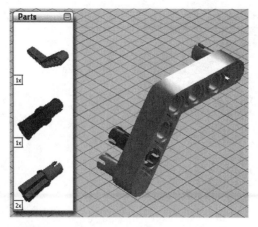

STEP 47. Add parts as shown..

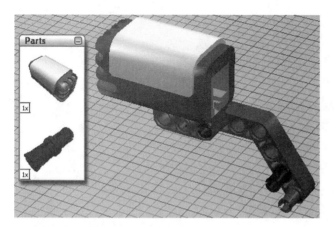

STEP 48. Add parts as shown.

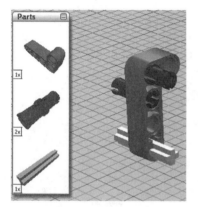

STEP 49. Add parts as shown.

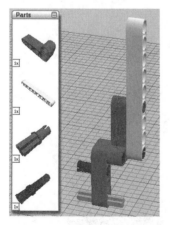

STEP 50. Add parts as shown.

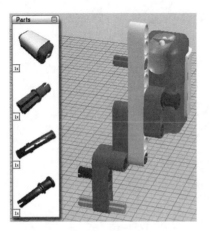

STEP 51. Add parts as shown.

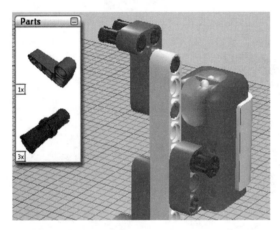

STEP 52. Add parts as shown.

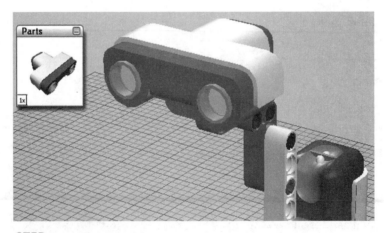

STEP 53. Add parts as shown.

Putting it all together

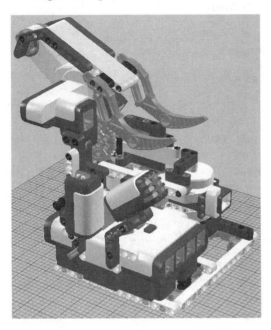

STEP 54. Combine the assembly from step 48 and the assembly from step 53 with the assembly from step 46 as shown. Place one of the large plastic balls in the launcher. This completes the security system.

The ultrasonic sensor connects to port 4. It attaches vertically to two of the three holes on the side of the NXT near the top using the black pin and the gray axle. The sound sensor connects to port 3. It attaches to two of the three vertical holes on the side of the NXT near the sensor ports. The light sensor connects to port 2. Finally, connect the cannon motor to output B and the turret motor to output C.

Security System

Our program runs in a loop, continually checking to see if any of the three sensors detect an abnormal state. If any of them notice something out of the ordinary, an indicator is tripped that tells the alarm system an intruder is detected. The alarm is raised and the indicator is reset so that after the alarm is completed the system returns to the scanning mode.

The program implements a very simple state machine where you are either in the alarm state or the scanning state. It is important that it doesn't sound the alarm at the same time as it checks for intruders. That would cause the alarm to sound continually because of the speaker sound detected rather than because of an intruder still present.

Let's have a look at the NXC source code, starting with the main task.

```
 1. task main() {
 2.    InitializeSecuritySystem();
 3.    start lightWatcher;
 4.    start soundWatcher;
 5.    start distanceWatcher;
 6.    while (true) {
 7.      until(gIntruderPresent);
 8.      start elevateCannon;
 9.      start fireCannon;
10.    start soundAlarm;
11.    gIntruderPresent = false;
12.  }
13.}
```

The program performs some initialization each time it runs. It must configure the sensors to the correct types and determine average sensor readings for the distance, sound, and light sensors. The main task then starts three sensor watcher tasks and enters an infinite loop where it waits until one of the watcher tasks shifts the state machine into the "intruder is present" state. Once in that state, the main task starts three tasks for the alarm. Then the task shifts the state machine back into the "intruder is not present" state and continues looping.

Let's look at the initialization code. This code includes the functions for configuring our sensors and calculating an average sensor reading. It also includes all of our preprocessor definitions, global constants, and global variables, which are at the very beginning of the program.

It's a good idea to create meaningful names for sensor and motor ports. Use them instead of the port number or letter constants. If you reconfigure the system you will only need to change the port setting in one place rather than throughout your code.

```
 1. #define lightPort   S3
 2. #define lightLevel  SENSOR_3
 3. #define soundPort   S2
 4. #define soundLevel  SENSOR_2
 5. #define distPort    S4
 6. #define distValue   SensorUS(distPort)
 7.
 8. #define cannonPort   OUT_B
 9. #define turretPort   OUT_C
10.
11. #define LIGHT     0
12. #define SOUND     1
13. #define DISTANCE  2
14.
15. const int nSamples        = 10;
16. const int nLightThreshold = 200;
17. const int nSoundThreshold = 200;
18. const int nDistThreshold  = 50;
19.
```

```
20.  // global variables
21.  int gAverageLightLevel;
22.  int gAverageSoundLevel;
23.  int gAverageDistValue;
24.
25.  bool gIntruderPresent = FALSE;
26.
27.  void SetSensorLightInactive(const int port) {
28.     SetSensorType(port, IN_TYPE_LIGHT_INACTIVE);
29.     SetSensorMode(port, IN_MODE_RAW);
30.     ResetSensor(port);
31.  }
32.
33.  void SetSensorSoundDBA(const int port) {
34.     SetSensorType(port, IN_TYPE_SOUND_DBA);
35.     SetSensorMode(port, IN_MODE_RAW);
36.     ResetSensor(port);
37.  }
38.
39.  int CalcAverage(const int x) {
40.     int sensorTotal = 0;
41.     for(int i=0; i<nSamples; i++) {
42.       if (x == LIGHT)
43.         sensorTotal += lightLevel;
44.       else if (x == SOUND)
45.         sensorTotal += soundLevel;
46.       else
47.         sensorTotal += distValue;
48.       Wait(50);
49.     }
50.     return sensorTotal / nSamples;
51.  }
52.
53.  void InitializeSecuritySystem() {
54.     // configure our sensors
55.     SetSensorLightInactive(lightPort);
56.     SetSensorSoundDBA(soundPort);
57.     SetSensorLowspeed(distPort);
58.     // initialize average sensor values
59.     gAverageLightLevel = CalcAverage(LIGHT);
60.     gAverageSoundLevel = CalcAverage(SOUND);
61.     gAverageDistValue  = CalcAverage(DISTANCE);
62.  }
```

Lines 16, 17, and 18 define threshold constants for each of the three sensors. You will need to tweak these values to adjust the sensitivity of the security system. If you set them too low, the alarm may go off too frequently. Setting too high a threshold will make the system unresponsive to real intrusions.

The three sensor watcher tasks are all very simple. They loop forever, reading the current sensor value and comparing it to the average, plus or minus the sensor threshold value. If the threshold is exceeded then the watcher shifts the state machine into the "intruder is present state" and then waits for ten seconds.

```
1.  task lightWatcher() {
2.     while(true) {
3.         if (lightLevel > gAverageLightLevel+ nLightThreshold) {
4.             gIntruderPresent = true;
5.             Wait(10000);
6.         }
7.     }
8.  }
9.
10. task soundWatcher() {
11.     while(true) {
12.         if (soundLevel > gAverageSoundLevel+ nSoundThreshold) {
13.             gIntruderPresent = true;
14.             Wait(10000);
15.         }
16.     }
17. }
18.
19. task distanceWatcher() {
20.     while(true) {
21.         if (distValue < gAverageDistValue- nDistThreshold) {
22.             gIntruderPresent = true;
23.             Wait(10000);
24.         }
25.     }
26. }
```

The alarm tasks could have been written as subroutines but the goal is for all three things to happen at the same time rather than sequentially. By making them tasks that do not run forever, they will all run at the same time and then simply end.

```
1.  task fireCannon() {
2.     RotateMotor(cannonPort, 100, 135);
3.     RotateMotor(cannonPort, 100, -135);
4.  }
5.
6.  task elevateCannon() {
7.     RotateMotor(turretPort, 80,  1080);
8.     Wait(1000);
9.     RotateMotor(turretPort, 80, -1080);
10. }
11.
12. task soundAlarm() {
13.     // play sound file and tones
```

```
14.     PlayFile("Shout.rso");
15.     Wait(2000);
16.     PlayFile("Stop.rso");
17.  }
```

When you put the code together, place the main task at the end of the file. Since it refers to each of the other tasks, they need to be defined before the main task. Also make sure that when you start the program the turret is lowered so that the ball launcher is roughly horizontal. Copy the "Shout.rso" and "Stop.rso" sound files to the NXT using BricxCC's NXT Explorer, the LEGO MINDSTORMS NXT software, or any one of the downloading tools for Mac and Linux platforms.

Get your parents' permission before you start launching plastic balls at bedroom intruders! Otherwise you may end up with worse problems than unwanted guests in your room. If they don't mind you running the launcher, you will need to keep your batteries fully charged so that the motor has enough strength to launch a ball.

 TRY IT: *Add the ability to log intrusions to a file on the NXT. New log entries should be appended to the existing intrusion log.*

Check out mindsensors' real time clock I²C device. It gives you the ability to read the current date and time. Try using this device to log not only a generic text message when an intruder enters your room but the exact date and time that it occurs. The ReadSensorMSRTClock NXC API function returns values that you can use to create a date-time string. You can also schedule your intruder alarm to scan at certain times of the day.

FIGURE 14-1. Realtime clock I²C device

 WEBSITE: *Find out more information about the Realtime Clock for the NXT and how to order it at www.mindsensors.com. Click on the NXT Sensors & Interfaces link and then click on the Realtime Clock link.*

Games People Play

Topics in this Chapter

- Tic-Tac-Toe
- Nxt-o-sketch
- Pong

Chapter 15

L ike many of us, I've always enjoyed playing electronic games. I fondly recall playing simple Atari 2600 games that hooked up to a television set when I was a kid. I'd drop a Rush album onto a record player and play breakout or pong for hours while listening to the music. Now it's a Nintendo Wii and an iPod, but *plus ca change, plus c'est la meme chose.* The more things change, the more they stay the same. In this chapter we'll look at the source code for three simple games for your NXT.

Tic-Tac-Toe

We'll get started with the simple game of tic-tac-toe. The NXT buttons are the only inputs, while the LCD screen and the sound system are the only outputs. The computer is your opponent and he's fairly smart. You will need to play carefully if you want to beat him. Ross Crawford wrote the original version of this tic-tac-toe game for the NXT using NBC.

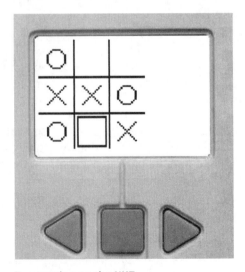

Figure 15-1. Tic-Tac-Toe running on the NXT

WEBSITE: *Visit* http://www.br-eng.info/My-MOCs/Current/ *for more of Ross Crawford's excellent LEGO projects.*

Here is the main task from the original code.

```
1.  task main() {
2.    while (true) {
3.      InitBoard();
4.      while (!bGameOver) {
5.        GetUserMove();
6.        CalcMove();
7.      }
8.      DrawBoard();
9.      TextOut(65, LCD_LINE2, "Game");
10.     TextOut(65, LCD_LINE3, "Over!");
11.     TextOut(65, LCD_LINE5, sResult);
12.     TextOut(65, LCD_LINE6, "Win!");
13.     Wait(3000);
14.     ClearScreen();
15.     TextOut(10, LCD_LINE5, "Another game?");
16.     TextOut( 0, LCD_LINE8, "Yes");
17.     TextOut(80, LCD_LINE8, "No");
18.     do {
19.       WaitButton();
20.       if (pressedButton == BTNLEFT)
21.         break;
22.     } while (pressedButton != BTNRIGHT);
23.     if (pressedButton == BTNRIGHT)
24.       break;
25.   }
26.   TextOut(20, LCD_LINE5, "Goodbye!", true);
27.   Wait(1500);
28. }
```

The main task contains three loops. The primary loop runs until you press the right arrow button when asked if you want to play another game. The main game loop starts on line four after the call to InitBoard. Once a single game has ended, the final do-while loop lets you choose whether to start a new game or end the program.

At the beginning of the source file, there are several global variables and preprocessor declarations. The Board array stores the moves you make along with the moves the computer makes. Initially the board is filled with zeros indicating empty spaces. Your moves are recorded as positive one in the array, while the NXT moves are negative one.

```
1.  #define USER_MOVE 1
2.  #define COMPUTER_MOVE -1
3.  #define ScoreForUserWin 3
4.  #define ScoreForComputerWin -2
5.  #define ScoreForUserBlock 2
6.  #define DX 2
7.  #define DY 2
```

```
 8. #define MAX_MOVES 9
 9.
10. byte ScoringRows[] = {0,1,2, 3,4,5,
11.                        6,7,8, 0,3,6,
12.                        1,4,7, 2,5,8,
13.                        0,4,8, 2,4,6};
14. byte Next[] = {2,3,4,1,1,5,8,7,6};
15. byte Order[] = {4,0,1,2,5,8,7,6,3};
16. string sResult;
17. byte Board[];
18. char Move;
19. byte pressedButton;
20. bool bGameOver;
21. char dbX, dbY;
```

The InitBoard routine simply fills the Board with zeros and sets the game-over boolean variable to false. Following that function is the GetDeltas function, which converts the Move position from a value that ranges from 0 through 8 into a pair of X and Y coordinates indicating the physical board position on the NXT screen. Another helper function is the WaitButton function. It waits for a button press and release. It then sets the global pressedButton variable to the button index of whatever button you press. It gives you some feedback as you press the buttons by playing a brief tone.

```
 1. void InitBoard() {
 2.   ArrayInit(Board, 0, MAX_MOVES);
 3.   bGameOver = false;
 4. }
 5.
 6. void GetDeltas(const char i) {
 7.   char temp;
 8.   dbY  = i / 3;
 9.   temp = dbY * 3;
10.   dbX  = i - temp;
11.   dbX *= 20;
12.   dbY *= 20;
13. }
14.
15. void WaitButton() {
16.   byte wbIndex = BTNRIGHT;
17.   while (wbIndex <= BTNCENTER) {
18.     if (ButtonPressed(wbIndex, FALSE) != 0)
19.       break;
20.     wbIndex++;
21.     if (wbIndex > BTNCENTER)
22.       wbIndex = BTNRIGHT;
23.   }
24.   until(!ButtonPressed(wbIndex, FALSE));
25.   pressedButton = wbIndex;
26.   PlayToneEx(440, 50, 1, false);
27. }
```

At first the DrawBoard routine draws the standard tic-tac-toe grid on the screen. It then iterates through all the Board positions and draws an X or an O if the Board array contains a non-zero value. A positive value draws an X while a negative value draws an O.

```
1.  void DrawBoard() {
2.     byte dbX1, dbX2, dbY1, dbY2;
3.     char i, dbv;
4.     ClearScreen();
5.     LineOut( 0+DX, 20+DY, 60+DX, 20+DY);
6.     LineOut( 0+DX, 40+DY, 60+DX, 40+DY);
7.     LineOut(20+DX,  0+DY, 20+DX, 60+DY);
8.     LineOut(40+DX,  0+DY, 40+DX, 60+DY);
9.     i = 0;
10.    while (i < MAX_MOVES) {
11.       GetDeltas(i);
12.       dbv = Board[i];
13.       if (dbv != 0) {
14.          if (dbv > 0) {
15.             dbX1 = dbX+5+DX;
16.             dbY1 = dbY+5+DY;
17.             dbX2 = dbX1+10;
18.             dbY2 = dbY1+10;
19.             LineOut(dbX1, dbY1, dbX2, dbY2);
20.             LineOut(dbX2, dbY1, dbX1, dbY2);
21.          }
22.          else {
23.             dbX1 = dbX+10+DX;
24.             dbY1 = dbY+10+DY;
25.             CircleOut(dbX1, dbY1, 5);
26.          }
27.       }
28.       i++;
29.    }
30. }
```

The next function is the longest of them all. It is the routine that the NXT uses to score the current game state and decide the best move. There are eight possible win states. The CalcMove function examines each state and calculates a score along those rows, columns, and diagonals. If the score equals 3, it knows you have won the game. If the score is -2, it means there are two NXT moves in a row, column, or diagonal and an empty space in the same line. That means the NXT has won the game. If the score is 2 then the NXT knows it has to block in the empty space so that you do not win.

```
1. void CalcMove() {
2.    char i, j, k;
3.    char v, t;
4.    char cms[];
5.
```

```
 6.    ArrayInit(cms, 0, 8);
 7.    i = 0;
 8.    j = 0;
 9.    // first check to see if the user has won
10.    while (i < 8) {
11.      v = ScoringRows[j];
12.      t = Board[v];
13.      j++;
14.      v = ScoringRows[j];
15.      v = Board[v];
16.      t += v;
17.      j++;
18.      v = ScoringRows[j];
19.      v = Board[v];
20.      t += v;
21.      cms[i] = t;
22.      if (t == ScoreForUserWin) {
23.        bGameOver= true;
24.        sResult = "You" ;
25.        Move = -1;
26.        return;
27.      }
28.      j++;
29.    i++;
30.  }
31.
32.  // now check to see if the computer has won
33.  i = 0;
34.  while (i < 8) {
35.    t = cms[i];
36.    if (t == ScoreForComputerWin) {
37.      bGameOver= true;
38.      sResult = "I" ;
39.      Move = -1;
40.      return;
41.    }
42.    i++;
43.  }
44.
45.  // nobody has won yet so
46.  // pick the next computer move
47.  // check if we need to block a user win
48.  i = 0;
49.  while (i < 8) {
50.    t = cms[i];
51.    if (t == ScoreForUserBlock) {
52.      j = i*3;
53.      i = ScoringRows[j];
54.      v = Board[i];
55.      if (v == 0) {
```

```
56.        Move = i;
57.        Board[Move] = COMPUTER_MOVE;
58.      }
59.      else {
60.        j++;
61.        i = ScoringRows[j];
62.        v = Board[i];
63.        if (v == 0) {
64.          Move = i;
65.          Board[Move] = COMPUTER_MOVE;
66.        }
67.        else {
68.          j++;
69.          i = ScoringRows[j];
70.          Move = i;
71.          Board[Move] = COMPUTER_MOVE;
72.        }
73.      }
74.      return;
75.    }
76.    i++;
77.  }
78.  // Always check centre square first
79.  if (Board[4] == 0) {
80.    Move = 4;
81.    Board[4] = COMPUTER_MOVE;
82.    return;
83.  }
84.  // Then try corners
85.  if (Board[0] == 0) {
86.    Move = 0;
87.    Board[0] = COMPUTER_MOVE;
88.    return;
89.  }
90.  if (Board[2] == 0) {
91.    Move = 2;
92.    Board[2] = COMPUTER_MOVE;
93.    return;
94.  }
95.  if (Board[6] == 0) {
96.    Move = 6;
97.    Board[6] = COMPUTER_MOVE;
98.    return;
99.  }
100. if (Board[8] == 0) {
101.   Move = 8;
102.   Board[8] = COMPUTER_MOVE;
103.   return;
104. }
105. // Then others in order
```

```
106.   i = Next[Move];
107.   k = i;
108.   while (true) {
109.      j = Order[i];
110.      v = Board[j];
111.      if (v == 0) {
112.         Move = j;
113.         Board[Move] = COMPUTER_MOVE;
114.         return;
115.      }
116.      i++;
117.      if (i == k) {
118.         bGameOver = true;
119.         sResult = "No";
120.         Move = -1;
121.         return;
122.      }
123.      if (i >= MAX_MOVES)
124.         i = 0;
125.   }
126. }
```

The last function in the program is GetUserMove. It is where you pick your move on the tic-tac-toe board. Use the left and right arrow buttons to scroll through all the empty spaces on the board. Once you have positioned the cursor to the space where you want to place your X, press the orange enter button.

```
1. void GetUserMove() {
2.    byte moveTaken, newX, newY;
3.    Move = 0;
4.    while (Move < MAX_MOVES) {
5.       moveTaken = Board[Move];
6.       if (moveTaken == 0)
7.          break;
8.       Move++;
9.    }
10.   if (Move < MAX_MOVES) {
11.      while (true) {
12.         DrawBoard();
13.         // Draw cursor
14.         GetDeltas(Move);
15.         newX = dbX+2+DX;
16.         newY = dbY+2+DY;
17.         RectOut(newX, newY, 16, 16);
18.
19.         WaitButton();
20.         if (pressedButton == BTNRIGHT) {
21.            do {
22.               Move++;
23.               if (Move >= MAX_MOVES)
```

```
24.              Move = 0;
25.            moveTaken = Board[Move];
26.          } while (moveTaken != 0);
27.        }
28.        else if (pressedButton == BTNLEFT) {
29.          do {
30.            if (Move <= 0)
31.              Move = MAX_MOVES;
32.            Move--;
33.            moveTaken = Board[Move];
34.          } while (moveTaken != 0);
35.        }
36.        else if (pressedButton == BTNCENTER) {
37.          Board[Move] = USER_MOVE;
38.          break;
39.        }
40.      }
41.    }
42.    else {
43.      Move = -1;
44.      bGameOver = true;
45.    }
46. }
```

That completes the tic-tac-toe code. You can extend the program in a number of interesting ways.

 TRY IT: *See if you can add a score-keeping system to the program where the NXT saves the wins, losses, and draws to a file.*

Enhance the game presentation by adding different sound effects that play depending on whether you win, lose, or draw.

NXT-o-Sketch

The next project is a simple on-screen drawing program. It requires a little bit of building first. We need a structure with two motors that will act as our controllers on the NXT-o-Sketch. Once you have built the device we'll dive into the program code.

Building the NXT-o-Sketch

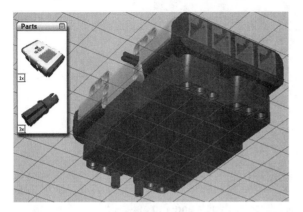

STEP 1. Add parts as shown.

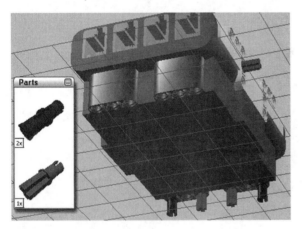

STEP 2. Add parts as shown.

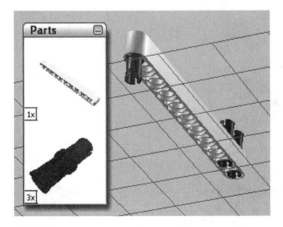

STEP 3. Add parts as shown.

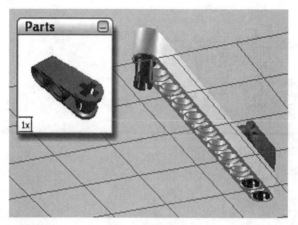

STEP 4. Add parts as shown. Repeat steps 3 and 4 to build a second copy of this assembly.

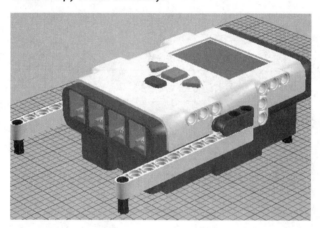

STEP 5. Combine the assemblies from step 4 with the assembly from step 2 as shown.

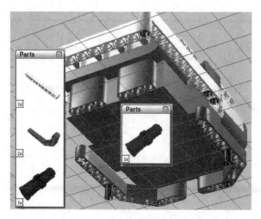

STEP 6. Add parts as shown.

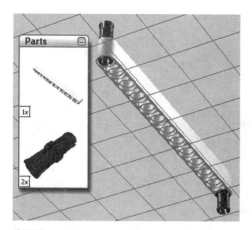

STEP 7. Add parts as shown. Make a second copy of this assembly.

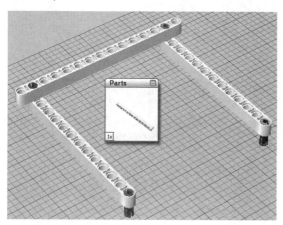

STEP 8. Add part as shown to the two assemblies from step 7.

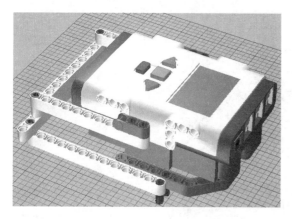

STEP 9. Combine the assembly from step 8 with the assembly from step 6 as shown.

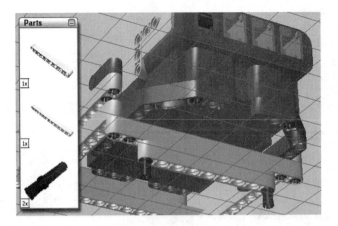

STEP 10. Add parts as shown.

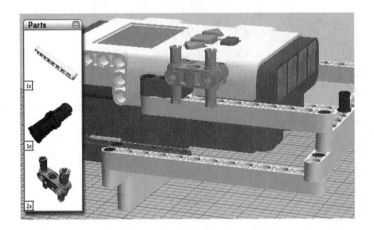

STEP 11. Add parts as shown.

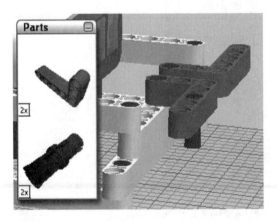

STEP 12. Add parts as shown.

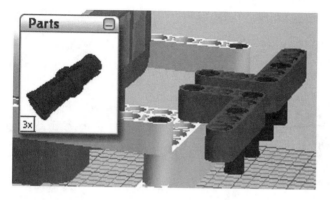

STEP 13. Add parts as shown.

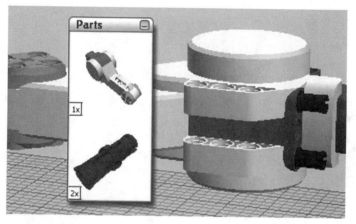

STEP 14. Add parts as shown.

STEP 15. Combine the assembly from step 14 with the assembly from step 13 as shown.

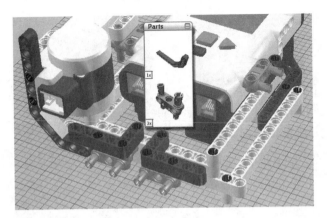

STEP 16. Add parts as shown.

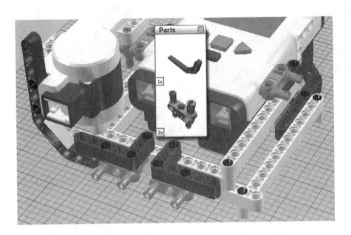

STEP 17. Add parts as shown.

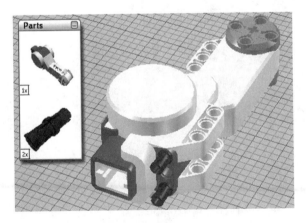

STEP 18. Add parts as shown.

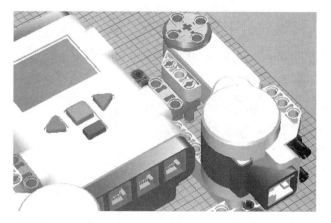

STEP 19. Combine the assembly from step 18 with the assembly from step 17 as shown.

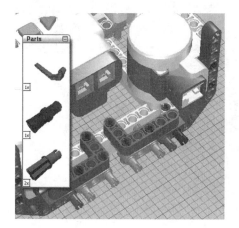

STEP 20. Add parts as shown.

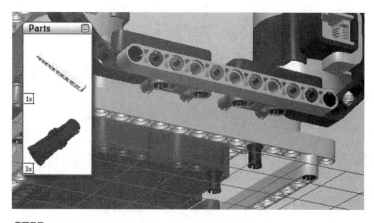

STEP 21. Add parts as shown.

STEP 22. Add parts as shown.

STEP 23. Add parts as shown.

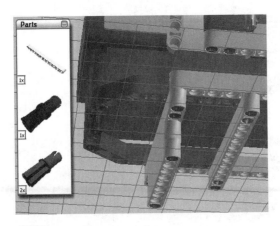

STEP 24. Add parts as shown.

STEP 25. Add parts as shown.

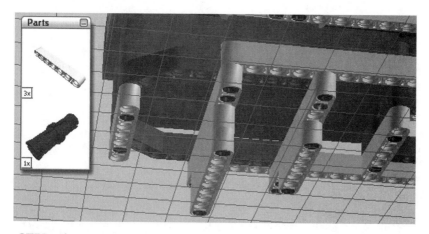

STEP 26. Add parts as shown.

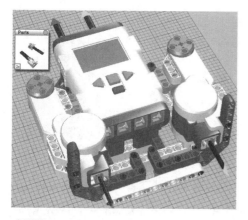

STEP 27. Connect medium cables from the left motor to port A and the right motor to port C.

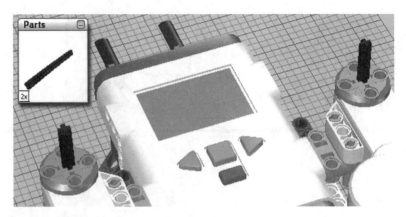

STEP 28. Add parts as shown. The axles are 6 units long.

STEP 29. Add parts as shown.

This is the complete NXT-o-Sketch robot. Now we can write a program that lets us draw on the NXT screen using the two motors as the x and y axis controllers.

The Program
The NXT-o-Sketch code uses the enhanced NBC/NXC firmware's ability to read and write more than 64 bytes at a time in an IOMapRead or IOMapWrite system call. So you will need to have the enhanced firmware installed on your NXT to run the program.

My brother Karl Hansen wrote the NXT-o-Sketch code. Two weeks after introducing him to the NXC programming language and the NXT, he had designed the NXT-o-Sketch platform and finished two separate games for the NXT. It just shows how easy it is to program using NXC!

Now let's look at the code that lets you draw your own sketches on the LCD screen. Figure 15-2 shows a drawing I made on my NXT while running the NXT-o-Sketch program. I'm sure you will make much prettier drawings than I can.

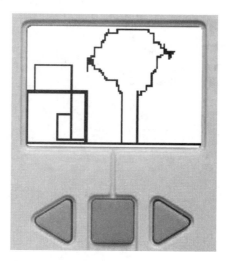

Figure 15-2. NXT-o-Sketch running on the NXT

The main task of NxtOSketch.nxc is very simple. It clears the screen, initializes the sketch state structure, configures the motors for their starting condition, and starts the Sketcher task. The Sketcher task is where most of the work of the program is accomplished. We'll have a look at it shortly. When we put all the source code together, the main task will go at the very end of the file.

```
1. task main() {
2.    ClearScreen();
3.    SketchState.X    = 0;
4.    SketchState.Y    = 0;
5.    SketchState.XSens = 5;
6.    SketchState.YSens = 5;
7.    SketchState.XDir  = 1;
8.    SketchState.YDir  = 1;
9.    ResetRotationCount(OUT_ABC);
10.   Float(OUT_ABC);
11.   start Sketcher;
12. } // main()
```

There are a number of preprocessor macros, a structure definition, and a few global variable declarations at the beginning of the source file. Most of the macros are used to define states for the two-level state machine that is implemented in the program. While the program is running, it is in either the STATE_SKETCHING or STATE_SETTING state. If you enter the STATE_SETTING state you can be in 6 different sub-states as you can see from the code below. The L1..L8 arrays are used to save and restore the image when you switch back and forth between the sketching state and the setting state, which is a menu system that lets you adjust several options.

```
1.  #define PIX_HEIGHT       64
2.  #define PIX_WIDTH        100
3.  #define STATE_SKETCHING  0
4.  #define STATE_SETTING    1
5.  #define SETTER_X_SENS    0
6.  #define SETTER_Y_SENS    1
7.  #define SETTER_X_DIR     2
8.  #define SETTER_Y_DIR     3
9.  #define SETTER_SAVE      4
10. #define SETTER_RETURN    5
11.
12. struct tStateSketch {
13.    int    X;
14.    int    Y;
15.    int    XSens;
16.    int    YSens;
17.    int    XDir;
18.    int    YDir;
19. };
20.
21. tStateSketch SketchState;
22.
23. byte  L1[PIX_WIDTH];
24. byte  L2[PIX_WIDTH];
25. byte  L3[PIX_WIDTH];
26. byte  L4[PIX_WIDTH];
27. byte  L5[PIX_WIDTH];
28. byte  L6[PIX_WIDTH];
29. byte  L7[PIX_WIDTH];
30. byte  L8[PIX_WIDTH];
31.
32. int   ProgramState;
33. int   SetterState;
```

The next function in the source code is RestoreSketch. This function is used to redraw the image that existed before you brought up the configuration menu system. The tones are used to provide feedback indicating that you are uploading an in-memory image of the sketch to the LCD display buffer.

```
1.  void RestoreSketch() {
2.     PlayTone(800, 250);
3.     SetDisplayNormal(0, 0, PIX_WIDTH, L1);
4.     SetDisplayNormal(0, 1, PIX_WIDTH, L2);
5.     SetDisplayNormal(0, 2, PIX_WIDTH, L3);
```

```
6.    SetDisplayNormal(0, 3, PIX_WIDTH, L4);
7.    SetDisplayNormal(0, 4, PIX_WIDTH, L5);
8.    SetDisplayNormal(0, 5, PIX_WIDTH, L6);
9.    SetDisplayNormal(0, 6, PIX_WIDTH, L7);
10.   SetDisplayNormal(0, 7, PIX_WIDTH, L8);
11.   Wait(250);
12.   PlayTone(1000,  500);
13.   Wait(500);
14. } // RestoreSketch()
```

The SaveSketch function is the opposite of RestoreSketch. It copies data from the LCD screen buffer into the eight line arrays so that it can restore the image later. The tones are played in reverse order to indicate that the screen is downloading into memory.

```
1. void SaveSketch() {
2.    PlayTone(1000,  250);
3.    GetDisplayNormal(0, 0, PIX_WIDTH, L1);
4.    GetDisplayNormal(0, 1, PIX_WIDTH, L2);
5.    GetDisplayNormal(0, 2, PIX_WIDTH, L3);
6.    GetDisplayNormal(0, 3, PIX_WIDTH, L4);
7.    GetDisplayNormal(0, 4, PIX_WIDTH, L5);
8.    GetDisplayNormal(0, 5, PIX_WIDTH, L6);
9.    GetDisplayNormal(0, 6, PIX_WIDTH, L7);
10.   GetDisplayNormal(0, 7, PIX_WIDTH, L8);
11.   Wait(250);
12.   PlayTone(800,  500);
13.   Wait(500);
14. } // SaveSketch()
```

The next two functions work together to provide tactile feedback when you try to move the pen off the edge of the screen. The program will shake the offending motor by calling the Shake function. It uses ShakeHelper to do part of its job.

```
1. void ShakeHelper(const int ID, const int pwr) {
2.    OnFwd(ID,  pwr);
3.    Wait(30);
4.    Off(ID);
5.    Wait(5);
6. } // ShakeHelper()
7.
8. void Shake(const int ID) {
9.    for (int i = 0; i < 5; i++) {
10.      ShakeHelper(ID, 100);
11.      ShakeHelper(ID, -100);
12.   }
13.   Wait(25);
14.   Coast(ID);
15.   Wait(25);
16. } // Shake()
```

The next three routines work together to adjust the X and Y position of the sketch pen. The bulk of the work here is in UpdateAxis. It is not used directly, however. Instead other functions call either UpdateX or UpdateY and they pass the appropriate values into the UpdateAxis routine.

```
1.  void SetterMenu() {
2.    ClearScreen();
3.    switch (SetterState) {
4.      case SETTER_X_SENS:
5.        TextOut(0, LCD_LINE1, "X Sensitivity");
6.        TextOut(0, LCD_LINE3, "to adjust value");
7.        NumOut( 0, LCD_LINE5, SketchState.XSens);
8.        break;
9.      case SETTER_Y_SENS:
10.       TextOut(0, LCD_LINE1, "Y Sensitivity");
11.       TextOut(0, LCD_LINE3, "to adjust value");
12.       NumOut( 0, LCD_LINE5, SketchState.YSens);
13.       break;
14.     case SETTER_X_DIR:
15.       TextOut(0, LCD_LINE1, "X Direction");
16.       TextOut(0, LCD_LINE3, "to toggle dir.");
17.       if (SketchState.XDir > 0)
18.         TextOut(0, LCD_LINE5, "CW");
19.       else
20.         TextOut(0, LCD_LINE5, "CCW");
21.       break;
22.     case SETTER_Y_DIR:
23.       TextOut(0, LCD_LINE1, "Y Direction");
24.       TextOut(0, LCD_LINE3, "to toggle dir.");
25.       if (SketchState.YDir > 0)
26.         TextOut(0, LCD_LINE5, "CW");
27.       else
28.         TextOut(0, LCD_LINE5, "CCW");
29.       break;
30.     case SETTER_SAVE:
31.       TextOut(0, LCD_LINE1, "Save/Clear");
32.       TextOut(0, LCD_LINE4, "L = save sketch");
33.       TextOut(0, LCD_LINE5, "R = clear sketch");
34.       break;
35.     case SETTER_RETURN:
36.       TextOut(0, LCD_LINE1, "Switch to Sketch Mode");
37.       TextOut(0, LCD_LINE3, "to sketch");
38.       break;
39.     }
40.   TextOut(0, LCD_LINE2, "Press L or R btn");
41.   TextOut(0, LCD_LINE7, "Press center btn");
42.   TextOut(0, LCD_LINE8, "for next option");
43.   Wait(250);
44. } // SetterMenu()
```

The setter task calls three helper functions as it shifts through the six setter states in response to the buttons you press. The first is ChangeSetterState. It updates the global SetterState variable, draws the correct menu by calling SetterMenu, and plays a pair of tones to provide you with feedback indicating the changing setter menu state.

The next two functions are executed if you choose one of the menu options that these functions are associated with. DoSaveImage is an unimplemented routine that is perfect for writing a program extension that lets you save the image data to a file. DoEraseImage is used to clear out the in-memory buffer so that you can start sketching with a blank canvas.

```
1.  void ChangeSetterState(const int NewState) {
2.      PlayTone(400, 100);
3.      SetterState = NewState;
4.      SetterMenu();
5.      Wait(100);
6.      PlayTone(600, 100);
7.      Wait(100);
8.  } // ChangeSetterState()
9.
10. void DoSaveImage() {
11.     ClearScreen();
12.     PlayTone(100, 1000);
13.     TextOut(0, LCD_LINE1, "Not currently");
14.     TextOut(0, LCD_LINE2, "implemented!");
15.     Wait(2000);
16. } // DoSaveImage()
17.
18. void DoEraseImage() {
19.     for (int i = 0; i < PIX_WIDTH; i++) {
20.        L1[i] = 0; L2[i] = 0; L3[i] = 0;
21.        L4[i] = 0; L5[i] = 0; L6[i] = 0;
22.        L7[i] = 0; L8[i] = 0;
23.        PlayTone(200+i*10, 1);
24.     }
25.     SketchState.X = 0;
26.     SketchState.Y = 0;
27. } // DoEraseImage()
```

Now lets look at the two tasks that remain. The Sketcher task comes first. It uses Setter and Setter uses Sketcher. This circular reference requires that we add a forward declaration of the Setter task as you can see in line 1.

The Sketcher task always starts by restoring the in-memory saved image if there is one. It updates the global ProgramState variable. The task enters a while-loop that continues so long as the program is in the sketching state. Inside the loop it updates the X and Y pen position, draws on the LCD appropriately, and checks to see if you pressed the orange button. If you press the enter button it exits the loop, saves the screen buffer, clears the screen, and then starts the Setter task.

```
1. task Setter();
2.
3. task Sketcher() {
4.    int CurX;
5.    int CurY;
6.    ProgramState = STATE_SKETCHING;
7.    RestoreSketch();
8.    while (ProgramState == STATE_SKETCHING) {
9.       CurX = SketchState.X;
10.      CurY = SketchState.Y;
11.      SketchState.X  = UpdateX();
12.      SketchState.Y  = UpdateY();
13.      if ((abs(CurX - SketchState.X) < 2) &&
14.          (abs(CurY - SketchState.Y) < 2))
15.        PointOut(SketchState.X, SketchState.Y);
16.      else
17.        LineOut(CurX, CurY,
18.                SketchState.X, SketchState.Y);
19.      if (ButtonCount(BTNCENTER, true) != 0)
20.        break;
21.      else
22.        Wait(20);
23.    }
24.    SaveSketch();
25.    ClearScreen();
26.    start Setter;
27. } // Sketcher()
```

The Setter task is long but simple. It is essentially a large while-loop that checks for button presses and steps through the setter states accordingly. As needed, it calls ChangeSetterState and SetterMenu to draw the next menu screen and update the setter state variable.

Each time through the while–loop, you either advance to the next setter state or adjust one of the SketchState structure members by using the left and right arrow buttons. The last menu in the setter state is where you can choose to either save the image or clear it. The state machine inside the loop is implemented via a switch statement containing six cases just like in SetterMenu.

```
1. task Setter() {
2.    int BtnU;
3.    int BtnL;
4.    int BtnR;
5.    int i;
6.    ProgramState = STATE_SETTING;
7.    ChangeSetterState(SETTER_X_SENS);
8.    while (ProgramState == STATE_SETTING) {
9.       Wait(20);
10.      BtnL = ButtonCount(BTNLEFT, true);
11.      BtnR = ButtonCount(BTNRIGHT, true);
```

```
12.      BtnU = ButtonCount(BTNCENTER, true);
13.      switch (SetterState) {
14.        case SETTER_X_SENS:
15.          if (BtnR) {
16.            if (SketchState.XSens++ >= 20)
17.              SketchState.XSens = 20;
18.            SetterMenu();
19.          }
20.          else if (BtnL) {
21.            if (SketchState.XSens-- <= 1)
22.              SketchState.XSens = 1;
23.            SetterMenu();
24.          }
25.          else if (BtnU)
26.            ChangeSetterState(SETTER_Y_SENS);
27.          break;
28.        case SETTER_Y_SENS:
29.          if (BtnR) {
30.            if (SketchState.YSens++ >= 20)
31.              SketchState.YSens = 20;
32.            SetterMenu();
33.          }
34.          else if (BtnL) {
35.            if (SketchState.YSens-- <= 1)
36.              SketchState.YSens = 1;
37.            SetterMenu();
38.          }
39.          else if (BtnU)
40.            ChangeSetterState(SETTER_X_DIR);
41.          break;
42.        case SETTER_X_DIR:
43.          if ((BtnR != 0) || (BtnL != 0)) {
44.            SketchState.XDir = -SketchState.XDir;
45.            SetterMenu();
46.          }
47.          else if (BtnU)
48.            ChangeSetterState(SETTER_Y_DIR);
49.          break;
50.        case SETTER_Y_DIR:
51.          if ((BtnR != 0) || (BtnL != 0)) {
52.            SketchState.YDir = -SketchState.YDir;
53.            SetterMenu();
54.          }
55.          else if (BtnU)
56.            ChangeSetterState(SETTER_SAVE);
57.          break;
58.        case SETTER_SAVE:
59.          if (BtnL) {
60.            DoSaveImage();
61.            SetterMenu();
```

```
62.            }
63.            else if (BtnR) {
64.               DoEraseImage();
65.               SetterMenu();
66.            }
67.            else if (BtnU)
68.               ChangeSetterState(SETTER_RETURN);
69.            break;
70.        case SETTER_RETURN:
71.            if (BtnR || BtnL)
72.               ProgramState = STATE_SKETCHING;
73.            else if (BtnU)
74.               ChangeSetterState(SETTER_X_SENS);
75.            break;
76.      }
77.   }
78.   start Sketcher;
79. } // Setter()
```

Put all the code together in a file using BricxCC or your favorite editor and give it a try. Remember to put main at the end of the file, since it refers to the Sketcher task and it uses the SketchState global variable.

TRY IT: *See if you can implement the ability to raise and lower the pen so that you can move to another screen location without having to follow an existing line.*

Add a sensor that you use to check if the NXT is upside down. If you turn the NXT upside down it clears the screen.

Add code to DoSaveImage to write the image data to a file using the RIC file format. That way you can draw your own RICs on the NXT.

If you are really ambitious, extend the code to support all the RIC op-codes for an on-brick NXT picture editor.

Pong

We started off the chapter with a reference to Pong, so it is fitting to end the chapter with a look at a fantastic implementation of that game written by Michael Andersen (Figure 15-3). Michael played an instrumental role in helping create the NeXT Byte Codes compiler and helped write the first NBC documentation.

To playing Pong on the NXT, you'll require a pair of motors with attached wheels. The motors act as the paddle controllers for the two pong players. Rather than build a completely new structure for this project, we will simply reuse the NXT-o-Sketch. Remove a few of the beams that tilted the NXT upward so that it will sit flat on a table. Relocate the right motor cable from port C to port B and you are ready to go. You really can't play this game on your own, so find a friend, crank up the tunes on your boom box, and let's start playing!

Figure 15-3. Pong on the NXT

The code for the main task is more involved in this program so we will look at it last. At the start of the file are the typical preprocessor macros just like we have seen in the other programs in this chapter.

```
 1. //
 2. // Pong for the LEGO Mindstorms NXT.
 3. //
 4. // Copyright (c) 2007, Michael Andersen
 5. // (mian@pobox.com)
 6. //
 7. #define HEIGHT       (64)
 8. #define WIDTH        (100)
 9. #define PADDLE_H     (paddlesize)
10. #define PADDLE_W     (1)
11. #define PADDLE_MIN   (0)
12. #define PADDLE_MAX   (HEIGHT-paddlesize)
13. #define BALL_DX      (4)
14. #define BALL_DY      (0)
```

The displayInfo function starts off the program with a title screen. It explains that the two motors should be attached to outputs A and B. It also displays Michael Andersen's copyright statement.

```
1. void displayInfo() {
2.   ClearScreen();
3.   TextOut( 0, LCD_LINE1, "Pong for the NXT");
4.   TextOut( 0, LCD_LINE3, "Attach motors to");
5.   TextOut(15, LCD_LINE4, "ports A & B");
6.   TextOut( 0, LCD_LINE6, "Copyright (c) by");
7.   TextOut( 6, LCD_LINE7, "mian@pobox.com");
8.   Wait(5000);
9. }
```

The selectValue function is used in the intial program configuration stage. You can pick the speed setting from 1 to 5 and the paddle size from 1 to 3. This function is used in both cases, with a string to adjust the displayed message appropriately. The min and max values are used to limit the allowed adjustments and set the initial default to the average of the two values. Inside the while-loop the function scans for button presses and either adjusts the resulting value up or down or exits the loop. There are also two wrapper functions here that encapsulate the selectValue call.

```
1.  int selectValue(string msg, int min, int max) {
2.     int result = (min+max)/2;
3.     ClearScreen();
4.     TextOut(12, LCD_LINE1, "SELECT "+msg);
5.     TextOut( 0, LCD_LINE3, "Press left/right");
6.     TextOut(18, LCD_LINE4, "to change.");
7.     TextOut(12, LCD_LINE5, "Middle = OK.");
8.     while (!ButtonPressed(BTNCENTER, false)) {
9.        TextOut(18, LCD_LINE7, msg+" =      ");
10.       NumOut(65, LCD_LINE7, result);
11.       if (ButtonPressed(BTNLEFT, false)) {
12.          while (ButtonPressed(BTNLEFT, false));
13.          result--;
14.       }
15.       if (ButtonPressed(BTNRIGHT, false)) {
16.          while (ButtonPressed(BTNRIGHT, false));
17.          result++;
18.       }
19.       if (result < min) result = min;
20.       if (result > max) result = max;
21.    }
22.    while (ButtonPressed(BTNCENTER, false));
23.    return result;
24. }
25.
26. int selectSpeed() {
27.    return selectValue("Speed", 1, 5));
28. }
29.
30. int selectPaddles() {
31.    return selectValue("Paddles", 1, 3));
32. }
```

Our pong program only has one task. The first thing that main does after the local variables are declared is to display the startup screen by calling displayInfo. Then it uses the selectValue function to configure the speed and paddle size values. It initializes the motor A and motor B positions next. Then it sets the ball and paddle positions and the ball size as well as the initial score.

The main game loop starts on line 34 and continues all the way down to line 123 below. Inside the loop the basic pattern is very straightforward. It updates both paddle positions. It updates the ball position and direction, accounting for bounces off either paddle or the top and bottom walls. Then it clears the screen and redraws the paddles, the ball, the net, and the two scores. On line 122 at the very end of the main loop it uses a simple for loop to control the game speed using the value you configured when the game started.

```
1. task main() {
2.    int i, j;
3.    int busywait;
4.    int paddlesize;
5.    int score_a, score_b;
6.    int motor_a_now, dA,
7.        motor_a_before, motor_a_init, a;
8.    int motor_b_now, dB,
9.        motor_b_before, motor_b_init, b;
10.   int ball_x, ball_y, ball_dx,
11.       ball_dy, ball_angle;
12.   displayInfo();
13.   busywait = 100 * (7 − selectSpeed());
14.   paddlesize = 4 * (1 + selectPaddles());
15.   // initial motor positions
16.   motor_a_init  = MotorTachoCount(OUT_A);
17.   motor_a_now   = motor_a_init;
18.   motor_a_before = motor_a_init;
19.   motor_b_init  = MotorTachoCount(OUT_B);
20.   motor_b_now   = motor_b_init;
21.   motor_b_before = motor_b_init;
22.   // center paddle positions
23.   a = HEIGHT/2-PADDLE_H/2;
24.   b = HEIGHT/2-PADDLE_H/2;
25.   // ball position
26.   ball_x = WIDTH/2;
27.   ball_y = HEIGHT/2;
28.   // ball speed
29.   ball_dx = BALL_DX;
30.   ball_dy = BALL_DY;
31.   // scores
32.   score_a = 0;
33.   score_b = 0;
34.   while (true) {
35.      // motor a
36.      motor_a_now =
37.         MotorTachoCount(OUT_A) - motor_a_init;
38.      if (motor_a_now <> motor_a_before) {
39.         dA = (motor_a_now − motor_a_before);
40.         a = a + dA / TPRATIO;
41.         if (a < PADDLE_MIN) a = PADDLE_MIN;
```

```
42.        if (a > PADDLE_MAX) a = PADDLE_MAX;
43.    }
44.    motor_a_before = motor_a_now;
45.    // motor b
46.    motor_b_now =
47.       MotorTachoCount(OUT_B) - motor_b_init;
48.    if (motor_b_now <> motor_b_before) {
49.       dB = (motor_b_now - motor_b_before);
50.       b = b + dB / TPRATIO;
51.       if (b < PADDLE_MIN) b = PADDLE_MIN;
52.       if (b > PADDLE_MAX) b = PADDLE_MAX;
53.    }
54.    motor_b_before = motor_b_now;
55.    // new position
56.    ball_x += ball_dx;
57.    ball_y += ball_dy;
58.    // adjust for A paddle bounce (left)
59.    if (ball_x < 1) {
60.        if ((a <= ball_y) &&
61.            (ball_y <= (a + PADDLE_H))) {
62.        // hit - calculate angle of return
63.        // range -4 to 4
64.        ball_dy =
65.            (ball_y - a - PADDLE_H/2)/2;
66.        ball_dx = 4;
67.        ball_x += ball_dx;
68.        }
69.        else {
70.        // missed
71.        score_b++;
72.        // ball position
73.        ball_x = WIDTH/2;
74.        ball_y = HEIGHT/2;
75.        // ball speed
76.        ball_dx = -BALL_DX;
77.        ball_dy = BALL_DY;
78.        }
79.    }
80.    // adjust for B paddle bounce (right)
81.    if (ball_x > 98) {
82.        if ((b <= ball_y) &&
83.            (ball_y <= (b + PADDLE_H))) {
84.        // hit - calculate angle of return
85.        // range -4 to 4
86.        ball_dy =
87.            (ball_y - b - PADDLE_H/2)/2;
88.        ball_dx = -4;
89.        ball_x += ball_dx;
90.        }
91.        else {
```

```
92.              // missed
93.              score_a++;
94.              // ball position
95.              ball_x = WIDTH/2;
96.              ball_y = HEIGHT/2;
97.              // ball speed
98.              ball_dx = BALL_DX;
99.              ball_dy = BALL_DY;
100.          }
101.      }
102.      // adjust for side bounce
103.      if ((ball_y <= 0) || (ball_y > 64)) {
104.          ball_y -= ball_dy;
105.          ball_dy = -ball_dy;
106.      }
107.      ClearScreen();
108.      // ball
109.      RectOut(ball_x-1, ball_y-1, 1, 1);
110.      RectOut(ball_x-2, ball_y-2, 3, 3);
111.      // paddles
112.      RectOut(A_X, a, PADDLE_W, PADDLE_H);
113.      RectOut(B_X, b, PADDLE_W, PADDLE_H);
114.      // net
115.      RectOut(49,3,1,8);
116.      RectOut(49,19,1,8);
117.      RectOut(49,35,1,8);
118.      RectOut(49,51,1,8);
119.      // score
120.      NumOut(35, LCD_LINE8, score_a);
121.      NumOut(60, LCD_LINE8, score_b);
122.      for (i = 0; i < busywait; i++);
123.    }
124. }
```

And that is all there is to pong on the NXT. As with the other game programs we've already seen, there are many areas where you can extend the program in interesting ways.

TRY IT: *Try adding a sound when the ball hits one of the paddles or bounces off the wall.*

Have the game incrementally speed up when a rally goes long. If a ball is missed, it should slow back down again.

While we are on the bouncing ball and paddle theme, try writing your own version of the classic Breakout game.

More Inspiration

One of the very first programs written for the NXT using a third-party programming language (NBC) was a falling-block game like Tetris. And Arno van der Vegt's Space Invader game is absolutely incredible.

You really need to check it out if you haven't already. Arno also contributed some amazing sample NBC programs that demonstrate how to manipulate 3D wire frame graphics on the NXT. His code is in the NBC samples area of the NXC/NBC website.

There simply isn't enough room in the book to show you all the games you can run on the NXT. In the sample program areas of the NXC/NBC website you can also find an NXT implementation of John Conway's famous artificial life simulator, written in NBC. There is a copy of a simple but still very fun driving game called RoadORouter, written in NXC by Karl Hansen. And you can find electronic copies of the three programs we looked at in this chapter as well. Be sure to check out these samples for inspiration in case you want to write your own NXT game.

Driving Around

Topics in this Chapter

- Direct Commands
- Remote Control Car

Chapter 16

This chapter is all about controlling an NXT remotely. There are lots of options when it comes to remote programming for the NXT. You can use some of the tools that were mentioned back in Chapter 2, such as the LEGO NXT Mobile Application that runs on a Bluetooth-enabled cell phone or a custom application using Bluetooth on a PDA. You can write your own programs on your platform of choice that use the Direct and System command protocols supported by the standard NXT firmware. Your programs can implement the protocol in your own code or you can leverage the Fantom SDK, which wraps this functionality in a higher level API on your PC.

You can even write a program on an NXT that controls another NXT remotely via Bluetooth messaging. These NXT-based programs can use the same Direct and System commands that you might use from a PC or they can rely on a custom protocol shared between the controlling program and the program running on the NXT slave. We'll examine the option of using Direct Commands from one NXT to another in this chapter. To get started, let's have a quick look at the direct commands.

Direct Commands

There are 20 direct commands supported by the NXT firmware. We'll look at each of them in this section before we start using a few of them to control our vehicle. Table 16-1 lists all the direct command names, their values, and their purpose.

Direct Command Name	Value	Purpose
StartProgram	0x00	Start running a program on the NXT
StopProgram	0x01	Stop any running program
PlaySoundFile	0x02	Play a sound or melody fil
PlayTone	0x03	Play a single tone

Direct Command Name	Value	Purpose
SetOutputState	0x04	Control an NXT motor
SetInputMode	0x05	Configure an NXT sensor port
GetOutputState	0x06	Read the state of a motor
GetInputValues	0x07	Read a sensor value
ResetInputScaledValue	0x08	Set a sensor's scaled value to zero
MessageWrite	0x09	Write a message to a mailbox on the NXT
ResetMotorPosition	0x0A	Reset a motor's position counter
GetBatteryLevel	0x0B	Read the current battery level
StopSoundPlayback	0x0C	Stop playing any sound or tone
KeepAlive	0x0D	Reset the sleep timer on the NXT
LSGetStatus	0x0E	Get the I2C status for one of the input ports
LSWrite	0x0F	Start an I2C transaction on one of the input ports
LSRead	0x10	Complete an I2C transaction on one of the input ports
GetCurrentProgramName	0x11	Read the name of the running program, if any
GetButtonState	0x12	Read the state of one of the NXT button
MessageRead	0x13	Read a message from an NXT mailbox

Table 16-1. NXT Direct Commands

Each of the direct commands has a specific format that is unique to the command. The details of the format for each direct command are found in the Bluetooth Developer Kit (BDK) on the LEGO MINDSTORMS NXT'reme page. While there are many high-level wrappers around the NXT protocol, the Fantom SDK does not hide the details of the direct commands from a user. It wraps all the system commands in high level functions but if you use Fantom directly you still need to build the direct command input buffers yourself and parse out the response data from the output buffer.

WEBSITE: *Download the BDK from the MINDSTORMS NXT'reme page at* http://mindstorms.lego.com/Overview/NXTreme.aspx.

Direct commands all start with a command type byte followed by the direct command byte itself. The command type byte tells the NXT two things: whether the command is a system or a direct command and whether a response is required or not. Direct commands always start with 0x00 if you want to receive a response or 0x80 if you do not want to receive a response. When you are communicating via Bluetooth, it is important to only request a response when you absolutely need one since it can take as much as 30 milliseconds for the NXT to switch from receiving Bluetooth messages to sending Bluetooth messages. Table 16-2 summarizes the command type byte information.

Command Type Byte	Value
Direct command with response	0x00
Direct command with no response	0x80
System command with response	0x01
System command with no response	0x81

Table 16-2. Command type byte

Each message is limited to a maximum of 64 bytes due to the use of USB bulk write operations for USB-based communication. If you are writing your own low-level code, when you are communicating via Bluetooth each message also needs to have two bytes added at the beginning. These are message length bytes that the NXT firmware requires in order for it to easily know how many bytes it needs to read before it can exit the Bluetooth message receiving state. Another limitation that you have to keep in mind is that filenames can't be longer than 15 bytes so that the filename plus the four-character file extension and the required null terminator fits within a maximum of 20 bytes.

If you request a response, the message you receive back from the NXT will start with the command reply byte, which is always 0x02. Following the reply byte is the original command byte that you sent to the brick. The third byte is the status byte. A status of 0x00 indicates a successful operation. The status can also have one of several positive values that indicate different types of errors. Table 16-3 lists all the direct command error status codes.

Code	Meaning
0x20	A communication transaction is already in progress
0x40	The specified mailbox queue is empty
0xBD	The specified file was not found
0xBE	Unknown command
0xBF	Insane packet
0xC0	Data contained values that were out of range
0xDD	An error occurred on the communication bus
0xDE	The communication buffer is out of memory
0xDF	The specified channel or connection is invalid
0xE0	The specified channel or connection is busy
0xEC	No program is active
0xED	The specified size is illegal
0xEE	The specified mailbox queue is invalid
0xEF	Error accessing invalid structure field
0xF0	The specified input or output is invalid
0xFB	Not enough memory available
0xFF	Invalid arguments

Table 16-23. Direct command error status codes

Direct driving

In this chapter we will use just a couple of the direct commands. The SetOutputState command is the only one that we must use so that we can control the motors in our vehicle. We will also use the PlayTone direct command so that we can honk the car's horn.

The SetOutputState direct command packet is twelve bytes long. The first two bytes are the command type and the command byte. Next is the port, ranging from zero to two. You can also use 0xFF, which means all three ports. Byte four is the power byte, which ranges from -100 to 100. Negative power levels are used to reverse the motor. The fifth byte sets the output mode. Use the same OUT_MODE_* constant values from NBC and NXC to set the motor mode. The sixth byte sets the motor regulation mode. It should be 0, 1, or 2 indicating no regulation, speed regulation, or motor synchronization. Byte seven is the turn ratio, ranging from -100 to 100. For our purposes this will always be zero. Byte eight is the output run state. Again, use the OUT_RUNSTATE_* constant values. We'll use 0x20, which is OUT_RUNSTATE_RUNNING. The last four bytes are the tachometer limit, in little-endian order. In this chapter we will not set a tachometer limit, so all four of these bytes will be zero.

The PlayTone direct command packet is six bytes long. As with all direct commands the first two bytes contain the command type and the direct command byte itself. The next two bytes are the tone frequency. The last two bytes are the tone duration. Both of these values are sent LSB first (little-endian).

Remote Control Car

Those of us who own two NXT sets are in luck. In this section we are going to use our base mobile robot as a radio controlled car and another NXT as the radio controller, so you will need two NXT sets to use this project as intended. The goal is to drive the mobile robot around and honk its horn using the second brick along with two motors and a touch sensor. Let's get started by building the remote control unit.

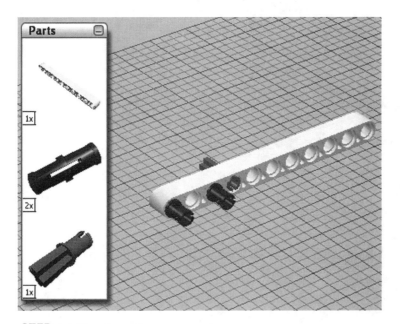

STEP 1. Add parts as shown.

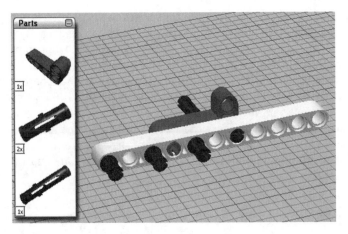

STEP 2. Add parts as shown.

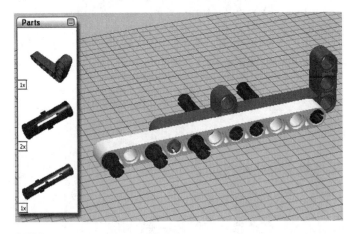

STEP 3. Add parts as shown.

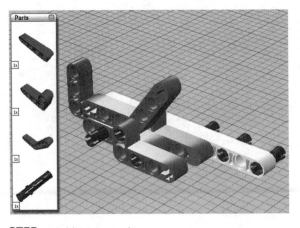

STEP 4. Add parts as shown.

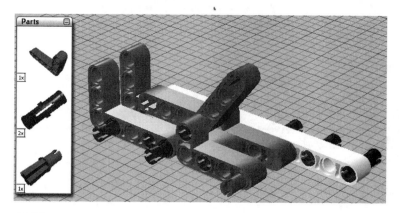

STEP 5. Add parts as shown.

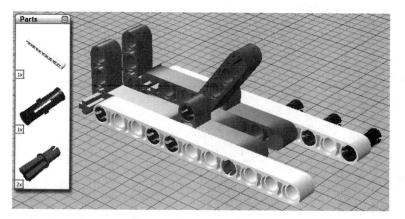

STEP 6. Add parts as shown.

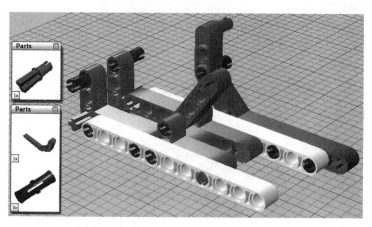

STEP 7. Add parts as shown.

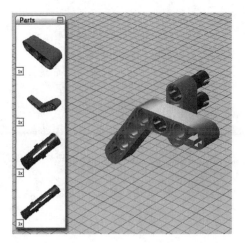

STEP 8. Add parts as shown.

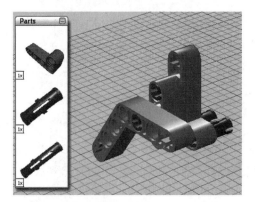

STEP 9. Add parts as shown.

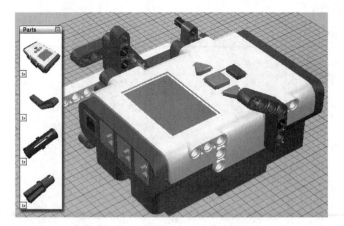

STEP 10. Clip the assembly from step 7 onto the lower-right side of the NXT brick. Then clip the assembly from step 9 onto the upper-right side. Add the remaining parts as shown.

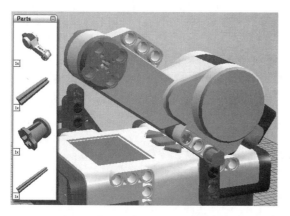

STEP 11. Add parts as shown. One axle is 3 units long and the other is 5 units long.

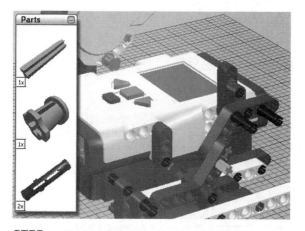

STEP 12. Add parts as shown. The axle is 3 units long.

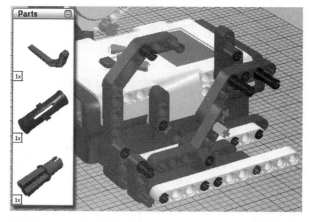

STEP 13. Add parts as shown.

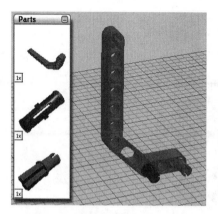

STEP 14. Add parts as shown.

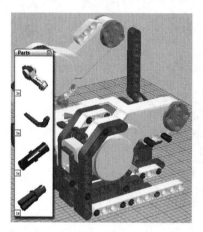

STEP 15. Combine the assembly from step 13 with the assembly from step 14. Attach the double bent beam to the orange motor face using the two pins as shown (pins are hidden).

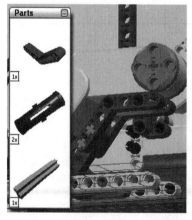

STEP 16. Add parts as shown. The axle is 3 units long.

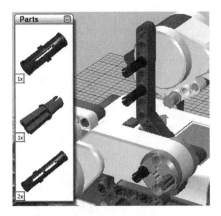

STEP 17. Add parts as shown.

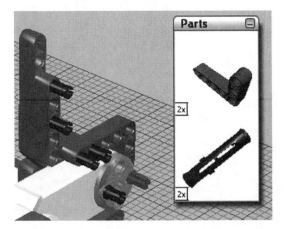

STEP 18. Add parts as shown.

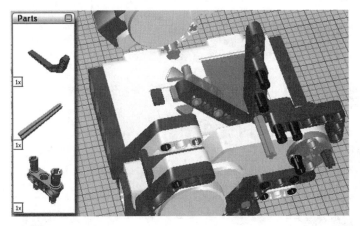

STEP 19. Attach the double bent beam to the black pins on the side of the NXT (hidden). One pin goes in the long slot. The axle is 5 units long.

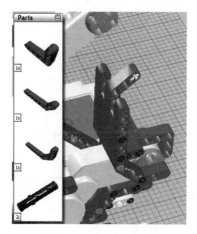

STEP 20. Add parts as shown.

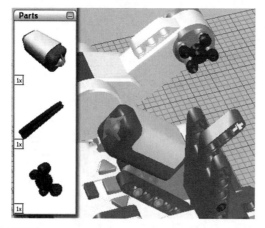

STEP 21. Add parts as shown. The axle is 4 units long.

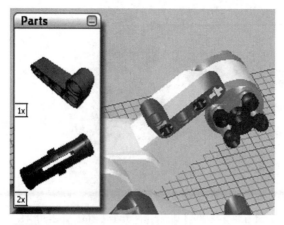

STEP 22. Add parts as shown. The axle is 3 units long.

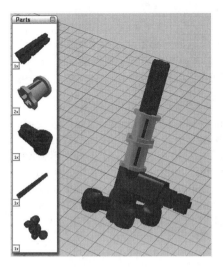

STEP 23. Add parts as shown. The axle is 6 units long.

STEP 24. Add parts as shown.

STEP 25. Combine the assembly from step 22 with the assembly from step 24 as shown. This completes our remote control cockpit. We can turn our steering wheel, control the throttle, and honk the horn.

Using Direct Commands

The program below shows how to send direct commands from one NXT to another via Bluetooth. The NXT in our remote control unit is the primary brick. Use the menu system on this brick to search for and pair with the mobile robot that we built back in Chapter 9.

The motor rotation sensors are used to calculate a pair of speed settings for the left and right motors on the mobile robot. You can see how this is done in GetSpeed starting at line 7 and GetTurnPercent starting at line 16. The CheckPairing routine at line 25 is used to make sure that the two NXT bricks are properly connected when the program starts running. If they are not then it displays an error message and stops the program. The inline function called WaitBluetooth is used to simplify the process of waiting until the last Bluetooth write has finished before commencing another one.

```
1.  #define HORN        SENSOR_1
2.  #define HORN_PORT S1
3.  #define BT_CONN    1
4.  #define MTR_SPEED OUT_C
5.  #define MTR_TURN   OUT_A
6.
7.  char GetSpeed() {
8.    TextOut(0, LCD_LINE1, "Speed = ");
9.    int speed = -2*MotorRotationCount(MTR_SPEED);
10. if (abs(speed) < 10)
11.    speed = 0;
12. NumOut(80, LCD_LINE1, speed);
13. return speed < 80 ? speed : 80;
14.}
15.
16.char GetTurnPercent() {
17. TextOut(0, LCD_LINE2, "Turn = ");
18. int turnpct = MotorRotationCount(MTR_TURN);
19. NumOut(80, LCD_LINE2, turnpct);
20. return ((turnpct < -100) ?
21.          -100 :
22.          (turnpct > 100 ? 100 : turnpct));
23.}
24.
25.void CheckPairing() {
26. int stat = BluetoothStatus(BT_CONN);
27. if (stat < 0) {
28.    TextOut(0, LCD_LINE1, "No connection");
29.    NumOut(0, LCD_LINE2, stat);
30.    Wait(4000);
31.    Stop(true);
32. }
33.}
34.
```

```
35. inline void WaitBluetooth() {
36.   while(BluetoothStatus(BT_CONN) != NO_ERR);
37. }
38.
39. task main() {
40.   CheckPairing();
41.   SetSensorTouch(HORN_PORT);
42.   ResetRotationCount(OUT_AC);
43.   while (!ButtonPressed(BTNCENTER, true)) {
44.     RemoteKeepAlive(BT_CONN);
45.     WaitBluetooth();
46.     byte speed = GetSpeed();
47.     char turnpct = GetTurnPercent();
48.     if (HORN) {
49.       RemotePlayTone(BT_CONN, 880, 1000);
50.       WaitBluetooth();
51.     }
52.     // negative turn means reduce OUT_C speed
53.     // positive turn means reduce OUT_A speed
54.     byte aspeed = speed-(turnpct/4);
55.     byte cspeed = speed+(turnpct/4);
56.     RemoteSetOutputState(BT_CONN, OUT_A, aspeed, OUT_
      MODE_MOTORON+OUT_MODE_BRAKE, OUT_REGMODE_IDLE, 0, OUT_
      RUNSTATE_RUNNING, 0);
57.     WaitBluetooth();
58.     RemoteSetOutputState(BT_CONN, OUT_C, cspeed, OUT_
      MODE_MOTORON+OUT_MODE_BRAKE, OUT_REGMODE_IDLE, 0, OUT_
      RUNSTATE_RUNNING, 0);
59.     WaitBluetooth();
60.   }
61.   RemoteSetOutputState(BT_CONN, 0xFF, 0, 0, 0, 0, 0);
62.   WaitBluetooth();
63. }
```

In the main and only task above the program performs the connection check and then configures the sensors before entering a while-loop that continues to run until the orange button (BTNCENTER) is pressed. Inside the loop, the program sends a keep alive message so that the robot doesn't fall asleep while it is driving — or being driven. The speed and turn percent values are gathered from the rotation sensors. Then the code checks for the touch sensor called HORN to see if it should send a PlayTone direct command to the vehicle. You could use RemotePlaySoundFile instead if you want to make the remote robot say "Watch Out" or some other sound every time you honk the horn.

Finally the code uses the turn percentage value along with the speed to calculate a motor A speed value and a motor C speed value for the two "engines" in our car. It sends these values to the robot using two separate direct commands. To end the program, simply press the orange enter button and the while-loop will terminate after the program sends one last direct command to stop both of the robot's motors. If you abort the program, rather than press the orange button, the motors on the robot will happily keep on turning.

 TRY IT: *Explore using the Fantom SDK and Direct Commands from your computer to control the NXT. Come up with the next great NXT utility and share it with the LEGO robotics community.*

Try using direct commands and standard Bluetooth NXT mailbox messaging between two or more NXT bricks in a collaborative robotics project.

NXC Quick Reference

Topics in this Appendix

- NXC Quick Reference Guide

Appendix A

Within these pages you'll find a quick guide to the NXC API. You can find more detailed information about NXC in the *NXC Programmers Guide*.

NXC API Functions

Many of the functions here are flexible in that they can take either constant or variable arguments. Some functions require that one or more of their arguments are constant values. These arguments are marked with the "const" keyword in the function argument list.

Task and Timing Functions

The functions in this section provide access to basic timing and task control features.

Function	Description	Example
Wait(time)	Wait for the specified amount of time in milliseconds. The time may be an expression or a constant.	Wait(x);
CurrentTick()	Return the current system timing value in milliseconds.	x = CurrentTick();
FirstTick()	Return the system timing value at the time the program began executing.	x = FirstTick();
SleepTime()	Return the number of minutes the NXT will remain on before it automatically shuts down.	x = SleepTime();
SleepTimer()	Return the number of minutes left before the NXT shuts down.	s = SleepTimer();

ResetSleepTimer()	Reset the sleep timer back to the SleepTime value.	ResetSleepTimer();
SetSleepTime(minutes)	Set the SleepTime value to the specified number of minutes.	SetSleepTime(8);
SetSleepTimer(minutes)	Set the SleepTimer value to the specified number of minutes.	SetSleepTimer(4);
Stop(bvalue)	Stop the running program if the specified value is true.	Stop(x==30);
StopAllTasks()	Stop all currently running tasks.	StopAllTasks();
StartTask(task)	Start the specified task.	StartTask(movement);
StopTask(task)	Stop the specified task. This function requires the enhanced NBC/NXC firmware.	StopTask(sound);
Acquire(mutex)	Acquire the specified mutex variable. If the mutex is already acquired then wait until the mutex is released.	Acquire(moveMotors);
Release(mutex)	Release the specified mutex variable.	Release(moveMotors);
Precedes(task1, task2, ..., taskN)	Schedule the specified tasks for execution once the current task has completed executing.	Precedes(moving, drawing, playing);
Follows(task1, task2, ..., taskN)	Schedule this task to follow the specified tasks so that it will execute once any of them finishes executing.	Follows(main);
ExitTo(task)	Immediately exit the current task and start executing the specified task.	ExitTo(nextTask);

String Functions

The functions in this section are for manipulating variables of the string data type.

Function	Description	Example
StrToNum(str)	Return the numeric value specified by the string passed to the function.	x = StrToNum("-10");
StrLen(str)	Return the length of the specified string (not including the null terminator).	x = StrLen(msg);
StrIndex(str, idx)	Return the numeric value of the character in the string at the specified index.	x = StrIndex(msg, 2);
NumToStr(value)	Return the string representation of the specified numeric value.	msg = NumToStr(-2);
FormatNum(fmtstr, value)	Return the formatted string using the format and value. Requires the enhanced NBC/ NXC firmware.	msg = FormatNum("value = %d", x);
.StrCat(str1, str2, ..., strN)	Return a string which is the combination of all the strings.	msg = StrCat("hi ", "there");
SubStr(string, idx, len)	Return a sub-string from the specified input string starting at idx and including len characters.	msg = SubStr(msg, 1, 2);
StrReplace(string, idx, newStr)	Return a string with part of the string replaced with the contents of the new string.	msg = StrReplace("testing", 3, "xx");
Flatten(value)	Return a string containing the byte representation of the specified value.	msg = Flatten(48);
FlattenVar(variable)	Like Flatten but you can use many different variable types such as structures, arrays, as well as signed and unsigned byte, int, and long.	msg = FlattenVar(myStruct);

UnflattenVar(string, out variable)	Reverses the operation performed by FlattenVar. The resulting value is output using the second function argument.	UnflattenVar(msg, myStruct);
ByteArrayToStr(arr)	Convert the array to a string by adding a zero to the end of the array.	str = ByteArrayToStr(arr);
ByteArrayToStrEx(arr, out str)	Convert the array to a string by adding a zero to the end of the array and return it via the second function argument.	ByteArrayToStrEx(arr, str);

Array Functions

The functions in this section are for manipulating array variables.

Function	Description	Example
StrToByteArray(str, out arr)	Convert the string to an array by removing the zero at the end of the string.	StrToByteArray(myStr, myArray);
ArrayLen(array)	Return the length of the specified array.	x = ArrayLen(myArray);
ArrayInit(array, value, count)	Initialize the array to have count elements equal to value.	ArrayInit(myArray, 0, 10);
ArraySubset(out aout, asrc, idx, len)	Copy a subset of asrc starting at idx and containing len elements into the output array.	ArraySubset(myArray, srcArray, 2, 5);
ArrayBuild(out aout, src1 [, src2, ..., srcN])	Build a new array from the specified source(s).	ArrayBuild(myArray, src1, src2);

Math Functions

The functions in this section are for performing math operations.

Function	Description	Example
Random(n)	Return an unsigned 16-bit random number between 0 and n (exclusive). N can be a constant or a variable.	x = Random(10);
Random()	Return a signed 16-bit random number.	x = Random();
Sqrt(x)	Return the square root of the specified value.	x = Sqrt(x);
Sin(degrees)	Return the sine of the specified degrees value. The result is 100 times the sine value (-100..100).	x = Sin(theta);
Cos(degrees)	Return the cosine of the specified degrees value. The result is 100 times the cosine value (-100..100).	x = Cos(y);
Asin(value)	Return the inverse sine of the specified value (-100..100). The result is degrees (-90..90).	deg = Asin(80);
Acos(value)	Return the inverse cosine of the specified value (-100..100). The result is degrees (0..180).	deg = Acos(0);
bcd2dec(bcd)	Convert a byte value in binary coded decimal (BCD) format into its equivalent decimal value.	dec = bcd2dec(bcdValue);

Input Functions

The functions in this section are for configuring, reading, and manipulating NXT inputs.

Function	Description	Example
SetSensor(port, const config)	Configure a sensor's type and mode.	SetSensor(S1, SENSOR_TOUCH);
SetSensorType(port, type)	Set a sensor's type	SetSensorType(S1, SENSOR_TYPE_TOUCH);
SetSensorMode(port, mode)	Set a sensor's mode	SetSensorMode(S1, SENSOR_MODE_RAW);
SetSensorLight(port)	Configure a sensor as an active light sensor	SetSensorLight(S1);
SetSensorSound(port)	Configure a sensor as a sound sensor with dB scaling.	SetSensorSound(S1);
SetSensorTouch(port)	Configure a sensor as a touch sensor.	SetSensorTouch(S1);
SetSensorLowspeed(port)	Configure a sensor as a lowspeed 9V sensor.	SetSensorLowspeed(S1);
SetInput(port, const field, value)	Set the specified field of the sensor on the specified port to the value provided.	SetInput(S1, Type, IN_TYPE_SOUND_DB);
ClearSensor(const port)	Clear the value of a sensor - only affects sensors that are configured to measure a cumulative quantity such as rotation or a pulse count.	ClearSensor(S1);
ResetSensor(port)	Reset the value of a sensor.	ResetSensor(S1);
SetCustomSensorZeroOffset (const p, value)	Sets the custom sensor zero offset value of a sensor.	SetCustomSensorZeroOffset (S1, 12);
SetCustomSensorPercent FullScale(const p, value)	Sets the custom sensor percent full scale value of a sensor.	SetCustomSensorPercent FullScale(S1, 100);

SetCustomSensorActive Status(const p, value)	Sets the custom sensor active status value of a sensor.	SetCustomSensorActive Status(S1, true);
SetSensorDigiPins Direction(const p, value)	Sets the digital pins direction value of a sensor.	SetSensorDigiPinsDirection (S1, 1);
SetSensorDigiPinsStatus (const p, value)	Sets the digital pins status value of a sensor.	SetSensorDigiPinsStatus (S1, false);
SetSensorDigiPinsOutput Level(const p, value)	Sets the digital pins output level value of a sensor.	SetSensorDigiPinsOutput Level(S1, 100);
SensorValue(p) or Sensor(p) or SensorScaled(p)	Return the processed or scaled value for a sensor on port p.	x = SensorValue(S1); x = SENSOR_1; x = SensorScaled(S3);
SensorUS(p)	Return the sensor reading for an ultrasonic sensor on port p, where p is 0, 1, 2, or 3 (or a sensor port name constant).	x = SensorUS(S1);
SensorType(p)	Return the configured type of a sensor on port p.	t = SensorType(S3);
SensorMode(p)	Return the configured mode of a sensor on port p.	m = SensorMode(S4);
SensorRaw(p)	Return the raw value for a sensor on port p.	raw = SensorRaw(S1);
SensorNormalized(p)	Return the normalized value for a sensor on port p.	n = SensorNormalized(S1);
SensorInvalid(p)	Return the value of the InvalidData flag of a sensor on port p.	invalid = SensorInvalid(S1);
SensorBoolean(const p)	Return the boolean value of a sensor on port p.	b = SensorBoolean(S1);
GetInput(p, const field)	Return the value of the specified field of a sensor on port p.	x = GetInput(S1, Type);
CustomSensorZeroOffset(const p)	Return the custom sensor zero offset value of a sensor.	x = CustomSensorZeroOffset (S1);
CustomSensorPercentFull Scale(const p)	Return the custom sensor percent full scale value of a sensor.	x = CustomSensorPercentFull Scale(S1);

CustomSensorActiveStatus (const p)	Return the custom sensor active status value of a sensor.	x = CustomSensorActive Status(S1);
SensorDigiPinsDirection (const p)	Return the digital pins direction value of a sensor.	x = SensorDigiPinsDirection (S1);
SensorDigiPinsStatus(const p)	Return the digital pins status value of a sensor.	x = SensorDigiPinsStatus (S1, false);
SensorDigiPinsOutputLevel (const p)	Return the digital pins output level value of a sensor.	x = SensorDigiPinsOutput Level(S1);

Output Functions

The functions in this section are for configuring and controlling NXT outputs.

Function	Description	Example
Off(outputs)	Turn the specified outputs off with braking. Use OUT_A, OUT_B, OUT_C, OUT_AB, OUT_AC, OUT_BC, or OUT_ABC for the outputs argument to motor functions. The non-Ex motor functions reset the Block and Tacho counters but they do not reset the Rotation counter.	Off(OUT_A);
OffEx(outputs, const reset)	Same as Off but you can specify which counters, if any, to reset.	OffEx(OUT_A, RESET_NONE);
Coast(outputs) or Float(outputs)	Turn off the outputs, making them coast to a stop.	Coast(OUT_A);
CoastEx(outputs, const reset)	Same as Coast but you can specify which counters, if any, to reset.	CoastEx(OUT_A, RESET_COUNT);
OnFwd(outputs, power)	Set outputs to forward direction and turn them on at the specified power level.	OnFwd(OUT_A, 75);
OnFwdEx(outputs, power, const reset)	Same as OnFwd but you can specify which counters, if any, to reset.	OnFwdEx(OUT_A, 75, RESET_BLOCK_COUNT);

OnRev(outputs, power)	Set outputs to reverse direction and turn them on at the specified power level.	OnRev(OUT_A, 75);
OnRevEx(outputs, power, const reset)	Same as OnRev but you can specify which counters, if any, to reset.	OnRevEx(OUT_A, 75, RESET_ALL);
OnFwdReg(outputs, power, regmode)	Run the outputs forward using the specified regulation mode.	OnFwdReg(OUT_A, 75, OUT_REGMODE_SPEED);
OnFwdRegEx(outputs, power, regmode, const reset)	Same as OnFwdReg but you can specify which counters, if any, to reset.	OnFwdRegEx(OUT_A, 75, OUT_REGMODE_SPEED, RESET_NONE);
OnRevReg(outputs, power, regmode)	Run the outputs in reverse using the specified regulation mode.	OnRevReg(OUT_A, 75, OUT_REGMODE_SPEED);
OnRevRegEx(outputs, power, regmode, const reset)	Same as OnRevReg but you can specify which counters, if any, to reset.	OnRevRegEx(OUT_A, 75, OUT_REGMODE_SPEED, RESET_NONE);
OnFwdSync(outputs, power, turnpct)	Run the specified outputs forward with regulated synchronization using the specified turn ratio.	OnFwdSync(OUT_A, 75, 50);
OnFwdSyncEx(outputs, power, turnpct, const reset)	Same as OnFwdSync but you can specify which counters, if any, to reset.	OnFwdSyncEx(OUT_A, 75, 50, RESET_NONE);
OnRevSync(outputs, power, turnpct)	Run the specified outputs in reverse with regulated synchronization using the specified turn ratio.	OnRevSync(OUT_A, 75, 50);
OnRevSyncEx(outputs, power, turnpct, const reset)	Same as OnRevSync but you can specify which counters, if any, to reset.	OnRevSyncEx(OUT_A, 75, 50, RESET_NONE);
RotateMotor(outputs, power, angle)	Run the outputs for the specified number of degrees at the specified power level. Calls to any of the RotateMotor* functions do not return until the motors reach the specified target or their power level is set to zero.	RotateMotor(OUT_AB, 75, 360);

RotateMotorPID(outputs, power, angle, p, i, d)	Same as RotateMotor but you can also control the proportional, integral, and derivative factors used by the firmware's motor control algorithm.	RotateMotorPID(OUT_BC, 50, 720, 50, 60, 60);
RotateMotorEx(outputs, power, angle, turnpct, bSync, bStop)	Same as RotateMotor but you can also set the turn percent, turn synchronization on or off, and indicate whether the motors should be stopped at the target angle.	RotateMotorEx(OUT_AB, 75, 360, 50, true, true);
RotateMotorExPID(outputs, power, angle, turnpct, bSync, bStop, p, i, d)	Same as RotateMotorEx but you can also control the proportional, integral, and derivative factors used by the firmware's motor control algorithm.	RotateMotorExPID(OUT_AB, 75, 360, 50, true, true, 30, 50, 90);
ResetTachoCount(outputs)	Reset the tachometer count and the tachometer limit goal for the outputs.	ResetTachoCount(OUT_AB);
ResetBlockTachoCount (outputs)	Reset the block-relative position counter for the outputs.	ResetBlockTachoCount (OUT_BC);
ResetRotationCount(outputs)	Reset the rotation sensor for the outputs.	ResetRotationCount (OUT_BC);
ResetAllTachoCounts(output)	Reset all three position counters and reset the current tachometer limit goal for the outputs.	ResetAllTachoCounts (OUT_ABC);
SetOutput(outputs, const field1, val1, ..., const fieldN, valN)	Set the specified field(s) of the outputs to the value(s) provided.	SetOutput(OUT_AB, TachoLimit, 720);
GetOutput(output, const field)	Get the value of the specified field for the output.	x = GetOutput(OUT_A, TachoLimit);
MotorMode(output)	Get the mode of the output.	x = MotorMode(OUT_A);
MotorPower(output)	Get the power level of the output.	x = MotorPower(OUT_A);
MotorActualSpeed(output)	Get the actual speed value of the output.	x = MotorActualSpeed (OUT_A);

MotorTachoCount(output)	Get the tachometer count value of the output.	x = MotorTachoCount (OUT_B);
MotorTachoLimit(output)	Get the tachometer limit value of the output.	x = MotorTachoLimit (OUT_B);
MotorRunState(output)	Get the run state value of the output.	x = MotorRunState(OUT_C);
MotorTurnRatio(output)	Get the turn ratio value of the output.	x = MotorTurnRatio(OUT_A);
MotorRegulation(output)	Get the regulation value of the output.	x = MotorRegulation(OUT_B);
MotorOverload(output)	Get the overload value of the output	x = MotorOverload(OUT_A);
MotorRegPValue(output)	Get the proportional PID value of the output.	p = MotorRegPValue(OUT_A);
MotorRegIValue(output)	Get the integral PID value of the output.	i = MotorRegIValue(OUT_A);
MotorRegDValue(output)	Get the derivative PID value of the output.	d = MotorRegDValue(OUT_A);
MotorBlockTachoCount (output)	Get the block-relative position counter value of the output.	x = MotorBlockTachoCount (OUT_A);
MotorRotationCount(output)	Get the rotation sensor value of the output	x = MotorRotationCount (OUT_A);
MotorPwnFreq()	Get the current motor pulse width modulation frequency	X = MotorPwnFreq();
SetMotorPwnFreq(value)	Set the current motor pulse width modulation frequency	SetMotorPwnFreq(x);

IOMA Functions

The functions in this section provide direct low-level access to input and output module values.

Function	Description	Example
IOMA(const n)	Get the specified IO Map Address value.	x = IOMA(InputIORawValue (S3));
SetIOMA(const n, value)	Set the specified IO Map Address to the value provided.	SetIOMA(OutputIOPower (OUT_A), x);

Sound Functions

The functions in this section are for controlling the sound module capabilities of the NXT.

Function	Description	Example
PlayTone(frequency, duration)	Play a single tone of the specified frequency and duration.	PlayTone(440, 500); Wait(500);
PlayToneEx(frequency, duration, volume, bLoop)	Play a single tone of the specified frequency, duration, and volume.	PlayToneEx(440, 500, 2, false);
PlayFile(filename)	Play the specified sound or melody file.	PlayFile("startup.rso");
PlayFileEx(filename, volume, bLoop)	Play the sound or melody file at the specified volume. Optionally loop at the end.	PlayFileEx("startup.rso", 2, false);
SoundFlags()	Return the current sound flags.	X = SoundFlags();
SetSoundFlags(flags)	Set the current sound flags.	SetSoundFlags (SOUND_FLAGS_UPDATE);
SoundState()	Return the current sound state.	X = SoundState();
SetSoundModuleState(state)	Set the current sound state.	SetSoundModuleState (SOUND_STATE_STOP);
SoundMode()	Return the current sound mode.	X = SoundMode();
SetSoundMode(mode)	Set the current sound mode.	SetSoundMode (SOUND_MODE_ONCE);
SoundFrequency()	Return the current sound frequency.	X = SoundFrequency();
SetSoundFrequency(freq)	Set the current sound frequency.	SetSoundFrequency(880);
SoundDuration()	Return the current sound duration.	X = SoundDuration();
SetSoundDuration(dur)	Set the current sound duration.	SetSoundDuration(500);

SoundSampleRate()	Return the current sound sample rate.	X = SoundSampleRate();
SetSoundSampleRate(rate)	Set the current sound sample rate.	SetSoundSampleRate(4000);
SoundVolume()	Return the current sound volume.	X = SoundVolume();
SetSoundVolume(volume)	Set the current sound volume.	SetSoundVolume(2);
StopSound()	Stop playing the current tone or file.	StopSound();

IO Control Functions

The functions in this section are for communicating with the AVR processor in the NXT.

Function	Description	Example
PowerDown() or SleepNow()	Turn off the NXT immediately.	PowerDown();
RebootInFirmwareMode()	Reboot the NXT in SAMBA or firmware download mode.	RebootInFirmwareMode();

Display Functions

The functions in this section are for controlling the display module capabilities of the NXT.

Function	Description	Example
NumOut(x, y, value, clear = false)	Draw a numeric value on the screen at the specified x and y location.	NumOut(0, LCD_LINE1, x);
TextOut(x, y, msg, clear = false)	Draw a text value on the screen at the specified x and y location.	TextOut(0, LCD_LINE8, msg_buffer, true);
GraphicOut(x, y, filename, clear = false)	Draw the specified graphic icon file on the screen at the specified x and y location.	GraphicOut(40, 40, "image.ric");

GraphicOutEx(x, y, filename, vars, clear = false)	Draw the specified graphic icon file on the screen at the specified x and y location. Use the values contained in the vars array to transform the drawing commands contained within the specified icon file.	GraphicOutEx(40, 40, "image. ric", variables);
CircleOut(x, y, radius, clear = false)	Draw a circle on the screen with its center at the specified x and y location, using the specified radius.	CircleOut(40, 40, 10);
LineOut(x1, y1, x2, y2, clear = false)	Draw a line on the screen from x1, y1 to x2, y2.	LineOut(40, 40, 10, 10);
PointOut(x, y, clear = false)	Draw a point on the screen at x, y.	PointOut(40, 40);
RectOut(x, y, width, height, clear = false)	Draw a rectangle on the screen at x, y with the specified width and height.	RectOut(40, 40, 30, 10);
ResetScreen()	Restore the standard NXT running program screen.	ResetScreen();
ClearScreen()	Clear the NXT LCD to a blank screen.	ClearScreen();
DisplayFlags()	Return the current display flags.	X = DisplayFlags();
SetDisplayFlags(n)	Set the current display flags.	SetDisplayFlags(x);
DisplayEraseMask()	Return the current display erase mask.	X = DisplayEraseMask();
SetDisplayEraseMask(n)	Set the current display erase mask.	SetDisplayEraseMask(x);
DisplayUpdateMask()	Return the current display update mask.	X = DisplayUpdateMask();
SetDisplayUpdateMask(n)	Set the current display update mask.	SetDisplayUpdateMask(x);
DisplayDisplay()	Return the current display memory address.	X = DisplayDisplay();
SetDisplayDisplay(n)	Set the current display memory address.	SetDisplayDisplay(x);

DisplayTextLinesCenter Flags()	Return the current display text lines center flags.	X = DisplayTextLinesCenter Flags();
SetDisplayTextLinesCenter Flags(n)	Set the current display text lines center flags.	SetDisplayTextLinesCenter Flags(x);
GetDisplayNormal(x, line, count, data)	Read "count" bytes from the normal display memory into the data array. Start reading from the specified x, line coordinate. Each byte of data read from screen memory is a vertical strip of 8 bits at the desired location. Each bit represents a single pixel on the LCD screen. Use TEXTLINE_1 through TEXTLINE_8 for the "line" parameter.	GetDisplayNormal(0, TEXTLINE_1, 8, ScreenMem);
SetDisplayNormal(x, line, count, data)	Write "count" bytes to the normal display memory from the data array. Start writing at the specified x, line coordinate.	SetDisplayNormal(0, TEXTLINE_1, 8, ScreenMem);
GetDisplayPopup(x, line, count, data)	Read "count" bytes from the popup display memory into the data array.	GetDisplayPopup(0, TEXTLINE_1, 8, PopupMem);
SetDisplayPopup(x, line, count, data)	Write "count" bytes to the popup display memory from the data array.	SetDisplayPopup(0, TEXTLINE_1, 8, PopupMem);

Loader Functions

The functions in this section are for controlling the loader module capabilities of the NXT. They include all forms of file reading and writing.

Function	Description	Example
FreeMemory()	Get the number of bytes of flash memory that are available for use.	x = FreeMemory();
CreateFile(filename, size, out handle)	Create a new file with the specified size for writing. The handle argument must be a variable.	result = CreateFile("data.txt", 1024, handle);

OpenFileAppend(filename, out size, out handle)	Open an existing file for writing. The size and handle arguments must be variables.	result = OpenFileAppend ("data.txt", fsize, handle);
OpenFileRead(filename, out size, out handle)	Open an existing file for reading. The size and handle arguments must be variables.	result = OpenFileRead ("data.txt", fsize, handle);
CloseFile(handle)	Close the file associated with the specified file handle.	result = CloseFile(handle);
ResolveHandle(filename, out handle, out bWriteable)	Resolve a file handle. The handle and bWriteable arguments must be variables.	result = ResolveHandle("data. txt", handle, bCanWrite);
RenameFile(oldfilename, newfilename)	Rename a file from the old filename to the new filename.	result = RenameFile("data. txt", "mydata.txt");
DeleteFile(filename)	Delete the specified file.	result = DeleteFile("data. txt");
ResizeFile(filename, newsize)	Resize the specified file to the new size.	result = ResizeFile("data.txt", 1024);
Read(handle, out value)	Read a numeric value from the file. The handle and value arguments must be variables. The type of the value argument determines how many bytes are read.	result = Read(handle, value);
ReadLn(handle, out value)	Read a numeric value from the file. It reads two more bytes from the file which should be CRLF.	result = ReadLn(handle, value);
ReadBytes(handle, in/out length, out buf)	Read the specified number of bytes from the file. The handle, length, and buf argumentss must be variables. The actual number of bytes read is returned in the length parameter.	result = ReadBytes(handle, len, buffer);
Write(handle, value)	Write a numeric value to a file. The handle parameter must be a variable. The type of the value argument determines how many bytes are written.	result = Write(handle, value);

WriteLn(handle, value)	Same as Write but also writes a CRLF to the file following the numeric data.	result = WriteLn(handle, value);
WriteString(handle, str, out count)	Write the string to a file. The handle and count arguments must be variables. The total number of bytes written is returned in count.	result = WriteString(handle, "testing", count);
WriteLnString(handle, str, out count)	Same as WriteString but also writes a CRLF to the file. result =	WriteLnString(handle, "testing", count);
WriteBytes(handle, data, out count)	Write the contents of the data array to a file. The total number of bytes written is returned in count.	result = WriteBytes(handle, buffer, count);
WriteBytesEx(handle, in/out length, buf)	Write the specified number of bytes to the file associated with the specified handle. The total number of bytes written is returned in count.	result = WriteBytesEx(handle, len, buffer);

Button Functions

The functions in this section are for controlling the button module capabilities of the NXT.

Function	Description	Example
ButtonCount(btn, reset)	Return the number of times the specified button has been pressed since the last time the button press count was reset. Optionally clear the count after reading it.	Value = ButtonCount(BTN1, true);
ButtonPressed(btn, reset)	Return whether the specified button is pressed. Optionally clear the press count.	Value = ButtonPressed(BTN1, true);
ReadButtonEx(btn, reset, out pressed, out count)	Read the specified button. Sets the pressed and count parameters with the current state of the button. Optionally reset the press count after reading it.	ReadButtonEx(BTN1, true, pressed, count);

ButtonPressCount(const btn)	Return the press count of the specified button.	Value = ButtonPressCount (BTN1);
ButtonLongPressCount (const btn)	Return the long press count of the specified button.	Value = ButtonLongPress Count(BTN1);
ButtonShortReleaseCount (const btn)	Return the short release count of the specified button.	Value = ButtonShortRelease Count(BTN1);
ButtonLongReleaseCount (const btn)	Return the long release count of the specified button.	Value = ButtonLongRelease Count(BTN1);
ButtonReleaseCount (const btn)	Return the release count of the specified button.	Value = ButtonReleaseCount (BTN1);
ButtonState(const btn)	Return the state of the specified button.	Value = ButtonState(BTN1);

User Interface Functions

The functions in this section are for controlling the user interface (UI) module capabilities of the NXT.

Function	Description	Example
Volume()	Return the user interface volume level.	x = Volume();
SetVolume(value)	Set the user interface volume level.	SetVolume(3);
BatteryLevel()	Return the battery level in millivolts.	x = BatteryLevel();
BluetoothState()	Return the Bluetooth state.	x = BluetoothState();
SetBluetoothState(value)	Set the Bluetooth state.	SetBluetoothState (UI_BT_STATE_OFF);
CommandFlags()	Return the command flags.	x = CommandFlags();
SetCommandFlags(value)	Set the command flags.	SetCommandFlags(UI_ FLAGS_REDRAW_STATUS);
UIState()	Return the user interface state.	x = UIState();
SetUIState(value)	Set the user interface state.	SetUIState(UI_STATE_ LOW_BATTERY);

UIButton()	Return user interface button information.	x = UIButton();
SetUIButton(value)	Set user interface button information.	SetUIButton(UI_BUTTON_ENTER);
VMRunState()	Return VM run state information.	x = VMRunState();
SetVMRunState(value)	Set VM run state information.	SetVMRunState(0);
BatteryState()	Return battery state information (0..4).	x = BatteryState();
RechargeableBattery()	Return whether the NXT has a rechargeable battery installed or not.	x = RechargeableBattery();
ForceOff(n)	Force the NXT to turn off if the specified value is greater than zero.	ForceOff(true);
UsbState()	Return USB state information (0=disconnected, 1=connected, 2=working).	x = UsbState();
OnBrickProgramPointer()	Return the current OBP (on-brick program) step.	x = OnBrickProgramPointer();
SetOnBrickProgramPointer (value)	Set the current OBP (on-brick program) step.	SetOnBrickProgramPointer (2);
LongAbort()	Return a boolean value indicating whether a long button press is required to abort a running program. This function requires the enhanced NBC/NXC firmware.	x = LongAbort();
SetLongAbort(value)	Set the current long abort configuration. This function requires the enhanced NBC/NXC firmware.	SetLongAbort(true);

Lowspeed I²C Functions

The functions in this section are for controlling the lowspeed module capabilities of the NXT.

Function	Description	Example
LowspeedWrite(port, returnlen, buffer) or I2CWrite(port, returnlen, buffer)	Start a transaction to write the bytes contained in the array buffer to the I2C device on the specified port.	X = LowspeedWrite(S1, 1, inbuffer);
LowspeedStatus(port, out bytesready) or I2CStatus(port, out bytesready)	Check the status of the I2C communication on the specified port and return the number of bytes ready.	X = LowspeedStatus(S1, nRead);
LowspeedCheckStatus(port) or I2CCheckStatus(port)	Check the status of the I2C communication on the specified port.	X = I2CCheckStatus(S1);
LowspeedBytesReady(port) or I2CBytesReady(port)	Check how many bytes are ready on the specified port.	nRead = I2CBytesReady(S1);
LowspeedRead(port, buflen, out buffer) or I2CRead(port, buflen, out buffer)	Read the specified number of bytes from the I2C device on the specified port and store the bytes read in the array buffer provided.	X = LowspeedRead(S1, 1, outbuffer);
I2CBytes(port, inbuf, in/out count, out outbuf)	This is a higher-level wrapper around the three main I2C functions. It also maintains a "last good read" buffer and returns values from that buffer if the I2C communication transaction fails.	X = I2CBytes(S4, writebuf, cnt, readbuf);
LSMode(const port)	Returns the mode of the lowspeed communication over the specified port.	X = LSMode(S1);
LSChannelState(const port)	Returns the channel state of the lowspeed communication over the specified port.	X = LSChannelState(S1);
LSErrorType(const port)	Returns the error type of the lowspeed communication over the specified port.	X = LSErrorType(S1);

LSState()	Returns the state of the lowspeed module.	X = LSState();
LSSpeed()	Returns the speed of the lowspeed module.	X = LSSpeed();

Bluetooth Functions

The functions in this section are for communicating via Bluetooth with other devices. Use the standard NXT communication protocol and mailbox system or NXT direct commands with these routines. You can also directly write Bluetooth messages which do not use the NXT protocols using API functions in this section.

Function	Description	Example
SendRemoteBool (connection, queue, bvalue)	Sends a boolean value to the device on the specified connection.	x = SendRemoteBool (1, queue, false);
SendRemoteNumber (connection, queue, value)	Sends a numeric value to the device on the specified connection.	x = SendRemoteNumber (1, queue, 123);
SendRemoteString (connection, queue, strval)	Sends a string value to the device on the specified connection.	x = SendRemoteString (1, queue, "hello world");
SendResponseBool(queue, bvalue)	Sends a boolean value as a response to a received message.	x = SendResponseBool (queue, false);
SendResponseNumber (queue, value)	Sends a numeric value as a response to a received message.	x = SendResponseNumber (queue, 123);
SendResponseString(queue, strval)	Sends a string value as a response to a received message.	x = SendResponseString (queue, "hello world");
ReceiveRemoteBool(queue, remove, out bvalue)	Use this function on a primary brick to receive a boolean value from a secondary device. Optionally remove the last read message from the message queue.	x = ReceiveRemoteBool (queue, true, bvalue);
ReceiveRemoteNumber (queue, remove, out value)	Same as ReceiveRemoteBool but for numeric messages.	x = ReceiveRemoteBool (queue, true, value);

ReceiveRemoteString(queue, remove, out strval)	Same as ReceiveRemoteBool but for string messages.	x = ReceiveRemoteString (queue, true, strval);
ReceiveRemoteMessageEx (queue, remove, out strval, out val, out bval)	Use this function to receive a string, boolean, or numeric value from another device.	x = ReceiveRemoteMessage Ex(queue, true, strval, val, bval);
SendMessage(queue, msg)	Write the message buffer contents to the specified mailbox or message queue.	x = SendMessage(mbox, data);
ReceiveMessage(queue, remove, out buffer)	Retrieve a message from the specified queue and write it to the buffer provided.	x = RecieveMessage(mbox, true, buffer);
BluetoothStatus(connection)	Returns the status of the specified Bluetooth connection. Avoid calling BluetoothWrite or any other API function that writes data over a Bluetooth connection while BluetoothStatus returns STAT_COMM_PENDING.	x = BluetoothStatus(1);
BluetoothWrite(connection, buffer)	Write the data in the buffer to the device on the specified Bluetooth connection.	x = BluetoothWrite(1, data);
RemoteMessageRead (connection, queue)	Send a MessageRead direct command to the device on the specified connection.	RemoteMessageRead(1, 5);
RemoteMessageWrite (connection, queue, msg)	Send a MessageWrite direct command to the device on the specified connection.	RemoteMessageWrite (1, 5, "test");
RemoteStartProgram (connection, filename)	Send a StartProgram direct command to the device on the specified connection.	RemoteStartProgram (1, "myprog.rxe");
RemoteStopProgram (connection)	Send a StopProgram direct command to the device on the specified connection.	RemoteStopProgram(1);
RemotePlaySoundFile (connection, filename, bLoop)	Send a PlaySoundFile direct command to the device on the specified connection.	RemotePlaySoundFile (1, "click.rso", false);

RemotePlayTone(connection, frequency, duration)	Send a PlayTone direct command to the device on the specified connection.	RemotePlayTone(1, 440, 1000);
RemoteStopSound (connection)	Send a StopSound direct command to the device on the specified connection.	RemoteStopSound(1);
RemoteKeepAlive (connection)	Send a KeepAlive direct command to the device on the specified connection.	RemoteKeepAlive(1);
RemoteResetScaledValue (connection, port)	Send a ResetScaledValue direct command to the device on the specified connection.	RemoteResetScaledValue (1, S1);
RemoteResetMotorPosition (connection, port, bRelative)	Send a ResetMotorPosition direct command to the device on the specified connection.	RemoteResetMotorPosition (1, OUT_A, true);
RemoteSetInputMode (connection, port, type, mode)	Send a SetInputMode direct command to the device on the specified connection.	RemoteSetInputMode (1, S1, IN_TYPE_LOWSPEED, IN_MODE_RAW);
RemoteSetOutputState (connection, port, speed, mode, regmode, turnpct, runstate, tacholimit)	Send a SetOutputState direct command to the device on the specified connection.	x = RemoteSetOutputState (1, OUT_A, 75, OUT_ MODE_MOTORON, OUT_REGMODE_IDLE, 0, OUT_RUNSTATE_RUNNING, 0);

USB Functions
The functions in this section are for communicating via USB with other devices.

Function	Description	Example
GetUSBInputBuffer(const offset, count, out data)	Read count bytes of data from the USB input buffer at the specified offset and write it to the buffer provided.	GetUSBInputBuffer (0, 10, buffer);
SetUSBInputBuffer(const offset, count, data)	Write count bytes of data to the USB input buffer at the specified offset.	SetUSBInputBuffer(0, 10, buffer);
SetUSBInputBufferInPtr(n)	Set the input pointer of the USB input buffer to the specified value.	SetUSBInputBufferInPtr(0);

USBInputBufferInPtr()	Return the value of the input pointer of the USB input buffer.	byte x = USBInputBuffer InPtr();
SetUSBInputBufferOutPtr(n)	Set the output pointer of the USB input buffer to the specified value.	SetUSBInputBufferOutPtr(o);
USBInputBufferOutPtr()	Return the value of the output pointer of the USB input buffer.	byte x = USBInputBufferOut Ptr();
GetUSBOutputBuffer(const offset, count, out data)	Read count bytes of data from the USB output buffer at the specified offset and write it to the buffer provided.	GetUSBOutputBuffer(0, 10, buffer);
SetUSBOutputBuffer(const offset, count, data)	Write count bytes of data to the USB output buffer at the specified offset.	SetUSBOutputBuffer(0, 10, buffer);
SetUSBOutputBufferInPtr(n)	Set the input pointer of the USB output buffer to the specified value.	SetUSBOutputBufferInPtr(o);
USBOutputBufferInPtr()	Return the value of the input pointer of the USB output buffer.	byte x = USBOutputBuffer InPtr();
SetUSBOutputBufferOutPtr(n)	Set the output pointer of the USB output buffer to the specified value.	SetUSBOutputBufferOutPtr(o);
USBOutputBufferOutPtr()	Return the value of the output pointer of the USB output buffer.	byte x = USBOutputBuffer OutPtr();
GetUSBPollBuffer(const offset, count, out data)	Read count bytes of data from the USB poll buffer and write it to the buffer provided.	GetUSBPollBuffer(0, 10, buffer);
SetUSBPollBuffer(const offset, count, data)	Write count bytes of data to the USB poll buffer at the specified offset.	SetUSBPollBuffer(0, 10, buffer);
SetUSBPollBufferInPtr(n)	Set the input pointer of the USB poll buffer to the specified value.	SetUSBPollBufferInPtr(o);

USBPollBufferInPtr()	Return the value of the input pointer of the USB poll buffer.	byte x = USBPollBufferInPtr();
SetUSBPollBufferOutPtr(n)	Set the output pointer of the USB poll buffer to the specified value.	SetUSBPollBufferOutPtr(o);
USBPollBufferOutPtr()	Return the value of the output pointer of the USB poll buffer.	byte x = USBPollBufferOut Ptr();
SetUSBState(n)	Set the USB state to the specified value.	SetUSBState(o);
USBState()	Return the USB state.	byte x = USBState();

High Speed Functions

The functions in this section are for communicating with other devices via the port 4 RS485 high-speed communication port.

Function	Description	Example
GetHSInputBuffer(const offset, count, out data)	Read count bytes of data from the High Speed input buffer and write it to the buffer provided.	GetHSInputBuffer(o, 10, buffer);
SetHSInputBuffer(const offset, count, data)	Write count bytes of data to the High Speed input buffer at the specified offset.	SetHSInputBuffer(o, 10, buffer);
SetHSInputBufferInPtr(n)	Set the input pointer of the High Speed input buffer to the specified value.	SetHSInputBufferInPtr(o);
HSInputBufferInPtr()	Return the value of the input pointer of the High Speed input buffer.	byte x = HSInputBufferInPtr();
SetHSInputBufferOutPtr(n)	Set the output pointer of the High Speed input buffer to the specified value.	SetHSInputBufferOutPtr(o);
HSInputBufferOutPtr()	Return the value of the output pointer of the High Speed input buffer.	byte x = HSInputBuffer OutPtr();

GetHSOutputBuffer(const offset, count, out data)	Read count bytes of data from the High Speed output buffer and write it to the buffer provided.	GetHSOutputBuffer(0, 10, buffer);
SetHSOutputBuffer(const offset, count, data)	Write count bytes of data to the High Speed output buffer at the specified offset.	SetHSOutputBuffer(0, 10, buffer);
SetHSOutputBufferInPtr(n)	Set the Output pointer of the High Speed output buffer to the specified value.	SetHSOutputBufferInPtr(0);
HSOutputBufferInPtr()	Return the value of the Output pointer of the High Speed output buffer.	byte x = HSOutputBuffer InPtr();
SetHSOutputBufferOutPtr(n)	Set the output pointer of the High Speed output buffer to the specified value.	SetHSOutputBufferOutPtr(0);
HSOutputBufferOutPtr()	Return the value of the output pointer of the High Speed output buffer.	byte x = HSOutputBuffer OutPtr();
SetHSFlags(n)	Set the High Speed flags to the specified value.	SetHSFlags(0);
HSFlags()	Return the value of the High Speed flags.	byte x = HSFlags();
SetHSSpeed(n)	Set the High Speed speed to the specified value.	SetHSSpeed(1);
HSSpeed()	Return the value of the High Speed speed.	byte x = HSSpeed();
SetHSState(n)	Set the High Speed state to the specified value.	SetHSState(1);
HSState()	Return the value of the High Speed state.	byte x = HSState();

Low-level Bluetooth Functions

The functions in this section provide direct low-level access to the Bluetooth input and output buffers as well as the brick data, Bluetooth device, and Bluetooth connection tables.

Function	Description	Example
GetBTInputBuffer(const offset, count, out data)	Read count bytes of data from the Bluetooth input buffer and write it to the buffer provided.	GetBTInputBuffer(0, 10, buffer);
SetBTInputBuffer(const offset, count, data)	Write count bytes of data to the Bluetooth input buffer at the specified offset.	SetBTInputBuffer(0, 10, buffer);
SetBTInputBufferInPtr(n)	Set the input pointer of the Bluetooth input buffer to the specified value.	SetBTInputBufferInPtr(0);
BTInputBufferInPtr()	Return the value of the input pointer of the Bluetooth input buffer.	byte x = BTInputBufferInPtr();
SetBTInputBufferOutPtr(n)	Set the output pointer of the Bluetooth input buffer to the specified value.	SetBTInputBufferOutPtr(0);
BTInputBufferOutPtr()	Return the value of the output pointer of the Bluetooth input buffer.	byte x = BTInputBufferOut Ptr();
GetBTOutputBuffer(const offset, count, out data)	Read count bytes of data from the Bluetooth output buffer and write it to the buffer provided.	GetBTOutputBuffer(0, 10, buffer);
SetBTOutputBuffer(const offset, count, data)	Write count bytes of data to the Bluetooth output buffer at the specified offset.	SetBTOutputBuffer(0, 10, buffer);
SetBTOutputBufferInPtr(n)	Set the input pointer of the Bluetooth output buffer to the specified value.	SetBTOutputBufferInPtr(0);
BTOutputBufferInPtr()	Return the value of the input pointer of the Bluetooth output buffer.	byte x = BTOutputBuffer InPtr();
SetBTOutputBufferOutPtr(n)	Set the output pointer of the Bluetooth output buffer to the specified value.	SetBTOutputBufferOutPtr(0);

BTOutputBufferOutPtr()	Return the value of the output pointer of the Bluetooth output buffer.	byte x = BTOutputBuffer OutPtr();
BTDeviceCount()	Return the number of devices defined within the Bluetooth device table.	byte x = BTDeviceCount();
BTDeviceNameCount()	Return the number of device names defined within the Bluetooth device table. This usually has the same value as BTDeviceCount but it can differ in some instances.	byte x = BTDeviceName Count();
BTDeviceName(const idx)	Return the name of the device at the specified index in the Bluetooth device table.	string name = BTDeviceName(o);
BTConnectionName(const idx)	Return the name of the device at the specified index in the Bluetooth connection table.	string name = BTConnectionName(o);
BTConnectionPinCode (const idx)	Return the pin code of the device at the specified index in the Bluetooth connection table.	string pincode = BTConnectionPinCode(o);
BrickDataName()	Return the name of the NXT.	string name = BrickDataName();
GetBTDeviceAddress(const idx, out data)	Read the address of the device at the specified index within the Bluetooth device table and store it in the data buffer provided.	GetBTDeviceAddress (o, buffer);
GetBTConnectionAddress(const idx, out data)	Read the address of the device at the specified index within the Bluetooth connection table and store it in the data buffer provided.	GetBTConnectionAddress(o, buffer);
GetBrickDataAddress(out data)	Read the address of the NXT and store it in the data buffer provided.	GetBrickDataAddress(buffer);
BTDeviceClass(const idx)	Return the class of the device at the specified index within the Bluetooth device table.	long class = BTDeviceClass(o);

BTDeviceStatus(const idx)	Return the status of the device at the specified index within the Bluetooth device table.	byte status = BTDevice Status(o);
BTConnectionClass(const idx)	Return the class of the device at the specified index within the Bluetooth connection table.	long class = BTConnection Class(o);
BTConnectionHandleNum (const idx)	Return the handle number of the device at the specified index within the Bluetooth connection table.	byte handlenum = BTConnectionHandleNum(o);
BTConnectionStreamStatus (const idx)	Return the stream status of the device at the specified index within the Bluetooth connection table.	byte streamstatus = BTConnectionStreamStatus (o);
BTConnectionLinkQuality (const idx)	Return the link quality of the device at the specified index within the Bluetooth connection table.	byte linkquality = BTConnectionLinkQuality(o);
BrickDataBluecoreVersion()	Return the bluecore version of the NXT.	int bv = BrickData BluecoreVersion();
BrickDataBtStateStatus()	Return the Bluetooth state status of the NXT.	int x = BrickDataBtState Status();
BrickDataBtHardwareStatus()	Return the Bluetooth hardware status of the NXT.	int x = BrickDataBtHardware Status();
BrickDataTimeoutValue()	Return the timeout value of the NXT.	int x = BrickDataTimeout Value();

System Call Functions
The functions in this section provide easy access to low-level NXT system calls.

Function	Description	Example

SysDrawText(DrawTextType & args)	Draw text on the NXT LCD given the DrawTextType structure.	DrawTextType dt;
		dt.Location.X = 0;
		dt.Location.Y = LCD_LINE1;
		dt.Text = "Please Work";
		dt.Options = 0x01;
		SysDrawText(dt);

SysDrawPoint (DrawPointType & args)	Draw or clear a pixel on the NXT LCD given the DrawPointType structure.	DrawPointType dp;
		dp.Location.X = 20;
		dp.Location.Y = 20;
		// clear this pixel
		dp.Options = 0x04;
		SysDrawPoint(dp);

SysDrawLine (DrawLineType & args)	Draw or clear a line on the NXT LCD given the DrawLineType structure.	DrawLineType dl;
		dl.StartLoc.X = 20;
		dl.StartLoc.Y = 20;
		dl.EndLoc.X = 60;
		dl.EndLoc.Y = 60;
		dl.Options = 0x01;
		SysDrawLine(dl);

SysDrawCircle (DrawCircleType & args)	Draw or clear a circle on the NXT LCD given the DrawCircleType structure.	DrawCircleType dc;
		dc.Center.X = 20;
		dc.Center.Y = 20;
		dc.Size = 10; // radius
		dc.Options = 0x01;
		SysDrawCircle(dc);

SysDrawRect (DrawRectType & args)	Draw or clear a rectangle on the NXT LCD given the DrawRectType structure.	DrawRectType dr;
		dr.Location.X = 20;
		dr.Location.Y = 20;
		dr.Size.Width = 20;
		dr.Size.Height = 10;
		dr.Options = 0x00;
		SysDrawRect(dr);

SysDrawGraphic (DrawGraphicType & args)	Draw an NXT picture on the NXT LCD given the DrawGraphicType structure.	DrawGraphicType dg;
		dg.Location.X = 20;
		dg.Location.Y = 20;
		dg.Filename = "image.ric";
		ArrayInit(dg.Variables, 0, 10);
		dg.Variables[0] = 12;
		dg.Variables[1] = 14;
		SysDrawGraphic(dg);

SysSetScreenMode
(SetScreenModeType & args)
 	
SCREEN_MODE_RESTORE
SCREEN_MODE_CLEAR
	Set the screen mode of the NXT
	LCD. The enhanced firmware
	lets you clear the screen in
	addition to restoring it to the
	default running screen.
		SetScreenModeType ssm;
		// restore default screen
		ssm.ScreenMode = 0x0;
		SysSetScreenMode(ssm);

SysSoundPlayFile
(SoundPlayFileType & args)
	Play a sound file given the
	SoundPlayFileType structure.
	The sound file can either be
	an RSO file or a melody file.
	SoundPlayFileType spf;
		spf.Filename = "hello.rso";
		spf.Loop = false;
		spf.SoundLevel = 3;
		SysSoundPlayFile(spf);

SysSoundPlayTone
(SoundPlayToneType & args)
	Play a tone given the
	SoundPlayToneType
	structure.
		SoundPlayToneType spt;
		spt.Frequency = 440;
		spt.Duration = 1000;
		spt.Loop = false;
		spt.SoundLevel = 3;
		SysSoundPlayTone(spt);

SysSoundGetState (SoundGetStateType & args)	Retrieve information about the sound module state via the SoundGetStateType structure.	SoundGetStateType sgs; SysSoundGetState(sgs); if (sgs.State == SOUND_ STATE_IDLE) {/* do stuff */}
SysSoundSetState (SoundSetStateType & args)	Set the sound module state via the SoundSetStateType structure.	SoundSetStateType sss; sss.State = SOUND_STATE_STOP; SysSoundSetState(sss);

SysReadButton
(ReadButtonType & args)
	Read button state information
	via the ReadButtonType
	structure.
		ReadButtonType rb;
		rb.Index = BTNRIGHT;
		SysReadButton(rb);
		if (rb.Pressed)
		{/* do something */}

SysRandomNumber (RandomNumberType & args)	Obtain a random number via the RandomNumberType structure.	RandomNumberType rn; SysRandomNumber(rn); int myRandomValue = rn.Result;
SysGetStartTick (GetStartTickType & args)	Obtain the tick value at the time your program began executing via the GetStartTickType structure.	GetStartTickType gst; SysGetStartTick(gst); unsigned long myStart = gst.Result;
SysKeepAlive(KeepAlive Type & args)	Reset the sleep timer via the KeepAliveType structure.	KeepAliveType ka; SysKeepAlive(ka);

SysFileOpenWrite (FileOpenType & args)	Create a file that you can write to using the FileOpenType structure.	FileOpenType fo; fo.Filename = "myfile.txt"; fo.Length = 256; SysFileOpenWrite(fo); if (fo.Result == NO_ERR) { // write to the file }
SysFileOpenAppend (FileOpenType & args)	Open an existing file that you can write to using the FileOpenType structure.	FileOpenType fo; fo.Filename = "myfile.txt"; SysFileOpenAppend(fo); if (fo.Result == NO_ERR) { // write to the file }
SysFileOpenRead (FileOpenType & args)	Open an existing file for reading using the FileOpenType structure.	FileOpenType fo; fo.Filename = "myfile.txt"; SysFileOpenRead(fo); // open the file for reading if (fo.Result == NO_ERR) { // read from file }
SysFileRead(FileRead WriteType & args)	Read from a file using the FileReadWriteType structure.	FileReadWriteType fr; fr.FileHandle = fo.FileHandle; fr.Length = 12; SysFileRead(fr); if (fr.Result == NO_ERR) TextOut(0, 0, fr.Buffer);
SysFileWrite(FileReadWrite Type & args)	Write to a file using the FileReadWriteType structure.	FileReadWriteType fw; fw.FileHandle = fo.FileHandle; fw.Buffer = "data"; SysFileWrite(fw);
SysFileClose(FileCloseType & args)	Close a file using the FileCloseType structure.	FileCloseType fc; fc.FileHandle = fo.FileHandle; SysFileClose(fc);
SysFileResolveHandle (FileResolveHandleType & args)	Resolve the handle of a file using the FileResolve HandleType structure.	FileResolveHandleType frh; frh.Filename = "myfile.txt"; SysFileResolveHandle(frh);
SysFileRename (FileRenameType & args)	Rename a file using the FileRenameType structure.	FileRenameType fr; fr.OldFilename = "myfile.txt"; fr.NewFilename = "myfile2.txt"; SysFileRename(fr);

SysFileDelete(FileDelete Type & args)	Delete a file using the FileDeleteType structure.	FileDeleteType fd; fd.Filename = "myfile.txt"; SysFileDelete(fd);

SysCommLSWrite (CommLSWriteType & args)	Write to a Lowspeed sensor using the CommLSWriteType structure.	CommLSWriteType args;
		args.Port = S1;
		args.Buffer = myBuf;
		args.ReturnLen = 8;
		SysCommLSWrite(args);

SysCommLSCheckStatus (CommLSCheckStatusType & args)	Check the status of a Lowspeed sensor transaction using the CommLSCheckStatusType structure.	CommLSCheckStatusType args; args.Port = S1; SysCommLSCheckStatus (args);

SysCommLSRead (CommLSReadType & args)	Read from a Lowspeed sensor using the CommLSReadType structure.	CommLSReadType args;
		args.Port = S1;
		args.Buffer = myBuf;
		args.BufferLen = 8;
		SysCommLSRead(args);

SysMessageWrite (MessageWriteType & args)	Write a message to a queue (aka mailbox) using the MessageWriteType structure.	MessageWriteType args; args.QueueID = MAILBOX1; args.Message = "testing"; SysMessageWrite(args);

SysMessageRead (MessageReadType & args)	Read a message from a queue (aka mailbox) using the MessageReadType structure.	MessageReadType args;
		args.QueueID = MAILBOX1;
		args.Remove = true;
		SysMessageRead(args);

SysCommBTWrite (CommBTWriteType & args)	Write to a Bluetooth connection using the CommBTWriteType structure.	CommBTWriteType args;
		args.Connection = 1;
		args.Buffer = myData;
		SysCommBTWrite(args);

SysCommBTCheckStatus (CommBTCheckStatusType & args)	Check the status of a Bluetooth connection using the CommBTCheckStatusType structure.	CommBTCheckStatusType args; args.Connection = 1; SysCommBTCheckStatus(args);

SysIOMapRead (IOMapReadType & args)	Read data from a firmware module IOMap using the IOMapReadType structure.	IOMapReadType args;
		args.ModuleName = CommandModuleName;
		args.Offset = CommandOffsetTick;
		args.Count = 4;
		SysIOMapRead(args);

SysIOMapWrite
(IOMapWriteType & args)
	Write data to a firmware module IOMap using the IOMapWriteType structure.	IOMapWriteType args;
		args.ModuleName = SoundModuleName;
		args.Offset = SoundOffsetSampleRate;
		args.Buffer = theData;
		SysIOMapWrite(args);

SysIOMapReadByID
(IOMapReadByIDType & args)
	Read data from a firmware module IOMap using the IOMapReadByIDType structure. This requires the enhanced NBC/NXC firmware.
	IOMapReadByIDType args;
	args.ModuleID =
		CommandModuleID;
		args.Offset = CommandOffsetTick;
		args.Count = 4;
		SysIOMapReadByID(args);

SysIOMapWriteByID
(IOMapWriteByIDType & args)
	Write data to a firmware module IOMap using the IOMapWriteByIDType structure. This requires the enhanced NBC/NXC firmware.
	IOMapWriteByIDType args;
		args.ModuleID = SoundModuleID;
		args.Offset = SoundOffsetSampleRate;
		args.Buffer = theData;
		SysIOMapWriteByID(args);

SysDisplayExecuteFunction (DisplayExecuteFunctionType & args)	Execute the Display module's primary drawing function using the DisplayExecuteFunctionType structure. This requires the enhanced NBC/NXC firmware.	DisplayExecuteFunctionType args; args.Cmd = DISPLAY_ERASE_ALL; SysDisplayExecuteFunction (args);
SysCommExecuteFunction (CommExecuteFunctionType & args)	Execute the Comm module's primary function using the CommExecuteFunctionType structure. This requires the enhanced NBC/NXC firmware.	CommExecuteFunctionType args; args.Cmd = INTF_BTOFF; SysCommExecuteFunction (args);

SysLoaderExecuteFunction
(LoaderExecuteFunctionType & args)
	Execute the Loader module's primary function using the LoaderExecuteFunctionType structure. This requires the enhanced NBC/NXC firmware.	LoaderExecuteFunctionType args;
		// delete user flash
		args.Cmd = 0xA0;
		SysLoaderExecuteFunction (args);

SysCall(const funcID, args)	A generic macro that can be used to call any system function.	DrawTextType dtArgs;
		dtArgs.Location.X = 0;
		dtArgs.Location.Y = 0;
		dtArgs.Text = "Please Work";
		SysCall(DrawText, dtArgs);

HiTechnic Functions

The functions in this section are for using various sensors and devices made by HiTechnic.

Function	Description	Example
SetSensorHTGyro(port)	Configure the sensor on the specified port as a HiTechnic Gyro sensor.	SetSensorHTGyro(S1);
SensorHTGyro(port, offset)	Return the HiTechnic Gyro sensor value using the specified offset value.	x = SensorHTGyro(S1, 10);
SensorHTCompass(port)	Return the HiTechnic Compass sensor value.	x = SensorHTCompass(S1);
SensorHTColorNum(port)	Return the HiTechnic Color sensor color number value.	x = SensorHTColorNum(S1);
SensorHTIRSeekerDir(port)	Return the HiTechnic IR Seeker direction value.	x = SensorHTIRSeekerDir(S1);
ReadSensorHTAccel(port, out x, out y, out z)	Read X, Y, and Z axis acceleration values from the HiTechnic Accelerometer sensor. Returns a boolean value indicating whether or not the operation completed successfully.	ReadSensorHTAccel(S1, x, y, z);
ReadSensorHTColor(port, out ColorNum, out Red, out Green, out Blue)	Read color number, red, green, and blue values from the HiTechnic Color sensor. Returns a boolean value indicating whether or not the operation completed successfully.	ReadSensorHTColor(S1, c, r, g, b);
ReadSensorHTIRSeeker(port, out dir, out s1, out s3, out s5, out s7, out s9)	Read direction, and five signal strength values from the HiTechnic IRSeeker sensor. Returns a boolean value indicating whether or not the operation completed successfully.	ReadSensorHTIRSeeker(port, dir, s1, s3, s5, s7, s9);
HTPowerFunctionCommand(port, channel, cmd1, cmd2)	Execute a pair of Power Function motor commands on the specified channel using the HiTechnic iRLink device. Commands are HTPF_CMD_ STOP, HTPF_CMD_REV, HTPF_CMD_FWD, and HTPF_ CMD_BRAKE. Valid channels are HTPF_CHANNEL_1 through HTPF_CHANNEL_4.	HTPowerFunctionComma nd(S1, HTPF_CHANNEL_1, HTPF_CMD_STOP, HTPF_CMD_FWD);

HiTechnic iRLink RCX Functions

The functions in this section are for controlling an RCX or Scout programmable brick using the HiTechnic iRLink device.

Function	Description	Example
HTRCXSetIRLinkPort(port)	Set the global port in advance of using the HTRCX* API functions for sending RCX messages over the HiTechnic iRLink device.	HTRCXSetIRLinkPort(S1);
HTRCXPoll(src, value)	Send the Poll command to an RCX to read a signed 2-byte value at the specified source and value combination.	X = HTRCXPoll(RCX_ VariableSrc, 0);
HTRCXBatteryLevel()	Send the BatteryLevel command to an RCX to read the current battery level.	X = HTRCXBatteryLevel();
HTRCXPing()	Send the Ping command to an RCX.	HTRCXPing();
HTRCXDeleteTasks()	Send the DeleteTasks command to an RCX.	HTRCXDeleteTasks();
HTRCXStopAllTasks()	Send the StopAllTasks command to an RCX.	HTRCXStopAllTasks();
HTRCXPBTurnOff()	Send the PBTurnOff command to an RCX.	HTRCXPBTurnOff();
HTRCXDeleteSubs()	Send the DeleteSubs command to an RCX.	HTRCXDeleteSubs();
HTRCXClearSound()	Send the ClearSound command to an RCX.	HTRCXClearSound();
HTRCXClearMsg()	Send the ClearMsg command to an RCX.	HTRCXClearMsg();
HTRCXMuteSound()	Send the MuteSound command to an RCX.	HTRCXMuteSound();
HTRCXUnmuteSound()	Send the UnmuteSound command to an RCX.	HTRCXUnmuteSound();
HTRCXClearAllEvents()	Send the ClearAllEvents command to an RCX.	HTRCXClearAllEvents();

HTRCXSetOutput(outputs, mode)	Send the SetOutput command to an RCX to configure the mode of the specified outputs	HTRCXSetOutput(RCX_OUT_ A, RCX_OUT_ON);
HTRCXSetDirection(outputs, dir)	Send the SetDirection command to an RCX to configure the direction of the specified outputs.	HTRCXSetDirection(RCX_ OUT_A, RCX_OUT_FWD);
HTRCXSetPower(outputs, pwrsrc, pwrval)	Send the SetPower command to an RCX to configure the power level of the specified outputs.	HTRCXSetPower(RCX_ OUT_A, RCX_ConstantSrc, RCX_OUT_FULL);
HTRCXOn(outputs)	Send commands to an RCX to turn on the specified outputs.	HTRCXOn(RCX_OUT_A);
HTRCXOff(outputs)	Send commands to an RCX to turn off the specified outputs.	HTRCXOff(RCX_OUT_A);
HTRCXFloat(outputs)	Send commands to an RCX to float the specified outputs.	HTRCXFloat(RCX_OUT_A);
HTRCXToggle(outputs)	Send commands to an RCX to toggle the direction of the specified outputs.	HTRCXToggle(RCX_OUT_A);
HTRCXFwd(outputs)	Send commands to an RCX to set the specified outputs to the forward direction.	HTRCXFwd(RCX_OUT_A);
HTRCXRev(outputs)	Send commands to an RCX to set the specified outputs to the reverse direction.	HTRCXRev(RCX_OUT_A);
HTRCXOnFwd(outputs)	Send commands to an RCX to turn on the specified outputs in the forward direction.	HTRCXOnFwd(RCX_OUT_A);
HTRCXOnRev(outputs)	Send commands to an RCX to turn on the specified outputs in the reverse direction.	HTRCXOnRev(RCX_OUT_A);
HTRCXOnFor(outputs, duration)	Send commands to an RCX to turn on the specified outputs in the forward direction for the specified duration.	HTRCXOnFor(RCX_OUT_A, 100);

HTRCXSetTxPower(pwr)	Send the SetTxPower command to an RCX.	HTRCXSetTxPower(0);
HTRCXPlaySound(snd)	Send the PlaySound command to an RCX.	HTRCXPlaySound(RCX_SOUND_UP);
HTRCXDeleteTask(n)	Send the DeleteTask command to an RCX.	HTRCXDeleteTask(3);
HTRCXStartTask(n)	Send the StartTask command to an RCX.	HTRCXStartTask(2);
HTRCXStopTask(n)	Send the StopTask command to an RCX.	HTRCXStopTask(1);
HTRCXSelectProgram(prog)	Send the SelectProgram command to an RCX.	HTRCXSelectProgram(3);
HTRCXClearTimer(timer)	Send the ClearTimer command to an RCX.	HTRCXClearTimer(0);
HTRCXSetSleepTime(t)	Send the SetSleepTime command to an RCX.	HTRCXSetSleepTime(4);
HTRCXDeleteSub(s)	Send the DeleteSub command to an RCX.	HTRCXDeleteSub(2);
HTRCXClearSensor(port)	Send the ClearSensor command to an RCX.	HTRCXClearSensor(S1);
HTRCXPlayToneVar(varnum, duration)	Send the PlayToneVar command to an RCX.	HTRCXPlayToneVar(0, 50);
HTRCXSetWatch(hours, minutes)	Send the SetWatch command to an RCX.	HTRCXSetWatch(3, 30);
HTRCXSetSensorType(port, type)	Send the SetSensorType command to an RCX.	HTRCXSetSensorType(S1, SENSOR_TYPE_TOUCH);
HTRCXSetSensorMode(port, mode)	Send the SetSensorMode command to an RCX.	HTRCXSetSensorMode(S1, SENSOR_MODE_BOOL);
HTRCXCreateDatalog(size)	Send the CreateDatalog command to an RCX.	HTRCXCreateDatalog(50);

HTRCXAddToDatalog(src, value)	Send the AddToDatalog command to an RCX.	HTRCXAddToDatalog(RCX_InputValueSrc, S1);
HTRCXSendSerial(first, count)	Send the SendSerial command to an RCX.	HTRCXSendSerial(0, 10);
HTRCXRemote(cmd)	Send the Remote command to an RCX.	HTRCXRemote(RCX_RemotePlayASound);
HTRCXEvent(src, value)	Send the Event command to an RCX.	HTRCXEvent(RCX_ConstantSrc, 2);
HTRCXPlayTone(freq, duration)	Send the PlayTone command to an RCX.	HTRCXPlayTone(440, 100);
HTRCXSelectDisplay (src, value)	Send the SelectDisplay command to an RCX.	HTRCXSelectDisplay (RCX_VariableSrc, 2);
HTRCXPollMemory(address, count)	Send the PollMemory command to an RCX.	HTRCXPollMemory(0, 10);
HTRCXSetEvent(evt, src, type)	Send the SetEvent command to an RCX.	HTRCXSetEvent(0, RCX_ConstantSrc, 5);
HTRCXSetGlobalOutput (outputs, mode)	Send the SetGlobalOutput command to an RCX.	HTRCXSetGlobalOutput (RCX_OUT_A, RCX_OUT_ON);
HTRCXSetGlobalDirection (outputs, dir)	Send the SetGlobalDirection command to an RCX.	HTRCXSetGlobalDirection (RCX_OUT_A, RCX_OUT_FWD);
HTRCXSetMaxPower(outputs, pwrsrc, pwrval)	Send the SetMaxPower command to an RCX.	HTRCXSetMaxPower(RCX_OUT_A, RCX_ConstantSrc, 5);
HTRCXEnableOutput(outputs)	Send the EnableOutput command to an RCX.	HTRCXEnableOutput (RCX_OUT_A);
HTRCXDisableOutput(outputs)	Send the DisableOutput command to an RCX.	HTRCXDisableOutput (RCX_OUT_A);
HTRCXInvertOutput(outputs)	Send the InvertOutput command to an RCX.	HTRCXInvertOutput (RCX_OUT_A);
HTRCXObvertOutput (outputs)	Send the ObvertOutput command to an RCX.	HTRCXObvertOutput (RCX_OUT_A);

HTRCXCalibrateEvent (evt, low, hi, hyst)	Send the CalibrateEvent command to an RCX.	HTRCXCalibrateEvent(0, 200, 500, 50);
HTRCXSetVar(varnum, src, value)	Send the SetVar command to an RCX.	HTRCXSetVar(0, RCX_VariableSrc, 1);
HTRCXSumVar(varnum, src, value)	Send the SumVar command to an RCX.	HTRCXSumVar(0, RCX_InputValueSrc, S1);
HTRCXSubVar(varnum, src, value)	Send the SubVar command to an RCX.	HTRCXSubVar(0, RCX_RandomSrc, 10);
HTRCXDivVar(varnum, src, value)	Send the DivVar command to an RCX.	HTRCXDivVar(0, RCX_ConstantSrc, 2);
HTRCXMulVar(varnum, src, value)	Send the MulVar command to an RCX.	HTRCXMulVar(0, RCX_VariableSrc, 4);
HTRCXSgnVar(varnum, src, value)	Send the SgnVar command to an RCX.	HTRCXSgnVar(0, RCX_VariableSrc, 0);
HTRCXAbsVar(varnum, src, value)	Send the AbsVar command to an RCX.	HTRCXAbsVar(0, RCX_VariableSrc, 0);
HTRCXAndVar(varnum, src, value)	Send the AndVar command to an RCX.	HTRCXAndVar(0, RCX_ConstantSrc, 0x7f);
HTRCXOrVar(varnum, src, value)	Send the OrVar command to an RCX.	HTRCXOrVar(0, RCX_ConstantSrc, 0xCC);
HTRCXSet(dstsrc, dstval, src, value)	Send the Set command to an RCX.	HTRCXSet(RCX_VariableSrc, 0, RCX_RandomSrc, 10000);
HTRCXUnlock()	Send the Unlock command to an RCX.	HTRCXUnlock();
HTRCXReset()	Send the Reset command to an RCX.	HTRCXReset();
HTRCXBoot()	Send the Boot command to an RCX.	HTRCXBoot();
HTRCXSetUserDisplay (src, value, precision)	Send the SetUserDisplay command to an RCX.	HTRCXSetUserDisplay (RCX_VariableSrc, 0, 2);
HTRCXIncCounter(counter)	Send the IncCounter command to an RCX.	HTRCXIncCounter(0);

HTRCXDecCounter(counter)	Send the DecCounter command to an RCX.	HTRCXDecCounter(0);
HTRCXClearCounter(counter)	Send the ClearCounter command to an RCX.	HTRCXClearCounter(0);
HTRCXSetPriority(p)	Send the SetPriority command to an RCX.	HTRCXSetPriority(2);
HTRCXSetMessage(msg)	Send the SetMessage command to an RCX.	HTRCXSetMessage(20);

HiTechnic iRLink Scout Functions

The functions in this section expose scout-specific commands for controlling the Scout programmable brick using the HiTechnic iRLink device.

Function	Description	Example
HTScoutCalibrateSensor()	Send the CalibrateSensor command to a Scout.	HTScoutCalibrateSensor();
HTScoutMuteSound()	Send the MuteSound command to a Scout.	HTScoutMuteSound();
HTScoutUnmuteSound()	Send the UnmuteSound command to a Scout.	HTScoutUnmuteSound();
HTScoutSelectSounds(group)	Send the SelectSounds command to a Scout.	HTScoutSelectSounds(0);
HTScoutSetLight(mode)	Send the SetLight command to a Scout.	HTScoutSetLight (SCOUT_LIGHT_ON);
HTScoutSetCounterLimit (counter, src, value)	Send the SetCounterLimit command to a Scout.	HTScoutSetCounterLimit(0, RCX_ConstantSrc, 2000);
HTScoutSetTimerLimit(timer, src, value)	Send the SetTimerLimit command to a Scout.	HTScoutSetTimerLimit(0, RCX_ConstantSrc, 10000);
HTScoutSetSensorClickTime (src, value)	Send the SetSensorClickTime command to a Scout.	HTScoutSetSensorClickTime (RCX_ConstantSrc, 200);
HTScoutSetSensorHysteresis (src, value)	Send the SetSensorHysteresis command to a Scout.	HTScoutSetSensorHysteresis (RCX_ConstantSrc, 50);

HTScoutSetSensorLower Limit(src, value)	Send the SetSensorLowerLimit command to a Scout.	HTScoutSetSensorLower Limit(RCX_ConstantSrc, 100);
HTScoutSetSensorUpper Limit(src, value)	Send the SetSensorUpperLimit command to a Scout.	HTScoutSetSensorUpper Limit(RCX_ConstantSrc, 400);
HTScoutSetEventFeedback (src, value)	Send the SetEventFeedback command to a Scout.	HTScoutSetEventFeedback (RCX_ConstantSrc, 10);
HTScoutSendVLL(src, value)	Send the SendVLL command to a Scout.	HTScoutSendVLL(RCX_ ConstantSrc, 0x30);
HTScoutSetScoutRules (motion, touch, light, time, effect)	Send the SetScoutRules command to a Scout.	HTScoutSetScoutRules(SCOUT_MR_FORWARD, SCOUT_TR_REVERSE, SCOUT_ LR_IGNORE, SCOUT_TGS_ SHORT, SCOUT_FXR_BUG);
HTScoutSetScoutMode (mode)	Send the SetScoutMode command to a Scout.	HTScoutSetScoutMode (SCOUT_MODE_POWER);

Mindsensors Functions

The functions in this section are for using various sensors made by mindsensors.com.

Function	Description	Example
ReadSensorMSRTClock (port, out sec, out min, out hrs, out dow, out date, out month, out year)	Read real-time clock values from the Mindsensors RTClock sensor. Returns a boolean value indicating whether or not the operation completed successfully.	ReadSensorMSRTClock (S1, ss, mm, hh, dow, dd, mon, yy);
SensorMSCompass(port)	Return the Mindsensors Compass sensor value.	x = SensorMSCompass(S1);

NBC Quick Reference

Topics in this Appendix

- NBC Quick Reference Guide

Appendix **B**

Within these pages you'll find a quick guide to the NBC API. You can find more detailed information about NBC in the *NBC Programmers Guide.*

NBC API Functions

Many of the functions here are flexible in that they can take either constant or variable arguments. Some functions require that one or more of their arguments are constant values. These arguments are marked with the "const" keyword in the function argument list.

Task and Timing Functions

The functions in this section provide access to basic timing and task control features.

Function	Description	Example
Wait(time)	Wait for the specified amount of time in milliseconds. The time may be an expression or a constant.	Wait(x)
GetFirstTick(out value)	Return the system timing value at the time the program began executing.	GetFirstTick(result)
GetSleepTime(out value)	Return the number of minutes the NXT will remain on before it automatically shuts down.	GetSleepTime(s)
GetSleepTimer(out value)	Return the number of minutes left before the NXT shuts down.	GetSleepTimer(s)
ResetSleepTimer	Reset the sleep timer back to the sleep timeout value.	ResetSleepTimer

SetSleepTimeout(minutes)	Set the sleep timeout value to the specified number of minutes.	SetSleepTimeout(8)
SetSleepTimer(minutes)	Set the sleep timer value to the specified number of minutes.	SetSleepTimer(4)

Math Functions

The functions in this section are for performing math operations.

Function	Description	Example
Random(out value, max)	Return an unsigned 16-bit random number between 0 and n (exclusive). N can be a constant or a variable.	Random(x, 10)
SignedRandom(out value)	Return a signed 16-bit random number.	SignedRandom(x)
Sqrt(x, out result)	Return the square root of the specified value.	Sqrt(x, result)
Sin(degrees, out result)	Return the sine of the specified degrees value. The result is 100 times the sine value (-100..100).	Sin(theta, x)
Cos(degrees, out result)	Return the cosine of the specified degrees value. The result is 100 times the cosine value (-100..100).	Cos(y, x)
Asin(value, out result)	Return the inverse sine of the specified value (-100..100). The result is degrees (-90..90).	Asin(80, deg)
Acos(value, out result)	Return the inverse cosine of the specified value (-100..100). The result is degrees (0..180).	Acos(0, deg)
bcd2dec(bcd, out dec)	Convert a byte value in binary coded decimal (BCD) format into its equivalent decimal value.	bcd2dec(bcdValue, dec)

Input Functions

The functions in this section are for configuring, reading, and manipulating NXT inputs.

Function	Description	Example
SetSensorType(port, type)	Set a sensor's type	SetSensorType(S1, SENSOR_TYPE_TOUCH)
SetSensorMode(port, mode)	Set a sensor's mode	SetSensorMode(S1, SENSOR_MODE_RAW)
SetSensorLight(port)	Configure a sensor as an active light sensor	SetSensorLight(S1)
SetSensorSound(port)	Configure a sensor as a sound sensor with dB scaling.	SetSensorSound(S1)
SetSensorTouch(port)	Configure a sensor as a touch sensor.	SetSensorTouch(S1)
SetSensorLowspeed(port) or SetSensorUltrasonic(port)	Configure a sensor as a lowspeed 9V sensor.	SetSensorLowspeed(S1)
ClearSensor(port)	Clear the value of a sensor – only affects sensors that are configured to measure a cumulative quantity such as rotation or a pulse count.	ClearSensor(S1)
ResetSensor(port)	Reset the value of a sensor.	ResetSensor(S1)
ReadSensor(p, out result)	Return the processed or scaled value for a sensor on port p (IN_1..IN_4).	ReadSensor(S1, scaledValue)
ReadSensorUS(p, out result)	Return the sensor reading for an ultrasonic sensor on port p.	ReadSensorUS(S1, dist)
SetInCustomZeroOffset (const p, value)	Sets the custom sensor zero offset value of a sensor on port p.	SetInCustomZeroOffset (S1, 12)
SetInCustomPercentFullScale (const p, value)	Sets the custom sensor percent full scale value of a sensor on port p.	SetInCustomPercentFull Scale(S1, 100)

SetInCustomActiveStatus (const p, value)	Sets the custom sensor active status value of a sensor on port p.	SetInCustomActiveStatus (S1, true)
SetInDigiPinsDirection (const p, value)	Sets the digital pins direction value of a sensor on port p.	SetInDigiPinsDirection(S1, 1)
SetInDigiPinsStatus (const p, value)	Sets the digital pins status value of a sensor on port p.	SetInDigiPinsStatus(S1, false)
SetInDigiPinsOutputLevel (const p, value)	Sets the digital pins output level value of a sensor on port p.	SetInDigiPinsOutput Level (S1, 100)
GetInSensorBoolean(const p, out value)	Return the boolean value of a sensor on port p.	GetInSensorBoolean(S1, b)
GetInCustomZeroOffset (const p, out value)	Return the custom sensor zero offset value of a sensor on port p.	GetInCustomZeroOffset (S1, zoff)
GetInCustomPercentFullScale (const p, out value)	Return the custom sensor percent full scale value of a sensor on port p.	GetInCustomPercentFullScale (S1, pct)
GetInCustomActiveStatus (const p, out value)	Return the custom sensor active status value of a sensor on port p.	GetInCustomActiveStatus (S1, status)
GetInDigiPinsDirection (const p, out value)	Return the digital pins direction value of a sensor on port p.	GetInDigiPinsDirection (S1, pdir)
GetInDigiPinsStatus (const p, out value)	Return the digital pins status value of a sensor on port p.	GetInDigiPinsStatus (S1, pstatus)
GetInDigiPinsOutputLevel (const p, out value)	Return the digital pins output level value of a sensor on port p.	GetInDigiPinsOutputLevel (S1, olevel)

Output Functions

The functions in this section are for configuring and controlling NXT outputs.

Function	Description	Example
Off(outputs)	Turn the specified outputs off with braking. Use OUT_A, OUT_B, OUT_C, OUT_AB, OUT_AC, OUT_BC, or OUT_ABC for the outputs argument to motor functions. The non-Ex motor functions reset the Block and Tacho counters but they do not reset the Rotation counter.	Off(OUT_A)
OffEx(outputs, const reset)	Same as Off but you can specify which counters, if any, to reset.	OffEx(OUT_A, RESET_NONE)
Coast(outputs) or Float(outputs)	Turn off the outputs, making them coast to a stop.	Coast(OUT_A)
CoastEx(outputs, const reset)	Same as Coast but you can specify which counters, if any, to reset.	CoastEx(OUT_A, RESET_COUNT)
OnFwd(outputs, power)	Set outputs to forward direction and turn them on at the specified power level.	OnFwd(OUT_A, 75)
OnFwdEx(outputs, power, const reset)	Same as OnFwd but you can specify which counters, if any, to reset.	OnFwdEx(OUT_A, 75, RESET_BLOCK_COUNT)
OnRev(outputs, power)	Set outputs to reverse direction and turn them on at the specified power level.	OnRev(OUT_A, 75)
OnRevEx(outputs, power, const reset)	Same as OnRev but you can specify which counters, if any, to reset.	OnRevEx(OUT_A, 75, RESET_ALL)
OnFwdReg(outputs, power, regmode)	Run the outputs forward using the specified regulation mode.	OnFwdReg(OUT_A, 75, OUT_REGMODE_SPEED)
OnFwdRegEx(outputs, power, regmode, const reset)	Same as OnFwdReg but you can specify which counters, if any, to reset.	OnFwdRegEx(OUT_A, 75, OUT_REGMODE_SPEED, RESET_NONE)

OnRevReg(outputs, power, regmode)	Run the outputs in reverse using the specified regulation mode.	OnRevReg(OUT_A, 75, OUT_REGMODE_SPEED)
OnRevRegEx(outputs, power, regmode, const reset)	Same as OnRevReg but you can specify which counters, if any, to reset.	OnRevRegEx(OUT_A, 75, OUT_REGMODE_SPEED, RESET_NONE)
OnFwdSync(outputs, power, turnpct)	Run the specified outputs forward with regulated synchronization using the specified turn ratio.	OnFwdSync(OUT_A, 75, 50)
OnFwdSyncEx(outputs, power, turnpct, const reset)	Same as OnFwdSync but you can specify which counters, if any, to reset.	OnFwdSyncEx(OUT_A, 75, 50, RESET_NONE)
OnRevSync(outputs, power, turnpct)	Run the specified outputs in reverse with regulated synchronization using the specified turn ratio.	OnRevSync(OUT_A, 75, 50)
OnRevSyncEx(outputs, power, turnpct, const reset)	Same as OnRevSync but you can specify which counters, if any, to reset.	OnRevSyncEx(OUT_A, 75, 50, RESET_NONE)
RotateMotor(outputs, power, angle)	Run the outputs for the specified number of degrees at the specified power level. Calls to any of the RotateMotor* functions do not return until the motors reach the specified target or their power level is set to zero.	RotateMotor(OUT_AB, 75, 360)
RotateMotorPID(outputs, power, angle, p, i, d)	Same as RotateMotor but you can also control the proportional, integral, and derivative factors used by the firmware's motor control algorithm.	RotateMotorPID(OUT_BC, 50, 720, 50, 60, 60)
RotateMotorEx(outputs, power, angle, turnpct, bSync, bStop)	Same as RotateMotor but you can also set the turn percent, turn synchronization on or off, and indicate whether the motors should be stopped at the target angle.	RotateMotorEx(OUT_AB, 75, 360, 50, true, true)

RotateMotorExPID(outputs, power, angle, turnpct, bSync, bStop, p, i, d)	Same as RotateMotorEx but you can also control the proportional, integral, and derivative factors used by the firmware's motor control algorithm.	RotateMotorExPID(OUT_AB, 75, 360, 50, true, true, 30, 50, 90)
ResetTachoCount(outputs)	Reset the tachometer count and the tachometer limit goal for the outputs.	ResetTachoCount(OUT_AB)
ResetBlockTachoCount (outputs)	Reset the block-relative position counter for the outputs.	ResetBlockTachoCount (OUT_BC)
ResetRotationCount(outputs)	Reset the rotation sensor for the outputs.	ResetRotationCount (OUT_BC)
ResetAllTachoCounts(output)	Reset all three position counters and reset the current tachometer limit goal for the outputs.	ResetAllTachoCounts (OUT_ABC)

Sound Functions

The functions in this section are for controlling the sound module capabilities of the NXT.

Function	Description	Example
PlayTone(frequency, duration)	Play a single tone of the specified frequency and duration.	PlayTone(440, 500) Wait(500)
PlayToneEx(frequency, duration, volume, bLoop)	Play a single tone of the specified frequency, duration, and volume.	PlayToneEx(440, 500, 2, false)
PlayFile(filename)	Play the specified sound or melody file.	PlayFile('startup.rso')
PlayFileEx(filename, volume, bLoop)	Play the sound or melody file at the specified volume. Optionally loop at the end.	PlayFileEx('startup.rso', 2, false)
GetSoundState(out state, out flags)	Return the current sound state.	GetSoundState(state, flags)
SetSoundState(state, flags, out result)	Set the current sound state.	SetSoundState(SOUND_ STATE_STOP, SOUND_FLAGS_ UPDATE, result)

SetSoundFlags(flags)	Set the current sound flags.	SetSoundFlags(SOUND_FLAGS_UPDATE)
SetSoundModuleState(state)	Set the current sound state.	SetSoundModuleState (SOUND_STATE_STOP)
GetSoundMode(out mode)	Return the current sound mode.	GetSoundMode(mode)
SetSoundMode(mode)	Set the current sound mode.	SetSoundMode(SOUND_MODE_ONCE)
GetSoundFrequency(out freq)	Return the current sound frequency.	GetSoundFrequency(freq)
SetSoundFrequency(freq)	Set the current sound frequency.	SetSoundFrequency(880)
GetSoundDuration(out dur)	Return the current sound duration.	GetSoundDuration(dur)
SetSoundDuration(dur)	Set the current sound duration.	SetSoundDuration(500)
GetSoundSampleRate (out rate)	Return the current sound sample rate.	GetSoundSampleRate(rate)
SetSoundSampleRate(rate)	Set the current sound sample rate.	SetSoundSampleRate(4000)
GetSoundVolume(out volume)	Return the current sound volume.	GetSoundVolume(volume)
SetSoundVolume(volume)	Set the current sound volume.	SetSoundVolume(2)

IO Control Functions

The functions in this section are for communicating with the AVR processor in the NXT.

Function	Description	Example
PowerDown	Turn off the NXT immediately.	PowerDown
RebootInFirmwareMode	Reboot the NXT in SAMBA or firmware download mode.	RebootInFirmwareMode

Display Functions

The functions in this section are for controlling the display module capabilities of the NXT.

Function	Description	Example
NumOut(x, y, value)	Draw a numeric value on the screen at the specified x and y location.	NumOut(0, LCD_LINE1, x)
NumOutEx(x, y, value, clear)	Draw a numeric value on the screen at the specified x and y location. Clear the screen first if clear equals TRUE.	NumOutEx(0, LCD_LINE1, x, TRUE)
TextOut(x, y, msg)	Draw a text value on the screen at the specified x and y location.	TextOut(0, LCD_LINE8, msg_buffer)
TextOutEx(x, y, msg, clear)	Draw a text value on the screen at the specified x and y location. Clear the screen first if clear equals TRUE.	TextOutEx(0, LCD_LINE8, msg_buffer, TRUE)
GraphicOut(x, y, filename)	Draw the specified graphic icon file on the screen at the specified x and y location.	GraphicOut(40, 40, 'image.ric')
GraphicOutEx(x, y, filename, vars, clear)	Draw the specified graphic icon file on the screen at the specified x and y location. Use the values contained in the vars array to transform the drawing commands contained within the specified icon file. Clear the screen first if clear equals TRUE.	GraphicOutEx(40, 40, 'image. ric', variables, TRUE)
CircleOut(x, y, radius)	Draw a circle on the screen with its center at the specified x and y location, using the specified radius.	CircleOut(40, 40, 10)
CircleOutEx(x, y, radius, clear)	Draw a circle on the screen with its center at the specified x and y location, using the specified radius. Clear the screen first if clear equals TRUE.	CircleOutEx(40, 40, 10, FALSE)

LineOut(x1, y1, x2, y2)	Draw a line on the screen from x1, y1 to x2, y2.	LineOut(40, 40, 10, 10)
LineOutEx(x1, y1, x2, y2, clear)	Draw a line on the screen from x1, y1 to x2, y2. Clear the screen first if clear equals TRUE.	LineOutEx(40, 40, 10, 10, TRUE)
PointOut(x, y)	Draw a point on the screen at x, y.	PointOut(40, 40)
PointOutEx(x, y, clear)	Draw a point on the screen at x, y. Clear the screen first if clear equals TRUE.	PointOutEx(40, 40, TRUE)
RectOut(x, y, width, height)	Draw a rectangle on the screen at x, y with the specified width and height.	RectOut(40, 40, 30, 10)
RectOutEx(x, y, width, height, clear)	Draw a rectangle on the screen at x, y with the specified width and height. Clear the screen first if clear equals TRUE.	RectOutEx(40, 40, 30, 10, FALSE)
ClearScreen()	Clear the NXT LCD to a blank screen.	ClearScreen()
GetDisplayFlags(out flags)	Return the current display flags.	GetDisplayFlags(flags)
SetDisplayFlags(n)	Set the current display flags.	SetDisplayFlags(x)
GetDisplayEraseMask (out mask)	Return the current display erase mask.	GetDisplayEraseMask(mask)
SetDisplayEraseMask(n)	Set the current display erase mask.	SetDisplayEraseMask(x)
GetDisplayUpdateMask (out mask)	Return the current display update mask.	GetDisplayUpdateMask (mask)
SetDisplayUpdateMask(n)	Set the current display update mask.	SetDisplayUpdateMask(x)
GetDisplayDisplay(out addr)	Return the current display memory address.	GetDisplayDisplay(addr)
SetDisplayDisplay(n)	Set the current display memory address.	SetDisplayDisplay(x)

GetDisplayTextLinesCenter Flags(out flags)	Return the current display text lines center flags.	GetDisplayTextLinesCenter Flags(flags)
SetDisplayTextLinesCenter Flags(n)	Set the current display text lines center flags.	SetDisplayTextLinesCenter Flags(x)
GetDisplayNormal(x, line, count, data)	Read "count" bytes from the normal display memory into the data array. Start reading from the specified x, line coordinate. Each byte of data read from screen memory is a vertical strip of 8 bits at the desired location. Each bit represents a single pixel on the LCD screen. Use TEXTLINE_1 through TEXTLINE_8 for the "line" parameter.	GetDisplayNormal(0, TEXTLINE_1, 8, ScreenMem)
SetDisplayNormal(x, line, count, data)	Write "count" bytes to the normal display memory from the data array. Start writing at the specified x, line coordinate.	SetDisplayNormal(0, TEXTLINE_1, 8, ScreenMem)
GetDisplayPopup(x, line, count, data)	Read "count" bytes from the popup display memory into the data array.	GetDisplayPopup(0, TEXTLINE_1, 8, PopupMem)
SetDisplayPopup(x, line, count, data)	Write "count" bytes to the popup display memory from the data array.	SetDisplayPopup(0, TEXTLINE_1, 8, PopupMem)

Loader Functions

The functions in this section are for controlling the loader module capabilities of the NXT. They include all forms of file reading and writing.

Function	Description	Example
GetFreeMemory(out mem)	Get the number of bytes of flash memory that are available for use.	GetFreeMemory(mem)
CreateFile(filename, size, out handle, out result)	Create a new file with the specified size for writing. The handle argument must be a variable.	CreateFile('data.txt', 1024, handle, result)

OpenFileAppend(filename, out size, out handle, out result)	Open an existing file for writing. The size and handle arguments must be variables.	OpenFileAppend('data.txt', fsize, handle, result)
OpenFileRead(filename, out size, out handle, out result)	Open an existing file for reading. The size and handle arguments must be variables.	OpenFileRead('data.txt', fsize, handle, result)
CloseFile(handle, out result)	Close the file associated with the specified file handle.	CloseFile(handle, result)
ResolveHandle(filename, out handle, out bWriteable, out result)	Resolve a file handle. The handle and bWriteable arguments must be variables.	ResolveHandle('data.txt', handle, bCanWrite, result)
RenameFile(oldfilename, newfilename, out result)	Rename a file from the old filename to the new filename.	RenameFile('data.txt', 'mydata.txt', result)
DeleteFile(filename, out result)	Delete the specified file.	DeleteFile('data.txt', result)
ResizeFile(filename, newsize, out result)	Resize the specified file to the new size.	ResizeFile('data.txt', 1024, result)
Read(handle, out value, out result)	Read a numeric value from the file. The handle and value arguments must be variables. The type of the value argument determines how many bytes are read.	Read(handle, value, result)
ReadLn(handle, out value, out result)	Read a numeric value from the file. It reads two more bytes from the file which should be CRLF.	ReadLn(handle, value, result)
ReadBytes(handle, in/out length, out buf, out result)	Read the specified number of bytes from the file. The handle, length, and buf argumentss must be variables. The actual number of bytes read is returned in the length parameter.	ReadBytes(handle, len, buffer, result)
Write(handle, value, out result)	Write a numeric value to a file. The handle parameter must be a variable. The type of the value argument determines how many bytes are written.	Write(handle, value, result)

WriteLn(handle, value, out result)	Same as Write but also writes a CRLF to the file following the numeric data.	WriteLn(handle, value, result)
WriteString(handle, str, out count, out result)	Write the string to a file. The handle and count arguments must be variables. The total number of bytes written is returned in count.	WriteString(handle, 'testing', count, result)
WriteLnString(handle, str, out count, out result)	Same as WriteString but also writes a CRLF to the file.	WriteLnString(handle, 'testing', count, result)
WriteBytes(handle, data, out count, out result)	Write the contents of the data array to a file. The total number of bytes written is returned in count.	WriteBytes(handle, buffer, count, result)
WriteBytesEx(handle, in/out length, buf, out result)	Write the specified number of bytes to the file associated with the specified handle. The total number of bytes written is returned in count.	WriteBytesEx(handle, len, buffer, result)

Button Functions

The functions in this section are for controlling the button module capabilities of the NXT.

Function	Description	Example
ReadButtonEx(btn, reset, out pressed, out count, out result)	Read the specified button. Sets the pressed and count parameters with the current state of the button. Optionally reset the press count after reading it.	ReadButtonEx(BTN1, true, pressed, count, result)
GetButtonPressCount(const btn, out result)	Return the press count of the specified button.	GetButtonPressCount(BTN1, result)
GetButtonLongPressCount (const btn, out result)	Return the long press count of the specified button.	GetButtonLongPressCount (BTN1, result)
GetButtonShortRelease Count(const btn, out result)	Return the short release count of the specified button.	GetButtonShortRelease Count(BTN1, result)

GetButtonLongRelease Count(const btn, out result)	Return the long release count of the specified button.	GetButtonLongRelease Count(BTN1, result)
GetButtonReleaseCount (const btn, out result)	Return the release count of the specified button.	GetButtonReleaseCount (BTN1, result)
GetButtonState(const btn, out result)	Return the state of the specified button.	GetButtonState(BTN1, result)

User Interface Functions

The functions in this section are for controlling the user interface (UI) module capabilities of the NXT.

Function	Description	Example
GetVolume(out value)	Return the user interface volume level.	GetVolume(x)
SetVolume(value)	Set the user interface volume level.	SetVolume(3)
GetBatteryLevel(out value)	Return the battery level in millivolts.	GetBatteryLevel(x)
GetBluetoothState(out value)	Return the Bluetooth state.	GetBluetoothState(x)
SetBluetoothState(value)	Set the Bluetooth state.	SetBluetoothState (UI_BT_STATE_OFF)
GetCommandFlags()	Return the command flags.	GetCommandFlags(x)
SetCommandFlags(value)	Set the command flags.	SetCommandFlags(UI_ FLAGS_REDRAW_STATUS)
GetUIState(out value)	Return the user interface state.	GetUIState(x)
SetUIState(value)	Set the user interface state.	SetUIState(UI_STATE_ LOW_BATTERY)
GetUIButton(out value)	Return user interface button information.	GetUIButton(x)
SetUIButton(value)	Set user interface button information.	SetUIButton(UI_BUTTON_ ENTER)

GetVMRunState(out value)	Return VM run state information.	GetVMRunState(x)
SetVMRunState(value)	Set VM run state information.	SetVMRunState(o)
GetBatteryState(out value)	Return battery state information (o..4).	GetBatteryState(value)
GetRechargeableBattery(out value)	Return whether the NXT has a rechargeable battery installed or not.	GetRechargeableBattery(value)
ForceOff(n)	Force the NXT to turn off if the specified value is greater than zero.	ForceOff(true)
GetUsbState(out value)	Return USB state information (o=disconnected, 1=connected, 2=working).	GetUsbState(value)
GetOnBrickProgramPointer(out value)	Return the current OBP (on-brick program) step.	GetOnBrickProgramPointer(value)
SetOnBrickProgramPointer(value)	Set the current OBP (on-brick program) step.	SetOnBrickProgramPointer(2)
GetLongAbort(out value)	Return a boolean value indicating whether a long button press is required to abort a running program. This function requires the enhanced NBC/NXC firmware.	GetLongAbort(x)
SetLongAbort(value)	Set the current long abort configuration. This function requires the enhanced NBC/NXC firmware.	SetLongAbort(true)

Lowspeed I²C Functions

The functions in this section are for controlling the lowspeed module capabilities of the NXT.

Function	Description	Example
LowspeedWrite(port, returnlen, buffer, out result)	Start a transaction to write the bytes contained in the array buffer to the I2C device on the specified port.	LowspeedWrite(S1, 1, inbuffer, X)
LowspeedStatus(port, out bytesready, out result)	Check the status of the I2C communication on the specified port and return the number of bytes ready.	LowspeedStatus(S1, nRead, X)
LowspeedCheckStatus(port, out result)	Check the status of the I2C communication on the specified port.	LowspeedCheckStatus(S1, X)
LowspeedBytesReady(port, out result)	Check how many bytes are ready on the specified port.	LowspeedBytesReady (S1, nReady)
LowspeedRead(port, buflen, out buffer, out result)	Read the specified number of bytes from the I2C device on the specified port and store the bytes read in the array buffer provided.	LowspeedRead(S1, 1, outbuffer, result)
ReadI2CBytes(port, inbuf, in/out count, out outbuf, out result)	This is a higher-level wrapper around the three main I2C functions. It also maintains a "last good read" buffer and returns values from that buffer if the I2C communication transaction fails.	ReadI2CBytes(S4, writebuf, cnt, readbuf, X)
GetLSMode(const port, out result)	Returns the mode of the lowspeed communication over the specified port.	GetLSMode(S1, X)
GetLSChannelState(const port, out result)	Returns the channel state of the lowspeed communication over the specified port.	GetLSChannelState(S1, X)
GetLSErrorType(const port, out result)	Returns the error type of the lowspeed communication over the specified port.	GetLSErrorType(S1, X)

GetLSState(out result)	Returns the state of the lowspeed module.	GetLSState(X)
GetLSSpeed(out result)	Returns the speed of the lowspeed module.	GetLSSpeed(X)

Bluetooth Functions

The functions in this section are for communicating via Bluetooth with other devices. Use the standard NXT communication protocol and mailbox system or NXT direct commands with these routines. You can also directly write Bluetooth messages which do not use the NXT protocols using API functions in this section.

Function	Description	Example
SendRemoteBool(connection, queue, bvalue, out result)	Sends a boolean value to the device on the specified connection.	SendRemoteBool(1, queue, false, x)
SendRemoteNumber (connection, queue, value, out result)	Sends a numeric value to the device on the specified connection.	SendRemoteNumber (1, queue, 123, x)
SendRemoteString (connection, queue, strval, out result)	Sends a string value to the device on the specified connection.	SendRemoteString(1, queue, 'hello world', x)
SendResponseBool(queue, bvalue, out result)	Sends a boolean value as a response to a received message.	SendResponseBool(queue, false, x)
SendResponseNumber (queue, value, out result)	Sends a numeric value as a response to a received message.	SendResponseNumber (queue, 123, x)
SendResponseString(queue, strval, out result)	Sends a string value as a response to a received message.	SendResponseString(queue, 'hello world', x)
ReceiveRemoteBool(queue, remove, out bvalue, out result)	Use this function on a primary brick to receive a boolean value from a secondary device. Optionally remove the last read message from the message queue.	ReceiveRemoteBool(queue, true, bvalue, x)
ReceiveRemoteNumber (queue, remove, out value, out result)	Same as ReceiveRemoteBool but for numeric messages.	ReceiveRemoteBool(queue, true, value, x)

ReceiveRemoteString(queue, remove, out strval, out result)	Same as ReceiveRemoteBool but for string messages.	ReceiveRemoteString(queue, true, strval, x)
ReceiveRemoteMessageEx (queue, remove, out strval, out val, out bval, out result)	Use this function to receive a string, boolean, or numeric value from another device.	ReceiveRemoteMessageEx (queue, true, strval, val, bval, x)
SendMessage(queue, msg, out result)	Write the message buffer contents to the specified mailbox or message queue.	SendMessage(mbox, data, x)
ReceiveMessage(queue, remove, out buffer, out result)	Retrieve a message from the specified queue and write it to the buffer provided.	RecieveMessage(mbox, true, buffer, x)
BluetoothStatus(connection, out result)	Returns the status of the specified Bluetooth connection. Avoid calling BluetoothWrite or any other API function that writes data over a Bluetooth connection while BluetoothStatus returns STAT_COMM_PENDING.	BluetoothStatus(1, x)
BluetoothWrite(connection, buffer, out result)	Write the data in the buffer to the device on the specified Bluetooth connection.	BluetoothWrite(1, data, x)
RemoteMessageRead (connection, queue, out result)	Send a MessageRead direct command to the device on the specified connection.	RemoteMessageRead(1, 5, x)
RemoteMessageWrite (connection, queue, msg, out result)	Send a MessageWrite direct command to the device on the specified connection.	RemoteMessageWrite(1, 5, 'test', x)
RemoteStartProgram (connection, filename, out result)	Send a StartProgram direct command to the device on the specified connection.	RemoteStartProgram (1, 'myprog.rxe', x)
RemoteStopProgram (connection, out result)	Send a StopProgram direct command to the device on the specified connection.	RemoteStopProgram(1, x)
RemotePlaySoundFile (connection, filename, bLoop, out result)	Send a PlaySoundFile direct command to the device on the specified connection.	RemotePlaySoundFile (1, 'click.rso', false, x)

RemotePlayTone(connection, frequency, duration, out result)	Send a PlayTone direct command to the device on the specified connection.	RemotePlayTone(1, 440, 1000, x)
RemoteStopSound (connection, out result)	Send a StopSound direct command to the device on the specified connection.	RemoteStopSound(1, x)
RemoteKeepAlive (connection, out result)	Send a KeepAlive direct command to the device on the specified connection.	RemoteKeepAlive(1, x)
RemoteResetScaledValue (connection, port, out result)	Send a ResetScaledValue direct command to the device on the specified connection.	RemoteResetScaledValue (1, S1, x)
RemoteResetMotorPosition (connection, port, bRelative, out result)	Send a ResetMotorPosition direct command to the device on the specified connection.	RemoteResetMotorPosition (1, OUT_A, true, x)
RemoteSetInputMode (connection, port, type, mode, out result)	Send a SetInputMode direct command to the device on the specified connection.	RemoteSetInputMode (1, S1, IN_TYPE_LOWSPEED, IN_MODE_RAW, x)
RemoteSetOutputState (connection, port, speed, mode, regmode, turnpct, runstate, tacholimit, out result)	Send a SetOutputState direct command to the device on the specified connection.	RemoteSetOutputState (1, OUT_A, 75, OUT_ MODE_MOTORON, OUT_REGMODE_IDLE, 0, OUT_RUNSTATE_RUNNING, 0, x)

USB Functions

The functions in this section are for communicating via USB with other devices.

Function	Description	Example
GetUSBInputBuffer(const offset, count, out data)	Read count bytes of data from the USB input buffer at the specified offset and write it to the buffer provided.	GetUSBInputBuffer (0, 10, buffer)
SetUSBInputBuffer(const offset, count, data)	Write count bytes of data to the USB input buffer at the specified offset.	SetUSBInputBuffer (0, 10, buffer)

SetUSBInputBufferInPtr(n)	Set the input pointer of the USB input buffer to the specified value.	SetUSBInputBufferInPtr(0)
GetUSBInputBufferInPtr (out result)	Return the value of the input pointer of the USB input buffer.	GetUSBInputBufferInPtr(x)
SetUSBInputBufferOutPtr(n)	Set the output pointer of the USB input buffer to the specified value.	SetUSBInputBufferOutPtr(0)
GetUSBInputBufferOutPtr (out result)	Return the value of the output pointer of the USB input buffer.	GetUSBInputBufferOutPtr(x)
GetUSBOutputBuffer(const offset, count, out data)	Read count bytes of data from the USB output buffer at the specified offset and write it to the buffer provided.	GetUSBOutputBuffer(0, 10, buffer)
SetUSBOutputBuffer(const offset, count, data)	Write count bytes of data to the USB output buffer at the specified offset.	SetUSBOutputBuffer(0, 10, buffer)
SetUSBOutputBufferInPtr(n)	Set the input pointer of the USB output buffer to the specified value.	SetUSBOutputBufferInPtr(0)
GetUSBOutputBufferInPtr (out result)	Return the value of the input pointer of the USB output buffer.	GetUSBOutputBufferInPtr(x)
SetUSBOutputBufferOutPtr(n)	Set the output pointer of the USB output buffer to the specified value.	SetUSBOutputBufferOutPtr(0)
GetUSBOutputBufferOutPtr (out result)	Return the value of the output pointer of the USB output buffer.	GetUSBOutputBufferOutPtr(x)
GetUSBPollBuffer(const offset, count, out data)	Read count bytes of data from the USB poll buffer and write it to the buffer provided.	GetUSBPollBuffer(0, 10, buffer)
SetUSBPollBuffer(const offset, count, data)	Write count bytes of data to the USB poll buffer at the specified offset.	SetUSBPollBuffer(0, 10, buffer)

SetUSBPollBufferInPtr(n)	Set the input pointer of the USB poll buffer to the specified value.	SetUSBPollBufferInPtr(o)
GetUSBPollBufferInPtr (out result)	Return the value of the input pointer of the USB poll buffer.	GetUSBPollBufferInPtr(x)
SetUSBPollBufferOutPtr(n)	Set the output pointer of the USB poll buffer to the specified value.	SetUSBPollBufferOutPtr(o)
GetUSBPollBufferOutPtr (out result)	Return the value of the output pointer of the USB poll buffer.	GetUSBPollBufferOutPtr(x)
SetUSBState(n)	Set the USB state to the specified value.	SetUSBState(o)
GetUSBState(out result)	Return the USB state.	GetUSBState(x)

High Speed Functions

The functions in this section are for communicating with other devices via the port 4 RS485 high-speed communication port.

Function	Description	Example
GetHSInputBuffer(const offset, count, out data)	Read count bytes of data from the High Speed input buffer and write it to the buffer provided.	GetHSInputBuffer(o, 10, buffer)
SetHSInputBuffer(const offset, count, data)	Write count bytes of data to the High Speed input buffer at the specified offset.	SetHSInputBuffer(o, 10, buffer)
SetHSInputBufferInPtr(n)	Set the input pointer of the High Speed input buffer to the specified value.	SetHSInputBufferInPtr(o)
GetHSInputBufferInPtr (out result)	Return the value of the input pointer of the High Speed input buffer.	GetHSInputBufferInPtr(x)
SetHSInputBufferOutPtr(n)	Set the output pointer of the High Speed input buffer to the specified value.	SetHSInputBufferOutPtr(o)

GetHSInputBufferOutPtr (out result)	Return the value of the output pointer of the High Speed input buffer.	GetHSInputBufferOutPtr(x)
GetHSOutputBuffer(const offset, count, out data)	Read count bytes of data from the High Speed output buffer and write it to the buffer provided.	GetHSOutputBuffer(0, 10, buffer)
SetHSOutputBuffer(const offset, count, data)	Write count bytes of data to the High Speed output buffer at the specified offset.	SetHSOutputBuffer(0, 10, buffer)
SetHSOutputBufferInPtr(n)	Set the Output pointer of the High Speed output buffer to the specified value.	SetHSOutputBufferInPtr(0)
GetHSOutputBufferInPtr (out result)	Return the value of the Output pointer of the High Speed output buffer.	GetHSOutputBufferInPtr(x)
SetHSOutputBufferOutPtr(n)	Set the output pointer of the High Speed output buffer to the specified value.	SetHSOutputBufferOutPtr(0)
GetHSOutputBufferOutPtr (out result)	Return the value of the output pointer of the High Speed output buffer.	GetHSOutputBufferOutPtr(x)
SetHSFlags(n)	Set the High Speed flags to the specified value.	SetHSFlags(0)
GetHSFlags(out result)	Return the value of the High Speed flags.	GetHSFlags(x)
SetHSSpeed(n)	Set the High Speed speed to the specified value.	SetHSSpeed(1)
GetHSSpeed(out result)	Return the value of the High Speed speed.	GetHSSpeed(x)
SetHSState(n)	Set the High Speed state to the specified value.	SetHSState(1)
GetHSState(out result)	Return the value of the High Speed state.	GetHSState(x)

Low-level Bluetooth Functions

The functions in this section provide direct low-level access to the Bluetooth input and output buffers as well as the brick data, Bluetooth device, and Bluetooth connection tables.

Function	Description	Example
GetBTInputBuffer(const offset, count, out data)	Read count bytes of data from the Bluetooth input buffer and write it to the buffer provided.	GetBTInputBuffer(0, 10, buffer)
SetBTInputBuffer(const offset, count, data)	Write count bytes of data to the Bluetooth input buffer at the specified offset.	SetBTInputBuffer(0, 10, buffer)
SetBTInputBufferInPtr(n)	Set the input pointer of the Bluetooth input buffer to the specified value.	SetBTInputBufferInPtr(0)
GetBTInputBufferInPtr(out result)	Return the value of the input pointer of the Bluetooth input buffer.	GetBTInputBufferInPtr(x)
SetBTInputBufferOutPtr(n)	Set the output pointer of the Bluetooth input buffer to the specified value.	SetBTInputBufferOutPtr(0)
GetBTInputBufferOutPtr (out result)	Return the value of the output pointer of the Bluetooth input buffer.	GetBTInputBufferOutPtr(x)
GetBTOutputBuffer(const offset, count, out data)	Read count bytes of data from the Bluetooth output buffer and write it to the buffer provided.	GetBTOutputBuffer(0, 10, buffer)
SetBTOutputBuffer(const offset, count, data)	Write count bytes of data to the Bluetooth output buffer at the specified offset.	SetBTOutputBuffer(0, 10, buffer)
SetBTOutputBufferInPtr(n)	Set the input pointer of the Bluetooth output buffer to the specified value.	SetBTOutputBufferInPtr(0)
GetBTOutputBufferInPtr (out result)	Return the value of the input pointer of the Bluetooth output buffer.	GetBTOutputBufferInPtr(x)

SetBTOutputBufferOutPtr(n)	Set the output pointer of the Bluetooth output buffer to the specified value.	SetBTOutputBufferOutPtr(o)
GetBTOutputBufferOutPtr (out result)	Return the value of the output pointer of the Bluetooth output buffer.	GetBTOutputBufferOutPtr(x)
GetBTDeviceCount(out result)	Return the number of devices defined within the Bluetooth device table.	GetBTDeviceCount(x)
GetBTDeviceNameCount (out result)	Return the number of device names defined within the Bluetooth device table. This usually has the same value as BTDeviceCount but it can differ in some instances.	GetBTDeviceNameCount(x)
GetBTDeviceName(const idx, out name)	Return the name of the device at the specified index in the Bluetooth device table.	GetBTDeviceName(o, name)
GetBTConnectionName(const idx, out name)	Return the name of the device at the specified index in the Bluetooth connection table.	GetBTConnectionName(o, name)
GetBTConnectionPinCode (const idx, out code)	Return the pin code of the device at the specified index in the Bluetooth connection table.	GetBTConnectionPinCode (o, code)
GetBrickDataName(out name)	Return the name of the NXT.	GetBrickDataName(name)
GetBTDeviceAddress(const idx, out data)	Read the address of the device at the specified index within the Bluetooth device table and store it in the data buffer provided.	GetBTDeviceAddress(o, buffer)
GetBTConnectionAddress (const idx, out data)	Read the address of the device at the specified index within the Bluetooth connection table and store it in the data buffer provided.	GetBTConnectionAddress(o, buffer)
GetBrickDataAddress (out data)	Read the address of the NXT and store it in the data buffer provided.	GetBrickDataAddress(buffer)

GetBTDeviceClass(const idx, out class)	Return the class of the device at the specified index within the Bluetooth device table.	GetBTDeviceClass(o, class)
GetBTDeviceStatus(const idx, out status)	Return the status of the device at the specified index within the Bluetooth device table.	GetBTDeviceStatus(o, status)
GetBTConnectionClass(const idx, out class)	Return the class of the device at the specified index within the Bluetooth connection table.	GetBTConnectionClass(o, class)
GetBTConnectionHandleNum (const idx, out handle)	Return the handle number of the device at the specified index within the Bluetooth connection table.	GetBTConnectionHandleNum (o, handle)
GetBTConnectionStream Status(const idx, out status)	Return the stream status of the device at the specified index within the Bluetooth connection table.	GetBTConnectionStream Status(o, status)
GetBTConnectionLink Quality(const idx, out quality)	Return the link quality of the device at the specified index within the Bluetooth connection table.	GetBTConnectionLinkQuality (o, quality)
GetBrickDataBluecore Version(out ver)	Return the bluecore version of the NXT.	GetBrickDataBluecore Version(bv)
GetBrickDataBtStateStatus (out value)	Return the Bluetooth state status of the NXT.	GetBrickDataBtStateStatus(x)
GetBrickDataBtHardware Status(out value)	Return the Bluetooth hardware status of the NXT.	GetBrickDataBtHardware Status(x)
GetBrickDataTimeoutValue (out value)	Return the timeout value of the NXT.	GetBrickDataTimeoutValue(x)

HiTechnic Functions

The functions in this section are for using various sensors and devices made by HiTechnic.

Function	Description	Example
SetSensorHTGyro(port)	Configure the sensor on the specified port as a HiTechnic Gyro sensor.	SetSensorHTGyro(S1)
ReadSensorHTGyro(port, offset, out result)	Return the HiTechnic Gyro sensor value using the specified offset value.	ReadSensorHTGyro(S1, 10, result)
ReadSensorHTCompass (port, out result)	Return the HiTechnic Compass sensor value.	ReadSensorHTCompass (S1, result)
ReadSensorHTColorNum (port, out result)	Return the HiTechnic Color sensor color number value.	ReadSensorHTColorNum (S1, result)
ReadSensorHTIRSeekerDir (port, out result)	Return the HiTechnic IR Seeker direction value.	ReadSensorHTIRSeekerDir (S1, result)
ReadSensorHTAccel(port, out x, out y, out z, out result)	Read X, Y, and Z axis acceleration values from the HiTechnic Accelerometer sensor. Also returns a boolean value indicating whether or not the operation completed successfully.	ReadSensorHTAccel(S1, x, y, z, result)
ReadSensorHTColor(port, out ColorNum, out Red, out Green, out Blue, out result)	Read color number, red, green, and blue values from the HiTechnic Color sensor. Also returns a boolean value indicating whether or not the operation completed successfully.	ReadSensorHTColor(S1, c, r, g, b, result)
ReadSensorHTIRSeeker(port, out dir, out s1, out s3, out s5, out s7, out s9, out result)	Read direction, and five signal strength values from the HiTechnic IRSeeker sensor. Returns a boolean value indicating whether or not the operation completed successfully.	ReadSensorHTIRSeeker(port, dir, s1, s3, s5, s7, s9, result)

HTPowerFunctionCommand (port, channel, cmd1, cmd2, out result)	Execute a pair of Power Function motor commands on the specified channel using the HiTechnic iRLink device. Commands are HTPF_CMD_ STOP, HTPF_CMD_REV, HTPF_CMD_FWD, and HTPF_ CMD_BRAKE. Valid channels are HTPF_CHANNEL_1 through HTPF_CHANNEL_4.	HTPowerFunctionCommand (S1, HTPF_CHANNEL_1, HTPF_ CMD_STOP, HTPF_CMD_FWD, result)

HiTechnic iRLink RCX Functions

The functions in this section are for controlling an RCX or Scout programmable brick using the HiTechnic iRLink device.

Function	Description	Example
HTRCXSetIRLinkPort(port)	Set the global port in advance of using the HTRCX* API functions for sending RCX messages over the HiTechnic iRLink device.	HTRCXSetIRLinkPort(S1)
HTRCXPoll(src, value, out result)	Send the Poll command to an RCX to read a signed 2-byte value at the specified source and value combination.	HTRCXPoll(RCX_VariableSrc, 0, x)
HTRCXBatteryLevel(out result)	Send the BatteryLevel command to an RCX to read the current battery level.	HTRCXBatteryLevel(x)
HTRCXPing()	Send the Ping command to an RCX.	HTRCXPing()
HTRCXDeleteTasks()	Send the DeleteTasks command to an RCX.	HTRCXDeleteTasks()
HTRCXStopAllTasks()	Send the StopAllTasks command to an RCX.	HTRCXStopAllTasks()
HTRCXPBTurnOff()	Send the PBTurnOff command to an RCX.	HTRCXPBTurnOff()
HTRCXDeleteSubs()	Send the DeleteSubs command to an RCX.	HTRCXDeleteSubs()
HTRCXClearSound()	Send the ClearSound command to an RCX.	HTRCXClearSound()

HTRCXClearMsg()	Send the ClearMsg command to an RCX.	HTRCXClearMsg()
HTRCXMuteSound()	Send the MuteSound command to an RCX.	HTRCXMuteSound()
HTRCXUnmuteSound()	Send the UnmuteSound command to an RCX.	HTRCXUnmuteSound()
HTRCXClearAllEvents()	Send the ClearAllEvents command to an RCX.	HTRCXClearAllEvents()
HTRCXSetOutput(outputs, mode)	Send the SetOutput command to an RCX to configure the mode of the specified outputs	HTRCXSetOutput(RCX_OUT_A, RCX_OUT_ON)
HTRCXSetDirection(outputs, dir)	Send the SetDirection command to an RCX to configure the direction of the specified outputs.	HTRCXSetDirection(RCX_OUT_A, RCX_OUT_FWD)
HTRCXSetPower(outputs, pwrsrc, pwrval)	Send the SetPower command to an RCX to configure the power level of the specified outputs.	HTRCXSetPower(RCX_OUT_A, RCX_ConstantSrc, RCX_OUT_FULL)
HTRCXOn(outputs)	Send commands to an RCX to turn on the specified outputs.	HTRCXOn(RCX_OUT_A)
HTRCXOff(outputs)	Send commands to an RCX to turn off the specified outputs.	HTRCXOff(RCX_OUT_A)
HTRCXFloat(outputs)	Send commands to an RCX to float the specified outputs.	HTRCXFloat(RCX_OUT_A)
HTRCXToggle(outputs)	Send commands to an RCX to toggle the direction of the specified outputs.	HTRCXToggle(RCX_OUT_A)
HTRCXFwd(outputs)	Send commands to an RCX to set the specified outputs to the forward direction.	HTRCXFwd(RCX_OUT_A)
HTRCXRev(outputs)	Send commands to an RCX to set the specified outputs to the reverse direction.	HTRCXRev(RCX_OUT_A)
HTRCXOnFwd(outputs)	Send commands to an RCX to turn on the specified outputs in the forward direction.	HTRCXOnFwd(RCX_OUT_A)

HTRCXOnRev(outputs)	Send commands to an RCX to turn on the specified outputs in the reverse direction.	HTRCXOnRev(RCX_OUT_A)
HTRCXOnFor(outputs, duration)	Send commands to an RCX to turn on the specified outputs in the forward direction for the specified duration.	HTRCXOnFor(RCX_OUT_A, 100)
HTRCXSetTxPower(pwr)	Send the SetTxPower command to an RCX.	HTRCXSetTxPower(0)
HTRCXPlaySound(snd)	Send the PlaySound command to an RCX.	HTRCXPlaySound (RCX_SOUND_UP)
HTRCXDeleteTask(n)	Send the DeleteTask command to an RCX.	HTRCXDeleteTask(3)
HTRCXStartTask(n)	Send the StartTask command to an RCX.	HTRCXStartTask(2)
HTRCXStopTask(n)	Send the StopTask command to an RCX.	HTRCXStopTask(1)
HTRCXSelectProgram(prog)	Send the SelectProgram command to an RCX.	HTRCXSelectProgram(3)
HTRCXClearTimer(timer)	Send the ClearTimer command to an RCX.	HTRCXClearTimer(0)
HTRCXSetSleepTime(t)	Send the SetSleepTime command to an RCX.	HTRCXSetSleepTime(4)
HTRCXDeleteSub(s)	Send the DeleteSub command to an RCX.	HTRCXDeleteSub(2)
HTRCXClearSensor(port)	Send the ClearSensor command to an RCX.	HTRCXClearSensor(S1)
HTRCXPlayToneVar(varnum, duration)	Send the PlayToneVar command to an RCX.	HTRCXPlayToneVar(0, 50)
HTRCXSetWatch(hours, minutes)	Send the SetWatch command to an RCX.	HTRCXSetWatch(3, 30)
HTRCXSetSensorType(port, type)	Send the SetSensorType command to an RCX.	HTRCXSetSensorType (S1, SENSOR_TYPE_TOUCH)
HTRCXSetSensorMode(port, mode)	Send the SetSensorMode command to an RCX.	HTRCXSetSensorMode (S1, SENSOR_MODE_BOOL)

HTRCXCreateDatalog(size)	Send the CreateDatalog command to an RCX.	HTRCXCreateDatalog(50)
HTRCXAddToDatalog(src, value)	Send the AddToDatalog command to an RCX.	HTRCXAddToDatalog (RCX_InputValueSrc, S1)
HTRCXSendSerial(first, count)	Send the SendSerial command to an RCX.	HTRCXSendSerial(0, 10)
HTRCXRemote(cmd)	Send the Remote command to an RCX.	HTRCXRemote(RCX_Remote PlayASound)
HTRCXEvent(src, value)	Send the Event command to an RCX.	HTRCXEvent(RCX_ ConstantSrc, 2)
HTRCXPlayTone(freq, duration)	Send the PlayTone command to an RCX.	HTRCXPlayTone(440, 100)
HTRCXSelectDisplay (src, value)	Send the SelectDisplay command to an RCX.	HTRCXSelectDisplay (RCX_VariableSrc, 2)
HTRCXPollMemory(address, count)	Send the PollMemory command to an RCX.	HTRCXPollMemory(0, 10)
HTRCXSetEvent(evt, src, type)	Send the SetEvent command to an RCX.	HTRCXSetEvent(0, RCX_ ConstantSrc, 5)
HTRCXSetGlobalOutput (outputs, mode)	Send the SetGlobalOutput command to an RCX.	HTRCXSetGlobalOutput (RCX_OUT_A, RCX_OUT_ON)
HTRCXSetGlobalDirection (outputs, dir)	Send the SetGlobalDirection command to an RCX.	HTRCXSetGlobalDirection (RCX_OUT_A, RCX_OUT_FWD)
HTRCXSetMaxPower(outputs, pwrsrc, pwrval)	Send the SetMaxPower command to an RCX.	HTRCXSetMaxPower(RCX_ OUT_A, RCX_ConstantSrc, 5)
HTRCXEnableOutput (outputs)	Send the EnableOutput command to an RCX.	HTRCXEnableOutput (RCX_OUT_A)
HTRCXDisableOutput (outputs)	Send the DisableOutput command to an RCX.	HTRCXDisableOutput (RCX_OUT_A)
HTRCXInvertOutput(outputs)	Send the InvertOutput command to an RCX.	HTRCXInvertOutput (RCX_OUT_A)
HTRCXObvertOutput(outputs)	Send the ObvertOutput command to an RCX.	HTRCXObvertOutput (RCX_OUT_A)
HTRCXCalibrateEvent(evt, low, hi, hyst)	Send the CalibrateEvent command to an RCX.	HTRCXCalibrateEvent(0, 200, 500, 50)

HTRCXSetVar(varnum, src, value)	Send the SetVar command to an RCX.	HTRCXSetVar(0, RCX_VariableSrc, 1)
HTRCXSumVar(varnum, src, value)	Send the SumVar command to an RCX.	HTRCXSumVar(0, RCX_InputValueSrc, S1)
HTRCXSubVar(varnum, src, value)	Send the SubVar command to an RCX.	HTRCXSubVar(0, RCX_RandomSrc, 10)
HTRCXDivVar(varnum, src, value)	Send the DivVar command to an RCX.	HTRCXDivVar(0, RCX_ConstantSrc, 2)
HTRCXMulVar(varnum, src, value)	Send the MulVar command to an RCX.	HTRCXMulVar(0, RCX_VariableSrc, 4)
HTRCXSgnVar(varnum, src, value)	Send the SgnVar command to an RCX.	HTRCXSgnVar(0, RCX_VariableSrc, 0)
HTRCXAbsVar(varnum, src, value)	Send the AbsVar command to an RCX.	HTRCXAbsVar(0, RCX_VariableSrc, 0)
HTRCXAndVar(varnum, src, value)	HTRCXAndVar(0, RCX_ConstantSrc, 0x7f)	Send the OrVar command to an RCX.
Send the AndVar command to an RCX.	HTRCXOrVar(varnum, src, value)	HTRCXOrVar(0, RCX_ConstantSrc, 0xCC)
HTRCXSet(dstsrc, dstval, src, value)	Send the Set command to an RCX.	HTRCXSet(RCX_VariableSrc, 0, RCX_RandomSrc, 10000)
HTRCXUnlock()	Send the Unlock command to an RCX.	HTRCXUnlock()
HTRCXReset()	Send the Reset command to an RCX.	HTRCXReset()
HTRCXBoot()	Send the Boot command to an RCX.	HTRCXBoot()
HTRCXSetUserDisplay(src, value, precision)	Send the SetUserDisplay command to an RCX.	HTRCXSetUserDisplay (RCX_VariableSrc, 0, 2)
HTRCXIncCounter(counter)	Send the IncCounter command to an RCX.	HTRCXIncCounter(0)
HTRCXDecCounter(counter)	Send the DecCounter command to an RCX.	HTRCXDecCounter(0)

HTRCXClearCounter(counter)	Send the ClearCounter command to an RCX.	HTRCXClearCounter(0)
HTRCXSetPriority(p)	Send the SetPriority command to an RCX.	HTRCXSetPriority(2)
HTRCXSetMessage(msg)	Send the SetMessage command to an RCX.	HTRCXSetMessage(20)

HiTechnic iRLink Scout Functions

The functions in this section expose scout-specific commands for controlling the Scout programmable brick using the HiTechnic iRLink device.

Function	Description	Example
HTScoutCalibrateSensor()	Send the CalibrateSensor command to a Scout.	HTScoutCalibrateSensor()
HTScoutMuteSound()	Send the MuteSound command to a Scout.	HTScoutMuteSound()
HTScoutUnmuteSound()	Send the UnmuteSound command to a Scout.	HTScoutUnmuteSound()
HTScoutSelectSounds(group)	Send the SelectSounds command to a Scout.	HTScoutSelectSounds(0)
HTScoutSetLight(mode)	Send the SetLight command to a Scout.	HTScoutSetLight (SCOUT_LIGHT_ON)
HTScoutSetCounterLimit (counter, src, value)	Send the SetCounterLimit command to a Scout.	HTScoutSetCounterLimit (0, RCX_ConstantSrc, 2000)
HTScoutSetTimerLimit(timer, src, value)	Send the SetTimerLimit command to a Scout.	HTScoutSetTimerLimit(0, RCX_ConstantSrc, 10000)
HTScoutSetSensorClickTime (src, value)	Send the SetSensorClickTime command to a Scout.	HTScoutSetSensorClickTime (RCX_ConstantSrc, 200)
HTScoutSetSensorHysteresis (src, value)	Send the SetSensorHysteresis command to a Scout.	HTScoutSetSensorHysteresis (RCX_ConstantSrc, 50)
HTScoutSetSensorLower Limit(src, value)	Send the SetSensorLower Limit command to a Scout.	HTScoutSetSensorLower Limit(RCX_ConstantSrc, 100)

HTScoutSetSensorUpper Limit(src, value)	Send the SetSensor UpperLimit command to a Scout.	HTScoutSetSensorUpper Limit(RCX_ConstantSrc, 400)
HTScoutSetEventFeedback (src, value)	Send the SetEventFeedback command to a Scout.	HTScoutSetEventFeedback (RCX_ConstantSrc, 10)
HTScoutSendVLL(src, value)	Send the SendVLL command to a Scout.	HTScoutSendVLL(RCX_ ConstantSrc, 0x30)
HTScoutSetScoutRules (motion, touch, light, time, effect)	Send the SetScoutRules command to a Scout.	HTScoutSetScoutRules (SCOUT_MR_FORWARD, SCOUT_TR_REVERSE, SCOUT_ LR_IGNORE, SCOUT_TGS_ SHORT, SCOUT_FXR_BUG)
HTScoutSetScoutMode (mode)	Send the SetScoutMode command to a Scout.	HTScoutSetScoutMode (SCOUT_MODE_POWER)

Mindsensors Functions

The functions in this section are for using various sensors made by mindsensors.com.

Function	Description	Example
ReadSensorMSRTClock(port, out sec, out min, out hrs, out dow, out date, out month, out year, out result)	Read real-time clock values from the Mindsensors RTClock sensor. Returns a boolean value indicating whether or not the operation completed successfully.	ReadSensorMSRTClock(S1, ss, mm, hh, dow, dd, mon, yy, result)
ReadSensorMSCompass (port, out result)	Return the Mindsensors Compass sensor value.	ReadSensorMSCompass (S1, result)

NXT Firmware Modules

Topics in this Appendix

- NXT Firmware Modules

Appendix C

If you want to know more about the inner workings of the firmware, look no further. The following appendix describes the modules that make up the NXT firmware.

Firmware Modules

The NXT firmware is built using several modules that interact and cooperate with each other, providing programmers with access to all that the NXT hardware has to offer. There are eleven modules in the standard firmware. Each has an important role to play in the operation of the firmware. In this section we'll have a brief look at each module.

Before we look at the modules, let's see what they all have in common. Each module in the firmware is accessed via a module header structure. Its definition is shown in Sample C-1.

```
typedef    struct
{
  ULONG    ModuleID;
  UBYTE    ModuleName[FILENAME_LENGTH + 1];
  void     (*cInit)(void* pHeader);
  void     (*cCtrl)(void);
  void     (*cExit)(void);
  void     *pIOMap;
  void     *pVars;
  UWORD    IOMapSize;
  UWORD    VarsSize;
  UWORD    ModuleSize;
} HEADER;
```

Sample C-1. Module header structure

Every module has a module ID, a module name, three function pointers (init, ctrl, and exit), an IOMap structure, and a Vars structure as well as fields that store the size of the IOMap and the Vars structures. The init, ctrl, and exit functions of every module are called in a loop by the firmware scheduler. The init and exit functions are only called when starting up or shutting down but the ctrl function is called repeatedly while the NXT is turned on.

The item we're most interested in happens to be the IOMap structure. We also need to know the full module names and their IDs. Table C-1 below lists all eleven modules, their IDs, and the IOMap module sizes.

Module Name	ID	IOMap Size
Command.mod	0x00010001	32820 bytes
Output.mod	0x00020001	100 bytes
Input.mod	0x00030001	80 bytes
Button.mod	0x00040001	36 bytes
Comm.mod	0x00050001	1896 bytes
IOCtrl.mod	0x00060001	2 bytes
Sound.mod	0x00080001	30 bytes
Loader.mod	0x00090001	8 bytes
Display.mod	0x000A0001	1720 bytes
Low Speed.mod	0x000B0001	167 bytes
Ui.mod	0x000C0001	44 bytes

Table C-1. Module names, IDs, and sizes

You can read values from and write values to the various module IOMaps from within your NBC or NXC programs using the IOMapRead and IOMapWrite system call functions. If you have the enhanced NBC/NXC firmware installed on your NXT, then you can read and write IOMap values much faster by using the IOMapReadByID and IOMapWriteByID system call functions. In both cases you not only need either the module name or its ID but you also need to know the right IOMap field offset constants that point to the desired field within the module IOMap structure. Those constants are listed in the IOMap offset tables in the subsections below.

The functionality of most of the firmware modules is not exposed solely via these IOMap structures. Usually there are higher-level API functions that hide the details of the IOMap from a programmer. But it is good to understand how things work under the hood so that you can take advantage of the extra power when you really need it.

Command Module

The command module provides support for the execution of user programs via the NXT virtual machine. It also implements the direct command protocol support that enables the NXT to respond to USB or Bluetooth requests from other devices such as a PC or another NXT brick. It is the largest module, since its IOMap structure includes 32k bytes of memory for use by running programs. The IOMap structure for the command module is shown in Sample C-2.

```
typedef struct
{
  UBYTE FormatString[VM_FORMAT_STRING_SIZE];
  UWORD (*pRCHandler)(UBYTE *,UBYTE *,UBYTE *);
  ULONG Tick;
  UWORD OffsetDS;
  UWORD OffsetDVA;
  PROGRAM_STATUS ProgStatus;
  UBYTE Awake;
  UBYTE ActivateFlag;
  UBYTE DeactivateFlag;
  UBYTE FileName[FILENAME_LENGTH + 1];
  ULONG MemoryPool[POOL_MAX_SIZE / 4];
} IOMAPCMD;
```

Sample C-2. Command module IOMap structure

From the perspective of a running program, the IOMap isn't incredibly interesting. You can read the Tick value but that is also available via other API functions. You can get information about the memory your program is using (via OffsetDS and OffsetDVA) but it isn't all that useful for making your robot do interesting things. The memory access provided by this IOMap is very useful from the standpoint of a PC-based development environment such as BricxCC. BricxCC uses this ability to let users view and modify variable values while a program is running on the NXT. You can access the fields in the command module IOMap using the offsets shown in Table C-2.

Command Module Offsets	Value	Size
CommandOffsetFormatString	0	16
CommandOffsetPRCHandler	16	4
CommandOffsetTick	20	4
CommandOffsetOffsetDS	24	2
CommandOffsetOffsetDVA	26	2
CommandOffsetProgStatus	28	1
CommandOffsetAwake	29	1
CommandOffsetActivateFlag	30	1
CommandOffsetDeactivateFlag	31	1
CommandOffsetFileName	32	20
CommandOffsetMemoryPool	52	32k

Table C-2. Command module IOMap offsets

The most important part of the Command module is the virtual machine, since that is what executes every line of code you write in your NBC or NXC program. It is where all the functionality accessible by a program is ultimately defined. We'll see more about the specific operations implemented by the virtual machine as well as the types of data that it can manipulate in chapter 8.

Output Module

The output module is the hub through which every form of motor control, as well as rotation sensor feedback, is routed. If you call any API function that controls an NXT motor, the ultimate result is a modification to one or more of the fields in the output module IOMap structure. It is only 100 bytes in size but it packs a lot of power into that small package. The IOMap structure for the output module is shown in Sample C-3.

```
typedef struct
{
  SLONG    TachoCount;
  SLONG    BlockTachoCount;
  SLONG    RotationCount;
  ULONG    TachoLimit;
  SWORD    MotorRPM; // unused
  UBYTE    Flags;
  UBYTE    Mode;
  SBYTE    Speed;
  SBYTE    ActualSpeed;
  UBYTE    RegPParameter;
  UBYTE    RegIParameter;
  UBYTE    RegDParameter;
  UBYTE    RunState;
  UBYTE    RegMode;
  UBYTE    Overloaded;
  SBYTE    SyncTurnParameter;
  UBYTE    SpareOne;
  UBYTE    SpareTwo;
  UBYTE    SpareThree;
} OUTPUT;

typedef struct
{
  OUTPUT   Outputs[NO_OF_OUTPUTS];
  UBYTE    PwnFreq;
} IOMAPOUTPUT;
```

Sample C-3. Output module IOMap structure

The IOMap structure contains 3 OUTPUT structures, one for each NXT output port. Each structure can be used to control a motor attached to the associated port. You can access the fields in the output module IOMap using the offsets shown in Table C-3. Use values for 'p' of 0 hrough 2 (outputs A through C).

Output Module Offsets	Value	Size
OutputOffsetTachoCount(p)	(((p)*32)+0)	4
OutputOffsetBlockTachoCount(p)	(((p)*32)+4)	4
OutputOffsetRotationCount(p)	(((p)*32)+8)	4
OutputOffsetTachoLimit(p)	(((p)*32)+12)	4
OutputOffsetMotorRPM(p)	(((p)*32)+16)	2
OutputOffsetFlags(p)	(((p)*32)+18)	1
OutputOffsetMode(p)	(((p)*32)+19)	1
OutputOffsetSpeed(p)	(((p)*32)+20)	1
OutputOffsetActualSpeed(p)	(((p)*32)+21)	1
OutputOffsetRegPParameter(p)	(((p)*32)+22)	1
OutputOffsetRegIParameter(p)	(((p)*32)+23)	1
OutputOffsetRegDParameter(p)	(((p)*32)+24)	1
OutputOffsetRunState(p)	(((p)*32)+25)	1
OutputOffsetRegMode(p)	(((p)*32)+26)	1
OutputOffsetOverloaded(p)	(((p)*32)+27)	1
OutputOffsetSyncTurnParameter(p)	(((p)*32)+28)	1
OutputOffsetPwnFreq	96	1

Table C-3. Output module IOMap offsets

The outputs each have several fields that define the current state of the output port. These fields are defined in Table C-4.

Field Constant	Type	Access	Range	Meaning
Flags	ubyte	Read/ Write	0, 255	This field can include any combination of the flag bits described in Table C-5.
				Use UF_UPDATE_MODE, UF_UPDATE_SPEED, UF_UPDATE_TACHO_LIMIT, and UF_UPDATE_PID_VAL-UES along with other fields to commit changes to the state of outputs. Set the appropriate flags after setting one or more of the output fields in order for the changes to actually go into affect.
Mode	ubyte	Read / Write	0, 255	This is a bitfield that can include any of the values listed in Table C-6.
				The OUT_MODE_MOTO-RON bit must be set in order for power to be applied to the motors. Add OUT_MODE_BRAKE to enable electronic brak-ing. Braking means that the output voltage is not allowed to float between active PWM pulses. It improves the accuracy of motor output but uses more battery power.
				To use motor regulation include OUT_MODE_REGU-LATED in the Mode value. Use UF_UPDATE_MODE with Flags to commit changes to this field.
Speed	sbyte	Read/ Write	-100, 100	Specify the power level of the output. The abso-lute value of speed is a percentage of the full power of the motor. The sign of Speed controls the rotation direction. Positive values tell the firmware

				to turn the motor forward, while negative values turn the motor backward. Use UF_UPDATE_SPEED with Flags to commit changes to this field.
ActualSpeed	sbyte	Read	-100, 100	Return the percent of full power the firmware is applying to the output. This may vary from the Speed value when auto-regulation code in the firmware responds to a load on the output
TachoCount	slong	Read	ful range of signed long	Return the internal position counter value for the specified output. The internal count is reset automatically when a new goal is set using the TachoLimit and the UF_UPDATE_TACHO_LIMIT flag. Set the UF_UPDATE_RESET_COUNT flag in Flags to reset TachoCount and cancel any TachoLimit. The sign of TachoCount indicates the motor rotation direction.
TachoLimit	ulong	Read/ Write	full range of unsigned long	Specify the number of degrees the motor should rotate. Use UF_UPDATE_TACHO_LIMIT with the Flags field to commit changes to the TachoLimit. The value of this field is a relative distance from the current motor position at the moment when the UF_UPDATE_TACHO_LIMIT flag is processed.
RunState	ubyte	Read/ Write	0..255	Use this field to specify the running state of an output. Set the RunState to OUT_RUNSTATE_RUNNING to enable power to any output.

				Use OUT_RUNSTATE_RAM-PUP to enable automatic ramping to a new Speed level greater than the current Speed level. Use OUT_RUNSTATE_RAMPDOWN to enable automatic ramping to a new Speed level less than the current Speed level.
				Both the rampup and ramp-down bits must be used in conjunction with appropriate TachoLimit and Speed values. In this case the firmware smoothly increases or decreases the actual power to the new Speed level over the total number of degrees of rotation specified in TachoLimit.
SyncTurn Parameter	sbyte	Read/ Write	-100, 100	Use this field to specify a proportional turning ratio. This field must be used in conjunction with other field values: Mode must include OUT_MODE_MOTORON and OUT_MODE_REGULATED, RegMode must be set to OUT_REGMODE_SYNC, RunState must not be OUT_RUNSTATE_IDLE, and Speed must be non-zero.
				There are only three valid combinations of left and right motors for use with SyncTurn-Parameter: OUT_AB, OUT_BC, and OUT_AC. In each of these three options the first motor listed is considered to be the left motor and the second motor is the right motor, regardless of the physical configuration of the robot.
				Negative SyncTurnParameter values shift power toward the left motor while positive

				values shift power toward the right motor. An absolute value of 50 usually results in one motor stopping. An absolute value of 100 usually results in two motors turning in opposite directions at equal power.
RegMode	ubyte	Read/Write	0..255	This field specifies the regulation mode to use with the specified port(s). It is ignored if the OUT_MODE_REGULATED bit is not set in the Mode field. Unlike the Mode field, RegMode is not a bitfield. Only one RegMode value can be set at a time. Valid RegMode values are listed in Table C-8.

Speed regulation means that the firmware tries to maintain a certain speed based on the Speed setting. The firmware adjusts the PWM duty cycle if the motor is affected by a physical load. This adjustment is reflected by the value of the ActualSpeed property. When using speed regulation, do not set Speed to its maximum value since the firmware cannot adjust to higher power levels in that situation.

Synchronization means the firmware tries to keep two motors in synch regardless of physical loads. Use this mode to maintain a straight path for a mobile robot automatically. Also use this mode with the SyncTurnParameter property to provide proportional turning.

Set OUT_REGMODE_SYNC on at least two motor ports in order for synchronization to function. Setting OUT_REGMODE_SYNC on all three motor ports will result in |

				only the first two (OUT_A and OUT_B) being synchronized.
Overload	ubyte	Read	0..1	This field will have a value of 1 (true) if the firmware speed regulation cannot overcome a physical load on the motor. In other words, the motor is turning more slowly than expected.
				If the motor speed can be maintained in spite of loading then this field value is zero (false).
				In order to use this field the motor must have a non-idle RunState, a Mode which includes OUT_MODE_MOTORON and OUT_MODE_REGULATED, and its RegMode must be set to OUT_REGMODE_SPEED.
RegPParameter	ubyte	Read/ Write	0..255	This field specifies the proportional term used in the internal proportional-integral-derivative (PID) control algorithm.
				Set UF_UPDATE_PID_VALUES to commit changes to RegPParameter, RegIParameter, and RegDParameter simultaneously.
RegIParameter	ubyte	Read/ Write	0..255	This field specifies the integral term used in the internal proportional-integral-derivative (PID) control algorithm.
				Set UF_UPDATE_PID_VALUES to commit changes to RegPParameter, RegIParameter, and RegDParameter simultaneously.
RegDParameter	ubyte	Read/ Write	0..255	This field specifies the derivative term used in the internal proportional-integral-derivative (PID) control algorithm.

				Set UF_UPDATE_PID_VALUES to commit changes to RegPParameter, RegIParameter, and RegDParameter simultaneously.
BlockTachoCount	slong	Read	full range of signed long	Return the block-relative position counter value for the specified port.
				Refer to the Flags description for information about how to use block-relative position counts.
				Set the UF_UPDATE_RESET_BLOCK_COUNT flag in Flags to request that the firmware reset the BlockTachoCount.
				The sign of BlockTachoCount indicates the direction of rotation. Positive values indicate forward rotation and negative values indicate reverse rotation. Forward and reverse depend on the orientation of the motor.
RotationCount	slong	Read	full range of signed long	Return the program-relative position counter value for the specified port.
				Refer to the Flags description for information about how to use program-relative position counts.
				Set the UF_UPDATE_RESET_ROTATION_COUNT flag in Flags to request that the firmware reset the RotationCount.
				The sign of RotationCount indicates the direction of rotation. Positive values indicate forward rotation and negative values indicate reverse rotation. Forward and reverse depend on the orientation of the motor.

Table C-4. Output IOMap field definitions

Valid values for the Flags field are described in Table C-5 below.

Flags Constants	Value	Meaning
UF_UPDATE_MODE	0X01	Commits changes to the Mode output property
UF_UPDATE_SPEED	0X02	Commits changes to the Speed output property
UF_UPDATE_TACHO_LIMIT	0X04	Commits changes to the TachoLimit output property
UF_UPDATE_RESET_COUNT	0X08	Resets all rotation counters, cancels the current goal, and resets the rotation error-correction system
UF_UPDATE_PID_VALUES	0X10	Commits changes to the PID motor regulation properties
UF_UPDATE_RESET_BLOCK_COUNT	0X20	Resets the block-relative rotation counter
UF_UPDATE_RESET_ROTATION_COUNT	0X40	Resets the program-relative rotation counter

Table C-5. Flags Constants

Valid values for the Mode field are described in Table C-6 below.

Mode Constants	Value	Meaning
OUT_MODE_COAST	0X00	No power and no braking so motors rotate freely
OUT_MODE_MOTORON	0X01	Enables PWM power to the outputs given the Power setting
OUT_MODE_BRAKE	0X02	Uses electronic braking to outputs
OUT_MODE_REGULATED	0X04	Enables active power regulation using the RegMode value
OUT_MODE_REGMETHOD	0Xf0	

Table C-6. Mode Constants

Valid values for the RunState field are described in Table C-7 below.

RunState Constants	Value	Meaning
UT_RUNSTATE_IDLE	0x00	Disable all power to motors.
OUT_RUNSTATE_RAMPUP	0x10	Enable ramping up from a current Speed to a new (higher) Speed over a specified TachoLimit goal.
OUT_RUNSTATE_RUNNING	0x20	Enable power to motors at the specified Speed level.
OUT_RUNSTATE_RAMPDOWN	0x40	Enable ramping down from a current Speed to a new (lower) Speed over a specified TachoLimit goal.

Table C-7. RunState Constants

Valid values for the RegMode field are described in Table C-8 below.

RegMode Constants	Value	Meaning
OUT_REGMODE_IDLE	0x00	No regulation
OUT_REGMODE_SPEED	0x01	Regulate a motor's Speed
OUT_REGMODE_SYNC	0x02	Synchronize the rotation of two motors

Table C-8. RegMode Constants

As with other modules, nearly all of the functionality provided by the output module is exposed at higher levels of API functions. You can control the motors directly via the IOMap but there are better and faster ways to do it. We'll look at the many output control routines provided by NBC and NXC in great detail in Chapters 7 and 8 as well as in many of the programs contained in this book.

Input Module

The NXT input module encompasses all sensor inputs except for digital I²C (LowSpeed) sensors. There are four sensors, which internally are numbered 0, 1, 2, and 3. This is potentially confusing since they are labeled on the NXT as sensors 1, 2, 3, and 4. To help mitigate this confusion, the sensor port names S1, S2, S3, and S4 are defined in the NXC API. These sensor names may be used in any function that requires a sensor port as an argument. Alternatively, the NBC port name constants IN_1, IN_2, IN_3, and IN_4 may also be used when a sensor port is required. The IOMap structure for the input module is shown in Sample C-4.

```
typedef    struct
{
  UWORD    CustomZeroOffset;
  UWORD    ADRaw;
  UWORD    SensorRaw;
  SWORD    SensorValue;
  UBYTE    SensorType;
  UBYTE    SensorMode;
  UBYTE    SensorBoolean;
  UBYTE    DigiPinsDir;
  UBYTE    DigiPinsIn;
  UBYTE    DigiPinsOut;
  UBYTE    CustomPctFullScale;
  UBYTE    CustomActiveStatus;
  UBYTE    InvalidData;
  UBYTE    Spare1;
  UBYTE    Spare2;
  UBYTE    Spare3;
} INPUT;

typedef    struct
{
  INPUT    Inputs[NO_OF_INPUTS];
} IOMAPINPUT;
```

Sample C-4. Input module IOMap structure

Much like the output module, the input module IOMap structure contains 4 INPUT structures, one for each NXT input port. Each structure can be used to configure and read values from a non-I^2C sensor attached to the associated port. Access the fields in the input module IOMap using the offsets shown in Table C-9. Use values for 'p' of 0 through 3 (inputs 1 through 4).

Input Module Offsets	Value	Size
InputOffsetCustomZeroOffset(p)	(((p)*20)+0)	2
InputOffsetADRaw(p)	(((p)*20)+2)	2
InputOffsetSensorRaw(p)	(((p)*20)+4)	2
InputOffsetSensorValue(p)	(((p)*20)+6)	2
InputOffsetSensorType(p)	(((p)*20)+8)	1
InputOffsetSensorMode(p)	(((p)*20)+9)	1
InputOffsetSensorBoolean(p)	(((p)*20)+10)	1
InputOffsetDigiPinsDir(p)	(((p)*20)+11)	1
InputOffsetDigiPinsIn(p)	(((p)*20)+12)	1
InputOffsetDigiPinsOut(p)	(((p)*20)+13)	1
InputOffsetCustomPctFullScale(p)	(((p)*20)+14)	1
InputOffsetCustomActiveStatus(p)	(((p)*20)+15)	1
InputOffsetInvalidData(p)	(((p)*20)+16)	1

Table C-9. Input module IOMap offsets

The sensor ports on the NXT are capable of interfacing to a variety of different sensors. It is up to the program to tell the NXT the type of sensor is attached to each port. Setting the SensorType field in the IOMap configures a sensor's type. There are 12 sensor types, each corresponding to a specific LEGO RCX or NXT sensor. A thirteenth type (SENSOR_TYPE_NONE) is used to indicate that no sensor has been configured.

In general, a program should configure the type to match the actual sensor. If a sensor port is configured as the wrong type, the NXT may not be able to read it accurately. Use either the NXC Sensor Type constants or the NBC Sensor Type constants as listed in Table C-10.

Sensor Type	NBC Sensor Type	Meaning
SENSOR_TYPE_NONE	IN_TYPE_NO_SENSOR	no sensor configured
SENSOR_TYPE_TOUCH	IN_TYPE_SWITCH	NXT or RCX touch sensor
SENSOR_TYPE_TEMPERATURE	IN_TYPE_TEMPERATURE	RCX temperature sensor
SENSOR_TYPE_LIGHT	IN_TYPE_REFLECTION	RCX light sensor
SENSOR_TYPE_ROTATION	IN_TYPE_ANGLE	RCX rotation sensor
SENSOR_TYPE_LIGHT_ACTIVE	IN_TYPE_LIGHT_ACTIVE	NXT light sensor with light
SENSOR_TYPE_LIGHT_INACTIVE	IN_TYPE_LIGHT_INACTIVE	NXT light sensor without light
SENSOR_TYPE_SOUND_DB	IN_TYPE_SOUND_DB	NXT sound sensor with dB scaling
SENSOR_TYPE_SOUND_DBA	IN_TYPE_SOUND_DBA	NXT sound sensor with dBA scaling
SENSOR_TYPE_CUSTOM	IN_TYPE_CUSTOM	Custom sensor (unused)
SENSOR_TYPE_LOWSPEED	IN_TYPE_LOWSPEED	I2C digital sensor
SENSOR_TYPE_LOWSPEED_9V	IN_TYPE_LOWSPEED_9V	I2C digital sensor (9V power)
SENSOR_TYPE_HIGHSPEED	IN_TYPE_HISPEED	Highspeed sensor (unused)

Table C-10. SensorType constants

The NXT allows a sensor to be configured in different modes. The sensor mode determines how a sensor's raw value is processed. Some modes only make sense for certain types of sensors, for example SENSOR_MODE_ROTATION is useful only with rotation sensors. You can set the SensorMode field of the input module IOMap to set the sensor mode. Valid SensorMode values are shown in Table C-11. Use either the NXC Sensor Mode constant or the NBC Sensor Mode constant.

Sensor Mode	NBC Sensor Mode	Meaning
SENSOR_MODE_RAW	IN_MODE_RAW	raw value from 0 to 1023
SENSOR_MODE_BOOL	IN_MODE_BOOLEAN	boolean value (0 or 1)
SENSOR_MODE_EDGE	IN_MODE_TRANSITIONCNT	counts number of boolean transitions
SENSOR_MODE_PULSE	IN_MODE_PERIODCOUNTER	counts number of boolean periods
SENSOR_MODE_PERCENT	IN_MODE_PCTFULLSCALE	value from 0 to 100
SENSOR_MODE_FAHRENHEIT	IN_MODE_FAHRENHEIT	degrees F
SENSOR_MODE_CELSIUS	IN_MODE_CELSIUS	degrees C
SENSOR_MODE_ROTATION	IN_MODE_ANGLESTEP	rotation (16 ticks per revolution)

Table C-11. SensorMode constants

The NXT provides a boolean conversion for all sensors - not just touch sensors. This boolean conversion is normally based on preset thresholds for the raw value. A "low" value (less than 460) is a boolean value of 1. A high value (greater than 562) is a boolean value of 0. This conversion can be modified: a *slope value* between 0 and 31 may be added to a sensor's mode when setting the SensorMode field. If the sensor's raw value changes more than the specified slope during a certain time (3ms), then the sensor's boolean value will change. This allows the boolean value to reflect rapid changes in the raw value. A rapid increase will result in a boolean value of 0, a rapid decrease is a boolean value of 1.

There are, as expected, many high level API functions in both NBC and NXC that hide the details of the input module IOMap. They are a lot easier to use and a bit faster than directly manipulating the IOMap. We'll explore these functions in several upcoming chapters. But there are several fields that are only accessible via the IOMap structure so there may very well be situations where you will need to know how to use the lower level functionality. Creating a custom sensor and configuring it for use in your NXC program is one example of when you will need this knowledge.

Button Module

The NXT button module encompasses support for checking the pressed state and press counts of the four buttons on the NXT brick. The IOMap structure for the button module is shown in Sample C-5.

```
typedef struct
{
  UBYTE PressedCnt;
  UBYTE LongPressCnt;
  UBYTE ShortRelCnt;
  UBYTE LongRelCnt;
  UBYTE RelCnt;
  UBYTE SpareOne;
  UBYTE SpareTwo;
  UBYTE SpareThree;
} BTNCNT;

typedef struct
{
  BTNCNT BtnCnt[4];
  UBYTE State[4];
} IOMAPBUTTON;
```

Sample C-5. Button module IOMap structure

Much like the output module and the input module, the button module IOMap structure contains four BUTTON structures, one for each NXT button. Each structure can be used to read values from one of the four NXT buttons. Access the fields in the button module IOMap using the offsets shown in Table C-12. Use values for 'p' of 0 through 3 (BTN1 through BTN4).

Button Module Offsets	Value	Size
ButtonOffsetPressedCnt(b)	(((b)*8)+0)	1
ButtonOffsetLongPressCnt(b)	(((b)*8)+1)	1
ButtonOffsetShortRelCnt(b)	(((b)*8)+2)	1
ButtonOffsetLongRelCnt(b)	(((b)*8)+3)	1
ButtonOffsetRelCnt(b)	(((b)*8)+4)	1
ButtonOffsetState(b)	((b)+32)	1*4

Table C-12. Button module IOMap offsets

Button constant values that can be used with the module offsets shown above are listed in Table C-13.

Button Constants	Value
BTN1, BTNEXIT	0
BTN2, BTNRIGHT	1
BTN3, BTNLEFT	2
BTN4, BTNCENTER	3
NO_OF_BTNS	4

Table C-13. Button constants

Valid values for the button module IOMap State field are listed in Table C-14.

State Constants	Value
BTNSTATE_PRESSED_EV	0x01
BTNSTATE_SHORT_RELEASED_EV	0x02
BTNSTATE_LONG_PRESSED_EV	0x04
BTNSTATE_LONG_RELEASED_EV	0x08
BTNSTATE_PRESSED_STATE	0x80

Table C-14. State Constants

The virtual machine in the standard NXT firmware intercepts button presses of the dark gray exit button and uses it to abort a running program. That makes it impossible to use this button as part of your own programs. The enhanced NBC/NXC firmware provides the option of telling the virtual machine that aborting requires a long press of the exit button. If we do so then all four buttons can be used within our program in whatever way we choose.

Comm Module

The NXT comm module encompasses support for all forms of Bluetooth, USB, and HiSpeed communication. You can use the Bluetooth communication methods to send information to other devices connected to the NXT brick. The NXT firmware also implements a message queuing or mailbox system which you can access using these methods.

The structures used to define the IOMap for this module are shown in Sample C-6.

```
typedef struct
{
  UBYTE Buf[62];
  UBYTE InPtr;
  UBYTE OutPtr;
  UBYTE Spare1;
  UBYTE Spare2;
} USBBUF;
typedef struct
{
  UBYTE Buf[128];
  UBYTE InPtr;
  UBYTE OutPtr;
  UBYTE Spare1;
  UBYTE Spare2;
} HSBUF;
typedef struct
{
  UBYTE Buf[128];
  UBYTE InPtr;
  UBYTE OutPtr;
  UBYTE Spare1;
  UBYTE Spare2;
} BTBUF;
typedef struct
{
  UBYTE Name[16];
  UBYTE ClassOfDevice[4];
  UBYTE BdAddr[7];
  UBYTE DeviceStatus;
  UBYTE Spare1;
  UBYTE Spare2;
  UBYTE Spare3;
} BDDEVICETABLE;
typedef struct
{
  UBYTE Name[16];
  UBYTE ClassOfDevice[4];
  UBYTE PinCode[16];
  UBYTE BdAddr[7];
  UBYTE HandleNr;
  UBYTE StreamStatus;
  UBYTE LinkQuality;
  UBYTE Spare;
} BDCONNECTTABLE;
typedef struct
{
  UBYTE Name[16];
```

```
    UBYTE BluecoreVersion[2];
    UBYTE BdAddr[7];
    UBYTE BtStateStatus;
    UBYTE BtHwStatus;
    UBYTE TimeOutValue;
    UBYTE Spare1;
    UBYTE Spare2;
    UBYTE Spare3;
} BRICKDATA;
typedef struct
{
      UWORD (*pFunc)(UBYTE, UBYTE, UBYTE, UBYTE, UBYTE*,
         UWORD*);
    void  (*pFunc2)(UBYTE*);
    // BT related entries
    BDDEVICETABLE BtDeviceTable[30];
    BDCONNECTTABLE BtConnectTable[4];
    //General brick data
    BRICKDATA BrickData;
    BTBUF BtInBuf;
    BTBUF BtOutBuf;
    // HI Speed related entries
    HSBUF HsInBuf;
    HSBUF HsOutBuf;
    // USB related entries
    USBBUF UsbInBuf;
    USBBUF UsbOutBuf;
    USBBUF UsbPollBuf;
    UBYTE BtDeviceCnt;
    UBYTE BtDeviceNameCnt;
    UBYTE HsFlags;
    UBYTE HsSpeed;
    UBYTE HsState;
    UBYTE UsbState;
} IOMAPCOMM;
```

Sample C-6. Comm module IOMap structure

There are a number of interesting items in the Comm module IOMap structure and its various sub-structures. Being able to use the USB, Bluetooth, and HiSpeed buffers is potentially useful. The Bluetooth device table and the Bluetooth connection table can also provide useful information. You can access each of the fields in the comm module IOMap using the offsets shown in Table C-15.

Comm Module Offsets	Value	Size
CommOffsetPFunc	0	4
CommOffsetPFuncTwo	4	4
CommOffsetBtDeviceTableName(p)	(((p)*31)+8)	16
CommOffsetBtDeviceTableClassOfDevice(p)	(((p)*31)+24)	4
CommOffsetBtDeviceTableBdAddr(p)	(((p)*31)+28)	7
CommOffsetBtDeviceTableDeviceStatus(p)	(((p)*31)+35)	1
CommOffsetBtConnectTableName(p)	(((p)*47)+938)	16
CommOffsetBtConnectTableClassOfDevice (p)	(((p)*47)+954)	4
CommOffsetBtConnectTablePinCode(p)	(((p)*47)+958)	16
CommOffsetBtConnectTableBdAddr(p)	(((p)*47)+974)	7
CommOffsetBtConnectTableHandleNr(p)	(((p)*47)+981)	1
CommOffsetBtConnectTableStreamStatus(p)	(((p)*47)+982)	1
CommOffsetBtConnectTableLinkQuality(p)	(((p)*47)+983)	1
CommOffsetBrickDataName	1126	16
CommOffsetBrickDataBluecoreVersion	1142	2
CommOffsetBrickDataBdAddr	1144	7
CommOffsetBrickDataBtStateStatus	1151	1
CommOffsetBrickDataBtHwStatus	1152	1
CommOffsetBrickDataTimeOutValue	1153	1
CommOffsetBtInBufBuf	1157	128
CommOffsetBtInBufInPtr	1285	1
CommOffsetBtInBufOutPtr	1286	1
CommOffsetBtOutBufBuf	1289	128
CommOffsetBtOutBufInPtr	1417	1
CommOffsetBtOutBufOutPtr	1418	1
CommOffsetHsInBufBuf	1421	128
CommOffsetHsInBufInPtr	1549	1
CommOffsetHsInBufOutPtr	1550	1
CommOffsetHsOutBufBuf	1553	128
CommOffsetHsOutBufInPtr	1681	1
CommOffsetHsOutBufOutPtr	1682	1

CommOffsetUsbInBufBuf	1685	64
CommOffsetUsbInBufInPtr	1749	1
CommOffsetUsbInBufOutPtr	1750	1
CommOffsetUsbOutBufBuf	1753	64
CommOffsetUsbOutBufInPtr	1817	1
CommOffsetUsbOutBufOutPtr	1818	1
CommOffsetUsbPollBufBuf	1821	64
CommOffsetUsbPollBufInPtr	1885	1
CommOffsetUsbPollBufOutPtr	1886	1
CommOffsetBtDeviceCnt	1889	1
CommOffsetBtDeviceNameCnt	1890	1
CommOffsetHsFlags	1891	1
CommOffsetHsSpeed	1892	1
CommOffsetHsState	1893	1
CommOffsetUsbState	1894	1

Table C-15. Comm module IOMap offsets

Communication via Bluetooth uses a master/slave connection system. One device must be designated as the master device before you run a program using Bluetooth. If the NXT is the master device then you can configure up to three slave devices using connections 1, 2, and 3 on the NXT brick. If your NXT is a slave device then connection 0 on the brick must be reserved for the master device.

Programs running on the master NXT brick can send packets of data to any connected slave devices using various high-level API methods. Slave devices write response packets to the message queuing system where they wait for the master device to poll for the response. Using the direct command protocol, a master device can send messages to slave NXT bricks in the form of text strings addressed to a particular mailbox.

Each mailbox on the slave NXT brick is a circular message queue holding up to five messages. Each message can be up to 58 bytes long. A slave NXT brick must be running a program when an incoming message packet is received, otherwise, the slave NXT brick ignores the message and the message is dropped.

Various comm module constant values are listed in Table C-16 below.

Miscellaneous Constants	Value
SIZE_OF_USBBUF	64
USB_PROTOCOL_OVERHEAD	2
SIZE_OF_USBDATA	62
SIZE_OF_HSBUF	128
SIZE_OF_BTBUF	128
BT_CMD_BYTE	1
SIZE_OF_BT_DEVICE_TABLE	30
SIZE_OF_BT_CONNECT_TABLE	4
SIZE_OF_BT_NAME	16
SIZE_OF_BRICK_NAME	8
SIZE_OF_CLASS_OF_DEVICE	4
SIZE_OF_BDADDR	7
MAX_BT_MSG_SIZE	60000
BT_DEFAULT_INQUIRY_MAX	0
BT_DEFAULT_INQUIRY_TIMEOUT_LO	15
LR_SUCCESS	0x50
LR_COULD_NOT_SAVE	0x51
LR_STORE_IS_FULL	0x52
LR_ENTRY_REMOVED	0x53
LR_UNKNOWN_ADDR	0x54
USB_CMD_READY	0x01
BT_CMD_READY	0x02
HS_CMD_READY	0x04

Table C-16. Miscellaneous Constants

Valid values for the comm module IOMap BtState field are listed in Table C-17 below.

BtState Constants	Value
BT_ARM_OFF	0
BT_ARM_CMD_MODE	1
BT_ARM_DATA_MODE	2

Table C-17. BtState Constants

Valid values for the comm module IOMap BtStateStatus field are listed in Table C-18 below.

BtStateStatus Constants	Value
BT_BRICK_VISIBILITY	0X01
BT_BRICK_PORT_OPEN	0X02
BT_CONNECTION_0_ENABLE	0X10
BT_CONNECTION_1_ENABLE	0X20
BT_CONNECTION_2_ENABLE	0X40
BT_CONNECTION_3_ENABLE	0x80

Table C-18. BtStateStatus Constants

Values for the comm module IOMap BtHwStatus field are listed in Table C-19 below, which represent the Bluetooth hardware status.

BtHwStatus Constants	Value
BT_ENABLE	0X00
BT_DISABLE	0X01

Table C-19. BtHwStatus Constants

Valid values for the comm module IOMap HsFlags field are listed in Table C-20 below.

HsFlags Constants	Value
HS_UPDATE	1

Table C-20. HsFlags Constants

Valid values for the comm module IOMap HsState field are listed in Table C-21 below.

HsState Constants	Value
HS_INITIALISE	1
HS_INIT_RECEIVER	2
HS_SEND_DATA	3
HS_DISABLE	4

Table C-21. HsState Constants

Valid values for the comm module IOMap DeviceStatus field are listed in Table C-22 below.

DeviceStatus Constants	Value
BT_DEVICE_EMPTY	0X00
BT_DEVICE_UNKNOWN	0X01
BT_DEVICE_KNOWN	0X02
BT_DEVICE_NAME	0X40
BT_DEVICE_AWAY	0x80

Table C-22. DeviceStatus Constants

Valid module interface values are listed in Table C-23 below.

Interface Constants	Value
INTF_SENDFILE	0
INTF_SEARCH	1
INTF_STOPSEARCH	2
INTF_CONNECT	3
INTF_DISCONNECT	4
INTF_DISCONNECTALL	5
INTF_REMOVEDEVICE	6
INTF_VISIBILITY	7
INTF_SETCMDMODE	8
INTF_OPENSTREAM	9
INTF_SENDDATA	10

INTF_FACTORYRESET	11
INTF_BTON	12
INTF_BTOFF	13
INTF_SETBTNAME	14
INTF_EXTREAD	15
INTF_PINREQ	16
INTF_CONNECTREQ	17

Table C-23. Interface Constants

The function pointer shown at the beginning of the IOMap structure cannot be accessed using the standard NXT firmware. The interface values shown in Table C-23 are the function commands that can be passed as the first argument to this function pointer. If you install the enhanced NBC/NXC firmware on your NXT then you will have access to a special system call function that exposes access to this important feature of the comm module. Using this new system call, you will be able to establish Bluetooth connections dynamically under your program's control rather than having to create all your connections using the NXT menu system.

IOCtrl Module

The NXT IOCtrl module manages the low-level communication between the two processors that control the NXT. The module's IOMap is a tiny 2 bytes in size. This means there isn't a lot we need to know about this module and there isn't much we will ever do in a program we write for the NXT that will directly interact with the ioctrl module. The IOMap structure for this module is shown in Sample C-7 below.

```
typedef struct
{
   UWORD PowerOn;
} IOMAPIOCTRL;
```

Sample C-7. IOCtrl module IOMap structure

The NBC and NXC APIs expose a pair of functions that can be used within your NXT program. They both directly manipulate the IOMap's PowerOn field to perform their function. The necessary offset information is shown in Table C-24 below.

IOCtrl Module Offsets	Value	Size
IOCtrlOffsetPowerOn	0	2

Table C-24. IOCtrl module IOMap offsets

You can set this field to two different values for which the NXT checks and to which it reacts. Use the IOCTRL_POWERDOWN constant to tell the NXT to power down immediately. Of course, this will terminate your program as well. The other value you can use is the IOCTRL_BOOT constant. It forces the NXT to reboot in firmware download mode causing the firmware to be erased as well as all your programs. Using this value within a running program is not recommended!

Sound Module

The sound module provides support for all the various forms of sound output. The NXT sound files (.rso) are like .wav files. They contain thousands of sound samples that digitally represent an analog waveform. With sound files the NXT can speak or play music or make just about any sound imaginable.

Melody files are like MIDI files. They contain multiple tones with each tone being defined by a frequency and duration pair. When played on the NXT, a melody file sounds like a pure sine-wave tone generator playing back a series of notes. While not as fancy as sound files, melody files are usually much smaller than sound files.

The IOMap structure for the sound module is shown in Sample C-8.

```
typedef struct
{
  UWORD    Freq; // Hz
  UWORD    Duration; // ms
  UWORD    SampleRate; // [2000..16000]
  UBYTE    SoundFilename[FILENAME_LENGTH + 1];
  UBYTE    Flags;
  UBYTE    State;
  UBYTE    Mode;
  UBYTE    Volume; // [0..4] 0 = off
} IOMAPSOUND;
```

Sample C-8. Sound module IOMap structure

Several of the fields in this IOMap structure are also accessible via other API functions that we will examine in later chapters. You can experiment with changing some of these values while a sound is active. Access the fields in the sound module IOMap using the offsets shown in Table C-25.

When a sound or a file is played on the NXT, execution of the program does not wait for the previous playback to complete. To play multiple tones or files sequentially, it is necessary to wait for the previous tone or file playback to complete first. This can be done via wait API functions or by using the sound state value within a while loop. You can use the sound module Flags, State, and Mode fields to check the configuration and current state of the sound module.

Sound Module Offsets	Value	Size
SoundOffsetFreq	0	2
SoundOffsetDuration	2	2
SoundOffsetSampleRate	4	2
SoundOffsetSoundFilename	6	20
SoundOffsetFlags	26	1
SoundOffsetState	27	1
SoundOffsetMode	28	1
SoundOffsetVolume	29	1

Table C-25. Sound module IOMap offsets

Valid values for the Flags field of the sound module IOMap structure are listed in Table C-26 below.

Flags Constants	Read/Write	Meaning
SOUND_FLAGS_IDLE	Read	Sound is idle
SOUND_FLAGS_UPDATE	Write	Make changes take effect
SOUND_FLAGS_RUNNING	Read	Processing a tone or file

Table C-26. Flags Constants

Valid values for the State field of the sound module IOMap structure are listed in Table C-27 below.

Sound State Constants	Read/Write	Meaning
SOUND_STATE_IDLE	Read	Idle, ready for start sound
SOUND_STATE_FILE	Read	Processing file of sound/ melody data
SOUND_STATE_TONE	Read	Processing play tone request
SOUND_STATE_STOP	Write	Stop sound immediately and close hardware

Table C-27. State Constants

Valid values for the Mode field of the sound module IOMap are listed in Table C-28 below.

Mode Constants	Read/Write	Meaning
SOUND_MODE_ONCE	Read	Only play file once
SOUND_MODE_LOOP	Read	Play file until writing SOUND_STATE_STOP into State.
SOUND_MODE_TONE	Read	Play tone specified in Frequency for Duration milliseconds

Table C-28. Mode constants

Miscellaneous sound module constants are listed in Table C-29 below.

Misc. Sound Constants	Value	Meaning
FREQUENCY_MIN	220	Minimum frequency in Hz.
FREQUENCY_MAX	14080	Maximum frequency in Hz.
SAMPLERATE_MIN	2000	Minimum sample rate supported by NXT
SAMPLERATE_DEFAULT	8000	Default sample rate
SAMPLERATE_MAX	16000	Maximum sample rate supported by NXT

Table C-29. Miscellaneous sound module constants

Both NXC and NBC define frequency and duration constants that may be used with higher-level API functions as well as with fields in the sound module IOMap. Frequency constants start with TONE_A3 (the 'A' pitch in octave 3) and go to TONE_B7 (the 'B' pitch in octave 7). Duration constants start with MS_1 (1 millisecond) and go up to MIN_1 (60000 milliseconds) with several constants in between. See Appendix A for the complete list of sound constants.

Loader Module

The NXT loader module provides support for the NXT file system. The NXT supports creating files, opening existing files, reading, writing, renaming, and deleting files. Files in the NXT file system must adhere to the 15.3 naming convention for a maximum filename length of 19 characters. While multiple files can be opened simultaneously, a maximum of 4 files can be open for writing at any given time.

The loader module has a very small IOMap structure. Its structure is shown in Sample C-9 below.

```
typedef struct
{
  UWORD (*pFunc)(UBYTE,UBYTE *,UBYTE *,ULONG *);
  ULONG FreeUserFlash;
} IOMAPLOADER;
```

Sample C-9. Loader module IOMap structure

The function pointer at the start of this IOMap structure cannot be accessed via the standard NXT firmware but the enhanced NBC/NXC firmware includes support for a new system call function that was designed to expose this function for use by programmers like us. The amount of free flash memory can be accessed directly via the FreeUserFlash field in the IOMap using the offsets listed in Table C-30 below.

Loader Module Offsets	Value	Size
LoaderOffsetPFunc	0	4
LoaderOffsetFreeUserFlash	4	4

Table C-30. Loader module IOMap offsets

When accessing files on the NXT, errors can occur. Robust code will always check for error conditions so that appropriate corrective actions can be taken. The NBC and NXC APIs provide several constants that define possible result codes. They are listed in Table C-31 below.

Loader Result Codes	Value
LDR_SUCCESS	0x0000
LDR_INPROGRESS	0x0001
LDR_REQPIN	0x0002
LDR_NOMOREHANDLES	0x8100
LDR_NOSPACE	0x8200
LDR_NOMOREFILES	0x8300
LDR_EOFEXPECTED	0x8400
LDR_ENDOFFILE	0x8500
LDR_NOTLINEARFILE	0x8600
LDR_FILENOTFOUND	0x8700
LDR_HANDLEALREADYCLOSED	0x8800

LDR_NOLINEARSPACE	0x8900
LDR_UNDEFINEDERROR	0x8A00
LDR_FILEISBUSY	0x8B00
LDR_NOWRITEBUFFERS	0x8C00
LDR_APPENDNOTPOSSIBLE	0x8D00
LDR_FILEISFULL	0x8E00
LDR_FILEEXISTS	0x8F00
LDR_MODULENOTFOUND	0x9000
LDR_OUTOFBOUNDARY	0x9100
LDR_ILLEGALFILENAME	0x9200
LDR_ILLEGALHANDLE	0x9300
LDR_BTBUSY	0x9400
LDR_BTCONNECTFAIL	0x9500
LDR_BTTIMEOUT	0x9600
LDR_FILETX_TIMEOUT	0x9700
LDR_FILETX_DSTEXISTS	0x9800
LDR_FILETX_SRCMISSING	0x9900
LDR_FILETX_STREAMERROR	0x9A00
LDR_FILETX_CLOSEERROR	0x9B00

Table C-31. Loader result codes

As with other modules the bulk of the functionality provided by the loader module is exposed to us via means other than the module's IOMap structure. We will examine all the NXC and NBC API functions supported by the loader module in several chapters coming up.

Display Module

The NXT display module encompasses support for drawing to the NXT LCD. The NXT supports drawing points, lines, rectangles, and circles on the LCD. It supports drawing NXT picture files on the screen as well as text and numbers. The LCD screen has its origin (0, 0) at the bottom left-hand corner of the screen with the positive Y-axis extending upward and the positive X-axis extending toward the right.

The NBC and NXC APIs provide constants that allow you to specify LCD line numbers between 1 and 8, with line 1 being at the top of the screen and line 8 being at the bottom of the screen. These constants (LCD_LINE1, LCD_LINE2, LCD_LINE3, LCD_LINE4, LCD_LINE5,

LCD_LINE6, LCD_LINE7, LCD_LINE8) can be used as the Y coordinate when calling NBC and NXC API functions. If you specify a Y coordinate for text and numeric output other than these constants, the value will be adjusted so that the text and numbers are on one of 8 fixed line positions.

The IOMap for the display module is larger than many of the other IOMap structures since it includes memory for storing two different screen images – normal and popup screens. Data written to these screen buffers is copied to the LCD screen. The IOMap structure for the display module is shown in Sample C-10.

```
typedef    struct
{
      void (*pFunc)(UBYTE, UBYTE, UBYTE, UBYTE, UBYTE, UBYTE);
   ULONG    EraseMask;
   ULONG    UpdateMask;
   FONT     *pFont;
   UBYTE    *pTextLines[TEXTLINES];
   UBYTE    *pStatusText;
   ICON     *pStatusIcons;
   BMPMAP   *pScreens[SCREENS];
   BMPMAP   *pBitmaps[BITMAPS];
   UBYTE    *pMenuText;
   UBYTE    *pMenuIcons[MENUICONS];
   ICON     *pStepIcons;
   UBYTE    *Display;
   UBYTE    StatusIcons[STATUSICONS];
   UBYTE    StepIcons[STEPICONS];
   UBYTE    Flags;
   UBYTE    TextLinesCenterFlags;
   UBYTE    Normal[8][100];
   UBYTE    Popup[8][100];
} IOMAPDISPLAY;
```

Sample C-10. Display module IOMap structure

Several of the fields in this IOMap structure are not usable within a running program since they involve pointers and the NXT virtual machine does not give us access to pointers. There are, however, many high level API drawing functions that we will examine in later chapters. They provide a full range of support for controlling the LCD display. And you can write directly to the Normal field's 800 bytes of screen memory if you wish. Whatever you write to that field will be drawn on the LCD screen. You can access the fields in the display module IOMap using the offsets shown in Table C-32.

Display Module Offsets	Value	Size
DisplayOffsetPFunc	0	4
DisplayOffsetEraseMask	4	4
DisplayOffsetUpdateMask	8	4
DisplayOffsetPFont	12	4
DisplayOffsetPTextLines(p)	(((p)*4)+16)	4*8
DisplayOffsetPStatusText	48	4
DisplayOffsetPStatusIcons	52	4
DisplayOffsetPScreens(p)	(((p)*4)+56)	4*3
DisplayOffsetPBitmaps(p)	(((p)*4)+68)	4*4
DisplayOffsetPMenuText	84	4
DisplayOffsetPMenuIcons(p)	(((p)*4)+88)	4*3
DisplayOffsetPStepIcons	100	4
DisplayOffsetDisplay	104	4
DisplayOffsetStatusIcons(p)	((p)+108)	1*4
DisplayOffsetStepIcons(p)	((p)+112)	1*5
DisplayOffsetFlags	117	1
DisplayOffsetTextLinesCenterFlags	118	1
DisplayOffsetNormal(l,w)	(((l)*100)+(w)+119)	800
DisplayOffsetPopup(l,w)	(((l)*100)+(w)+919)	800

Table C-32. Display module IOMap offsets

If you install the enhanced NBC/NXC standard NXT firmware on your brick then you will be able to use the display module's function pointer that you can see at the beginning of the IOMap structure. The details of how to use this function are found in chapter 8. The enhanced firmware also contains several bug fixes related to drawing on the LCD screen. The ability to clear pixels in addition to setting them is fully exposed via the enhanced firmware. And all of the drawing routines have been optimized so that your programs run faster when they have to draw to the LCD screen frequently.

Low Speed Module

The NXT low speed module encompasses support for digital I^2C sensor communication. Use the low speed (aka I^2C) communication methods to access devices that use the I^2C protocol on the NXT brick's four input ports. The IOMap for the low speed module contains buffers for I^2C

communication on each input port. The IOMap structure for the low speed module is shown in Sample C-11.

```
typedef    struct
{
  UBYTE    Buf[SIZE_OF_LSBUF];
  UBYTE    InPtr;
  UBYTE    OutPtr;
  UBYTE    BytesToRx;
} LSBUF;

typedef    struct
{
  LSBUF    InBuf[NO_OF_LSBUF];
  LSBUF    OutBuf[NO_OF_LSBUF];
  UBYTE    Mode[NO_OF_LSBUF];
  UBYTE     ChannelState[NO_OF_LSBUF];
  UBYTE    ErrorType[NO_OF_LSBUF];
  UBYTE    State;
  UBYTE    Speed;
  UBYTE    Spare1;
} IOMAPLOWSPEED;
```

Sample C-11. Low Speed module IOMap structure

This IOMap structure is fully accessible from within a running program, but the primary functionality of using I²C digital sensors cannot be fully utilized with just the IOMap. The NXT firmware and the Low Speed module provide the required system call functions to perform I²C write transactions to any digital sensor that implements the NXT I²C specification. To simplify access to the Low Speed functionality there are a number of API functions that help make accessing I²C sensors a breeze. You can access the fields in the Low Speed module IOMap in your own programs, if necessary, by using the offsets shown in Table C-33.

LowSpeed Module Offsets	Value	Size
LowSpeedOffsetInBufBuf(p)	(((p)*19)+0)	16
LowSpeedOffsetInBufInPtr(p)	(((p)*19)+16)	1
LowSpeedOffsetInBufOutPtr(p)	(((p)*19)+17)	1
LowSpeedOffsetInBufBytesToRx(p)	(((p)*19)+18)	58
LowSpeedOffsetOutBufBuf(p)	(((p)*19)+76)	16
LowSpeedOffsetOutBufInPtr(p)	(((p)*19)+92)	1
LowSpeedOffsetOutBufOutPtr(p)	(((p)*19)+93)	1
LowSpeedOffsetOutBufBytesToRx(p)	(((p)*19)+94)	58

LowSpeedOffsetMode(p)	((p)+152)	4
LowSpeedOffsetChannelState(p)	((p)+156)	4
LowSpeedOffsetErrorType(p)	((p)+160)	4
LowSpeedOffsetState	164	1
LowSpeedOffsetSpeed	165	1

Table C-33. Low Speed module IOMap offsets

You must set the input port's SensorType property to SENSOR_TYPE_LOWSPEED or SENSOR_TYPE_LOWSPEED_9V on a given port before using an I²C device on that port. Use SENSOR_TYPE_LOWSPEED_9V if your device requires 9V power from the NXT brick. Remember that you also need to set the input port's InvalidData property to true after setting a new SensorType, and then wait in a loop for the NXT firmware to set InvalidData back to false. This process ensures that the firmware has time to properly initialize the port, including the 9V power lines, if applicable. Some digital devices might need additional time to initialize after power up.

When communicating with I²C devices, the NXT firmware uses a master/slave setup in which the NXT brick is always the master device. This means that the firmware is responsible for controlling the write and read operations. The NXT firmware maintains write and read buffers for each port, and three main Lowspeed (I²C) methods described below enable you to access these buffers.

Note that any of these calls might return various status codes at any time. A status code of 0 means the port is idle and the last transaction (if any) did not result in any errors. Negative status codes and the positive status code 32 indicate errors. There are a few possible errors per call.

Valid low speed return values are listed in Table C-34 below.

Low Speed Return Constants	Value	Meaning
NO_ERR	0	The operation succeeded.
STAT_COMM_PENDING	32	The specified port is busy performing a communication transaction.
ERR_INVALID_SIZE	-19	The specified buffer or byte count exceeded the 16 byte limit.
ERR_COMM_CHAN_NOT_READY	-32	The specified port is busy or improperly configured.

ERR_COMM_CHAN_INVALID	-33	The specified port is invalid. It must be between 0 and 3.
ERR_COMM_BUS_ERR	-35	The last transaction failed, possibly due to a device failure.

Table C-34. Low speed (I2C) return value constants

Valid values for the low speed module IOMap State field are listed in Table C-35 below.

State Constants	Value
COM_CHANNEL_NONE_ACTIVE	0x00
COM_CHANNEL_ONE_ACTIVE	0x01
COM_CHANNEL_TWO_ACTIVE	0x02
COM_CHANNEL_THREE_ACTIVE	0x04
COM_CHANNEL_NONE_ACTIVE	0x08

Table C-35. State Constants

Valid values for the low speed module IOMap ChannelState field are listed in the following table.

ChannelState Constants	Value
LOWSPEED_IDLE	0
LOWSPEED_INIT	1
LOWSPEED_LOAD_BUFFER	2
LOWSPEED_COMMUNICATING	3
LOWSPEED_ERROR	4
LOWSPEED_DONE	5

Table C-36. ChannelState Constants

Valid values for the low speed module IOMap Mode field are listed in Table C-37 below.

Mode Constants	Value
LOWSPEED_TRANSMITTING	1
LOWSPEED_RECEIVING	2
LOWSPEED_DATA_RECEIVED	3

Table C-37. Mode Constants

Valid values for the low speed module IOMap ErrorType field are listed in Table C-38 below.

ErrorType Constants	Value
LOWSPEED_NO_ERROR	0
LOWSPEED_CH_NOT_READY	1
LOWSPEED_TX_ERROR	2
LOWSPEED_RX_ERROR	3

Table C-38. ErrorType Constants

A call to LowspeedWrite starts an asynchronous transaction between the NXT brick and a digital I^{2}C device. Your program continues to run while the firmware manages the process of sending bytes from the write buffer and reading the response bytes from the I^{2}C device. Because the NXT is the master device, you must also specify the number of bytes to expect from the device in response to each write operation. You can exchange up to 16 bytes in each direction per transaction.

After you start a write transaction with LowspeedWrite, use LowspeedStatus in a loop to check the status of the port. If LowspeedStatus returns a status code of 0 and a count of bytes available in the read buffer, the system is ready for you to use LowspeedRead to copy the data from the read buffer into the buffer you provide.

The details of using these API functions and the constants described in the tables above will be covered in several upcoming chapters including chapter 8. There will be many sample programs in future chapters that use I^{2}C sensors to help a robot respond to its environment in very interesting and clever ways.

UI Module
The NXT UI module encompasses support for many different aspects of the user interface for the NXT brick. The IOMap structure for the UI module is shown in Sample C-12.

```
typedef    struct
{
  MENU     *pMenu;
  UWORD    BatteryVoltage;
  UBYTE    LMSfilename[FILENAME_LENGTH + 1];
  UBYTE    Flags;
  UBYTE    State;
  UBYTE    Button;
  UBYTE    RunState;
  UBYTE    BatteryState;
  UBYTE    BluetoothState;
  UBYTE    UsbState;
```

```
    UBYTE     SleepTimeout;
    UBYTE     SleepTimer;
    UBYTE     Rechargeable;
    UBYTE     Volume;
    UBYTE     Error;
    UBYTE     OBPPointer;
    UBYTE     ForceOff;
    UBYTE     LongAbort;
} IOMAPUI;
```

Sample C-12. UI module IOMap structure

Nearly all of the fields in this IOMap structure are directly usable from within a running program. To simplify access to these fields there are several API wrapper functions that hide the complexity of accessing IOMap data. When necessary, you can access the fields in the UI module IOMap in your own programs by using the offsets shown in Table C-39.

UI Module Offsets	Value	Size
UIOffsetPMenu	0	4
UIOffsetBatteryVoltage	4	2
UIOffsetLMSfilename	6	20
UIOffsetFlags	26	1
UIOffsetState	27	1
UIOffsetButton	28	1
UIOffsetRunState	29	1
UIOffsetBatteryState	30	1
UIOffsetBluetoothState	31	1
UIOffsetUsbState	32	1
UIOffsetSleepTimeout	33	1
UIOffsetSleepTimer	34	1
UIOffsetRechargeable	35	1
UIOffsetVolume	36	1
UIOffsetError	37	1
UIOffsetOBPPointer	38	1
UIOffsetForceOff	39	1
UIOffsetLongAbort	40	1

Table C-39. UI module IOMap offsets

Valid values for the Flags IOMap field are listed in Table C-40 below.

Flags Constants	Value
UI_FLAGS_UPDATE	0x01
UI_FLAGS_DISABLE_LEFT_RIGHT_ENTER	0x02
UI_FLAGS_DISABLE_EXIT	0x04
UI_FLAGS_REDRAW_STATUS	0x08
UI_FLAGS_RESET_SLEEP_TIMER	0x10
UI_FLAGS_EXECUTE_LMS_FILE	0x20
UI_FLAGS_BUSY	0x40
UI_FLAGS_ENABLE_STATUS_UPDATE	0x80

Table C-40. Flags Constants

Valid values for the State IOMap field are listed in Table C-41 below.

State Constants	Value
UI_STATE_INIT_DISPLAY	0
UI_STATE_INIT_LOW_BATTERY	1
UI_STATE_INIT_INTRO	2
UI_STATE_INIT_WAIT	3
UI_STATE_INIT_MENU	4
UI_STATE_NEXT_MENU	5
UI_STATE_DRAW_MENU	6
UI_STATE_TEST_BUTTONS	7
UI_STATE_LEFT_PRESSED	8
UI_STATE_RIGHT_PRESSED	9
UI_STATE_ENTER_PRESSED	10
UI_STATE_EXIT_PRESSED	11
UI_STATE_CONNECT_REQUEST	12
UI_STATE_EXECUTE_FILE	13
UI_STATE_EXECUTING_FILE	14
UI_STATE_LOW_BATTERY	15
UI_STATE_BT_ERROR	16

Table C-41. State Constants

Valid values for the Button IOMap field are listed in Table C-42 below.

Button Constants	Value
UI_BUTTON_NONE	1
UI_BUTTON_LEFT	2
UI_BUTTON_ENTER	3
UI_BUTTON_RIGHT	4
UI_BUTTON_EXIT	5

Table C-42. Button Constants

Valid values for the BluetoothState IOMap field are listed in Table C-43 below.

BluetoothState Constants	Value
UI_BT_STATE_VISIBLE	0x01
UI_BT_STATE_CONNECTED	0x02
UI_BT_STATE_OFF	0x04
UI_BT_ERROR_ATTENTION	0x08
UI_BT_CONNECT_REQUEST	0x40
UI_BT_PIN_REQUEST	0x80

Table C-43. BluetoothState Constants

You can use the various fields in this IOMap structure directly to find out such things as the battery voltage remaining or whether the NXT has a rechargeable battery pack installed or not. All of these field values can also be accessed via high-level API functions in NBC and NXC that hide the IOMap details.

NXT Picture Format

Topics in this Appendix

Appendix D

RICscript allows you to create your own pictures using a scripting language. This appendix details the NXT picture format, allowing you to understand the internal structure of the picture format.

The NXT picture format

So what are the opcodes or commands that you can use to create an NXT picture? Table D-1 contains a complete list of all the opcodes. It also has each opcode's ID and the associated structure size.

Opcode	ID	Size
Description	0	10
Sprite	1	12+
VarMap	2	16+
CopyBits	3	20
Pixel	4	12
Line	5	14
Rectangle	6	14
Circle	7	12
NumBox	8	12

Table D-1. NXT Picture opcodes

Each of these commands can occur any number of times within a single picture file. When you draw a picture you are telling the firmware to sequentially execute one or more of the commands listed above. Each command is defined by its ID or opcode value, which is a number from 0 to 8, and it is followed by the opcode arguments. The opcodes in an RIC file take the form of a structure with various member fields that define the arguments of the opcode.

In order to make reading the opcodes from a picture simple, each opcode begins with an unsigned word value or two bytes that indicate how many more bytes in the file are used by the current command. This is the size of the opcode structure minus two, since two bytes for the size have already been read. Each opcode follows the unsigned word specifying the size with the opcode ID that is also an unsigned word value.

Description

The smallest opcode is the Description opcode. The structure size, in total, is 10 bytes. That means that in an NXT picture if you add a Description opcode to the file the opcode will start with 0x08 0x00. This is "little-endian" since the least significant byte (the low-order byte) is written first. As mentioned, the value for the structure size is two less than the actual size of the opcode, since it is there to tell the parser how many more bytes it needs to read before it gets to the end of the current opcode. Following the size is the Description opcode's ID, which is zero. So there are two bytes (an unsigned word written in little-endian format) of zero after the size to tell the NXT it is to execute a Description opcode.

The Description opcode has three word-sized arguments. The first is Options. It is meaningless in the current firmware. The second argument is the Width and the third argument is the Height. Both are unsigned words (two bytes each). They describe the total width and height of the picture in its entirety. If you use a Description command in a picture, it should correctly specify the drawing extents of every other drawing command within the picture. However, the Description opcode is actually a no-op when the NXT picture is executed by the firmware. It ignores all the values within this opcode so they can be whatever you want. Leave out the Description opcode altogether if you want to and you will save 10 bytes. The Description structure is shown in Sample D-1 below.

```
typedef struct
{
  UWORD OpSize;
  UWORD OpCode;
  UWORD Options;
  UWORD Width;
  UWORD Height;
} IMG_OP_DESCRIPTION;
```

Sample D-1. Description opcode structure

The one caveat that needs to be mentioned is that the LEGO MINDSTORMS NXT Software prefers that an NXT picture start with a Description opcode. It uses the information in the Description opcode to help it correctly draw the picture in the Display block preview window.

A picture that does not start with a Description opcode may not always draw correctly within the NXT software preview window but it will draw correctly when your program runs on the NXT.

Sprite

The most important opcode from the LEGO MINDSTORMS NXT Software perspective, and one of three opcodes that are used exclusively by the pictures that come with the NXT software, is the Sprite opcode. A Sprite opcode is basically a bitmap. Its arguments define the number of rows and columns of pixels that are to be set or cleared. The Sprite structure is shown in Sample D-2 below.

```
typedef struct
{
  UWORD OpSize;
  UWORD OpCode;
  UWORD DataAddr;
  UWORD Rows;
  UWORD RowBytes;
  UBYTE Bytes[2];
} IMG_OP_SPRITE;
```

Sample D-2. Sprite opcode structure

As you can see, the size of this opcode is not a constant value. It depends on how many pixels are defined within the Sprite. The pixel data in the Sprite is laid out by rows starting with the top row of the Sprite and continuing with each row below it in sequence. Each byte in the Bytes array represents eight horizontal pixels in a single row. So if your sprite image is 24 pixels wide then the RowBytes value is three. If your sprite image is anywhere from 25 pixels wide to 32 pixels wide then the RowBytes is four (since each byte accounts for 8 pixels). The height of your image is the Row value in the Sprite. If you examine a sprite whose width is 24 pixels then Bytes[0] through Bytes[2] will represent the top row of Sprite data, Bytes[3] through Bytes[5] will represent the second row, and so on until all the Rows in the Sprite are defined by data in the Bytes array.

The total size of the Bytes array is Rows*RowBytes. The size of the Bytes array must be even, however, so if Rows*RowBytes is odd, the Bytes array size will be Rows*RowsBytes+1 where the extra byte is equal to zero. The reason why the Bytes array size must be even is because the NXT firmware running on the ARM processor wants these structures to be word-aligned. It does not like structures to have an odd number of bytes.

If your sprite image has a height of one, the Rows value will be one. If your sprite image has a width of one, the RowBytes value will be one and the Bytes array will have only two bytes in it (one byte for the single pixel and one pad byte). In that case the total size of the Sprite opcode would be 12 bytes and the value of the OpSize field would be 10. If the byte has all eight bits (pixels) set then the corresponding bytes of data in the RIC file for the Sprite opcode would be like this:

```
0A 00 01 00 01 00 01 00 01 00 FF 00
```

The DataAddr argument to the Sprite opcode is very important. This address is an index into a global array of structure pointers. The array can store 10 entries numbered 1 through 10. Each Sprite in an NXT picture is stored at the index specified by this argument. This array is used not only by all the Sprite opcodes in the picture but also all the VarMap opcodes.

When the Sprite opcode is executed by the firmware, the only thing that occurs is that a pointer to the Sprite structure is copied into the data array at the specified index. No image data is drawn to the LCD at this point. What this means is if you reuse the same data address value in subsequent Sprite opcodes or VarMap opcodes, then any Sprite or VarMap data previously written to that address is no longer useable. All the pictures that are included with the NXT software use a single Sprite opcode and they all use one as the data address value.

VarMap

The least understood command that can be used within an NXT picture is the VarMap opcode. The structure size is not a constant value just like the Sprite opcode. The actual size depends on the number of entries in the MapElement array. The MapCount argument to this command specifies the actual number of elements in the array, so this value is used to calculate the total structure size. Another important requirement is that there must always be at least two elements in the array.

The VarMap opcode ID is 2. The structure for this opcode is shown in Sample D-3 below.

```
typedef struct
{
  UWORD OpSize;
  UWORD OpCode;
  UWORD DataAddr;
  UWORD MapCount;
  struct
  {
    UWORD Domain;
    UWORD Range;
  } MapElement[1];
} IMG_OP_VARMAP;
```

Sample D-3. VarMap opcode structure

If a VarMap command has the minimum number of elements in the MapElement array then its MapCount field is equal to two. In that case the total size of the VarMap opcode would be 16 bytes and the value of its OpSize field would be 14. If the DataAddr field happened to be two and the MapElement array structures were f(0) = 0 and f(64) = 64 then the bytes associated with this command in a picture file would like this:

```
0D 00 02 00 02 00 02 00 00 00 00 00 40 00 40 00
```

You may be wondering about the f(x) = y syntax that I used above to describe the elements in the MapElement array. Each element in the array is a structure containing a Domain field and a Range field. They represent an input x value and an output y value of a transformation function. The domain of a function is the set of input values and the range of a function is the set of output values. The VarMap opcode is a mechanism within the NXT picture format to define a transformation function that can be applied to values that are used within the picture to draw on the LCD screen. It does not actually draw anything on the screen by itself.

If you want to use a VarMap command within an NXT picture you have to do at least three things. First, you have to define the map function using this opcode as a series of domain/range pairs in the MapElement array. Then you also have to parameterize the arguments in other commands that you want to transform using the function in the VarMap. Finally, you have to use GraphicOutEx and pass into the function an array of values to use with the parameterized commands. You can also use GraphicOut with a parameterized NXT picture but if you do then all the parameters in the file will use zero as their value. We'll look at parameterized arguments in more detail and explain how VarMap commands work in concert with them after we examine the remaining picture opcodes.

CopyBits

The CopyBits opcode is the third command used by NXT pictures. Each of these pictures begins with a Description, followed by a Sprite, and concluding with a CopyBits opcode. The Display block's preview window can only draw RIC files that use a Sprite and a CopyBits opcode. It does not know how to preview pictures that have multiple Sprites and CopyBits opcodes or RIC files that use any of the other opcodes supported by the NXT firmware.

The CopyBits opcode ID is 3. Its size is fixed at 20 bytes. That means the OpSize value is 18 (0x12 hex). This opcode is defined using three structures as shown in Sample D-4 below.

```
typedef struct
{
  SWORD X, Y;
} IMG_PT;
```

```
typedef struct
{
  IMG_PT Pt;
  SWORD Width, Height;
} IMG_RECT;
typedef struct
{
  UWORD OpSize;
  UWORD OpCode;
  UWORD CopyOptions;
  UWORD DataAddr;
  IMG_RECT Src;
  IMG_PT Dst;
} IMG_OP_COPYBITS;
```

Sample D-4. CopyBits opcode structure

The CopyOptions argument of the CopyBits opcode sounds promising. It can be set to a value that was intended to mean Copy, CopyNot, Or, and BitClear (values 0 through 3). Unfortunately, the CopyOptions argument found in this opcode and in other opcodes that we'll examine shortly is unused in the NXT firmware. All the opcodes that have the CopyOptions argument operate as if the argument was always set to Copy (zero).

The CopyBits opcode is designed to copy data or bits from a previously loaded Sprite into the normal LCD display memory. Which Sprite data is copied is determined by the DataAddr argument to the CopyBits opcode. If you want to copy bits from the Sprite loaded into slot one then you need to set the same data address in your CopyBits opcode. The bits that you want to copy from the Sprite are specified by the source rectangle (left, bottom, width, height). The CopyBits opcode also specifies the destination point (left, bottom) where the copied bits of Sprite data should be drawn. This x, y location is usually set to 0, 0 but it can be used as an offset from the drawing origin specified in the call to GraphicOut or to the low level DrawGraphic system call.

If you want to copy all the bits in the Sprite to the NXT LCD, then the source rectangle values will be 0, 0, RowBytes*8, Rows. However, since you may have extra bits at the end of each row as a result of each byte in a row representing 8 pixels, you will probably want to use the actual desired sprite image width rather than RowBytes*8.

You can also use the CopyBits opcode to only draw a portion of the Sprite data on the screen by manipulating the source rectangle values. You could use a single Sprite opcode with multiple CopyBits opcodes that select different portions of the Sprite data and draw them to the screen or multiple CopyBits opcodes that select the same portion of Sprite data and draw it to different relative screen destinations. The data address value, the values in the source rectangle, and the values in the destination point can all be parameterized and possibly involve a VarMap transformation function if you want to use this functionality.

Pixel

The Pixel opcode is very simple. It has a fixed size of 12 bytes. The command's ID is equal to 4. It is defined by the two structures shown in Sample D-5 below. Use Pixel commands to set individual pixels on the NXT LCD. The X and Y values in the point structure, as well as the Value argument, can all be parameterized and replaced or transformed at runtime using the array of values in the GraphicOutEx API function call along with a VarMap command, if desired.

```
typedef struct
{
   SWORD X, Y;
} IMG_PT;
typedef struct
{
   UWORD OpSize;
   UWORD OpCode;
   UWORD CopyOptions;
   IMG_PT Pt;
   UWORD Value;
} IMG_OP_PIXEL;
```

Sample D-5. Pixel opcode structure

A bug in the NXT firmware, unfortunately, afflicts the Pixel opcode. The Pixel opcode's size is erroneously checked against the size of the Line opcode, which is 2 bytes larger. As a result, NXT pictures containing the Pixel opcode will not draw successfully on a brick running the NXT firmware. This problem has been fixed in the enhanced NBC/NXC firmware available on the BricxCC website. It is also possible to work around the standard firmware bug by padding Pixel opcodes in the picture file with two extra bytes and increasing the OpSize value appropriately.

Line

The Line opcode is used to draw lines from one point to another on the LCD screen. This simple command has a structure size of just 14 bytes. Its opcode ID is 5. The command is defined by the two structures shown in Sample D-6 below. Each of the X and Y values in the two point structures can be parameterized and replaced or transformed at runtime.

```
typedef struct
{
   SWORD X, Y;
} IMG_PT;
```

```
typedef struct
{
  UWORD OpSize;
  UWORD OpCode;
  UWORD CopyOptions;
  IMG_PT Pt1;
  IMG_PT Pt2;
} IMG_OP_LINE;
```

Sample D-6. Line opcode structure

Rectangle

Use the Rectangle opcode to draw rectangles on the LCD. This structure requires 14 bytes, just like the Line opcode. Its opcode ID is 6. The command is defined using the structures shown in Sample D-7 below. The X and Y values in the point structure as well as the Width and Height values can be parameterized and replaced or transformed at runtime.

```
typedef struct
{
  SWORD X, Y;
} IMG_PT;
typedef struct
{
  UWORD OpSize;
  UWORD OpCode;
  UWORD CopyOptions;
  IMG_PT Pt;
  SWORD Width, Height;
} IMG_OP_RECT;
```

Sample D-7. Rectangle opcode structure

Circle

Unfortunately, the Circle opcode cannot draw circles on the LCD when you are running the NXT firmware. This is because of a bug in the drawing code that left out the check for this command within a picture file. If you install the enhanced NBC/NXC firmware then this opcode will draw on the LCD screen as it was designed.

The Circle structure requires 12 bytes and its opcode ID is 7. The command is defined using the structures shown in Sample D-8 below. The X and Y values in the point structure as well as the Radius value can be parameterized and replaced or transformed at runtime.

```
typedef struct
{
  SWORD X, Y;
} IMG_PT;
typedef struct
{
```

```
    UWORD OpSize;
    UWORD OpCode;
    UWORD CopyOptions;
    IMG_PT Pt;
    UWORD Radius;
} IMG_OP_CIRCLE;
```

Sample D-8. Circle opcode structure

NumBox

The NumBox opcode draws a number using the normal NXT font on the LCD screen. The structure that defines this command is 12 bytes long and its opcode ID is 8. The command is defined using the structures shown in Sample D-9 below. The number that is drawn on the screen using this opcode is its Value. The X and Y values in the point structure and the Value argument can be parameterized and replaced or transformed at runtime.

```
typedef struct
{
    SWORD X, Y;
} IMG_PT;
typedef struct
{
    UWORD OpSize;
    UWORD OpCode;
    UWORD CopyOptions;
    IMG_PT Pt;
    UWORD Value;
} IMG_OP_NUMBOX;
```

Sample D-9. NumBox opcode structure

Ball Castor

Topics in this Appendix

- Ball Castor Instructions

Appendix **E**

Versa allows interchangeable modules to easily attach to the main platform. To further demonstrate this concept, this appendix presents an alternate castor module that uses a ball as opposed to a wheel.

The Ball Castor Module

The following steps show how you can build a ball castor module that you can use instead of the normal castor wheel module. Just follow these steps and attach the module to the base platform.

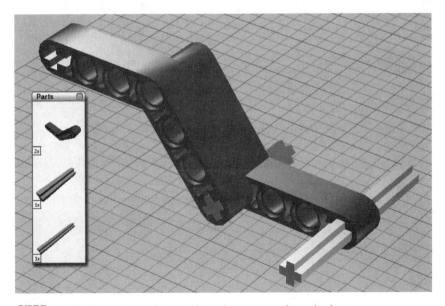

STEP 1. Add the parts as shown. The axles are 3 and 5 units long.

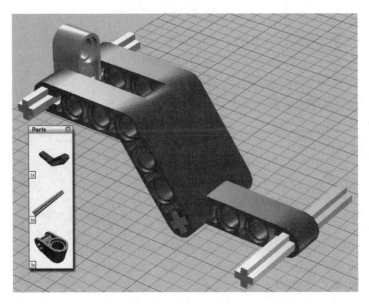

STEP 2. Add the parts as shown. The axle is 5 units long.

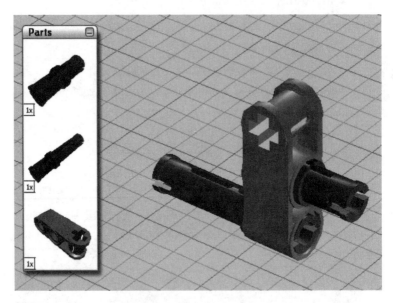

STEP 3. Add the parts as shown. Make two copies of this assembly.

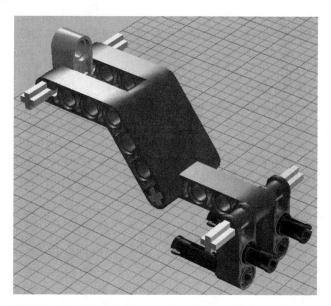

STEP 4. Combine the assemblies from step 3 with the assembly from step 2 as shown.

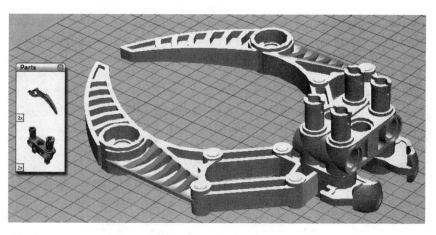

STEP 5. Add the parts as shown.

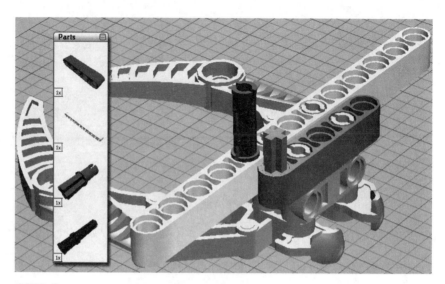

STEP 6. Add the parts as shown.

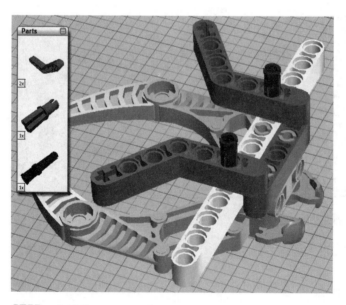

STEP 7. Add the parts as shown.

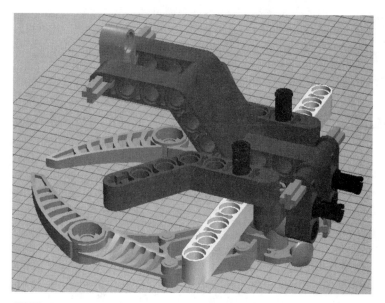

STEP 8. Combine the assembly from step 4 with the assembly from step 7 as shown.

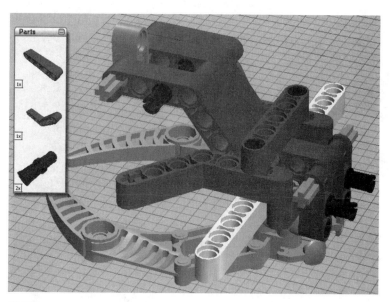

STEP 9. Add the parts as shown.

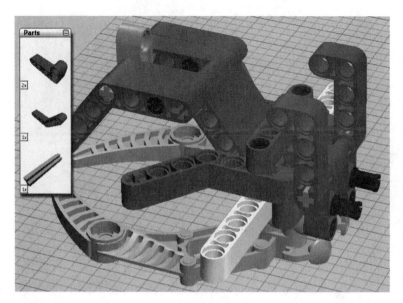

STEP 10. Add the parts as shown. The axle is 3 units long. Insert it through the three central beams at the top of the castor before adding the bent beam.

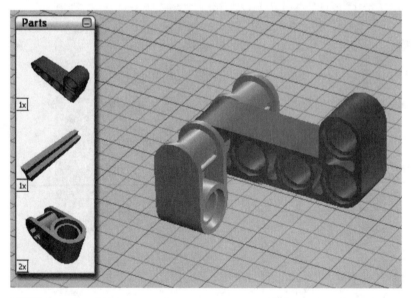

STEP 11. Add the parts as shown. The axle is 3 units long.

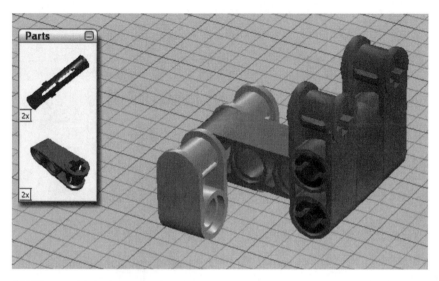

STEP 12. Add the parts as shown.

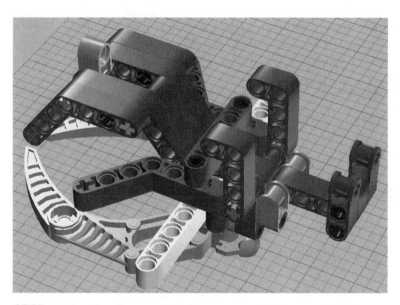

STEP 13. Combine the assembly from step 12 with the assembly from step 10 as shown. This completes the ball castor attachment.

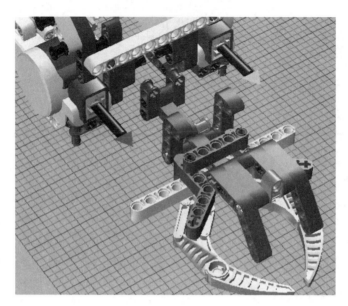

STEP 14. Line up the base platform with the ball castor module as shown.

STEP 15. Attach the two modules together as shown. Use the two 3 unit long axles attached to the black double cross block to lock the castor module into the base platform at the back. Attach the ball castor's white beam to the two blue axle pins with friction.

After you attach the ball castor module, take one of the two balls that came with the NXT set and push it gently up into the ball castor module. It should fit snuggly in between the Bionicle parts at the base of the module.

Index